Antenna Design for 5G and Beyond

Antenna Design for 5G and Beyond

Editor

Naser Ojaroudi Parch

MDPI • Basel • Beijing • Wuhan • Barcelona • Belgrade • Manchester • Tokyo • Cluj • Tianjin

Editor
Naser Ojaroudi Parch
Edinburgh Napier University
UK

Editorial Office
MDPI
St. Alban-Anlage 66
4052 Basel, Switzerland

This is a reprint of articles from the Special Issue published online in the open access journal *Sensors* (ISSN 1424-8220) (available at: https://www.mdpi.com/journal/sensors/special_issues/antenna_5G).

For citation purposes, cite each article independently as indicated on the article page online and as indicated below:

LastName, A.A.; LastName, B.B.; LastName, C.C. Article Title. *Journal Name* **Year**, *Volume Number*, Page Range.

ISBN 978-3-0365-3531-9 (Hbk)
ISBN 978-3-0365-3532-6 (PDF)

Contents

About the Editor

Naser Ojaroudi Parchin (Assistant Professor/Lecturer) is currently working at Edinburgh Napier University, UK. He obtained his Ph.D. in Electrical Engineering from the University of Bradford, UK, where he was a Postdoctoral Research Assistant at the Faculty of Engineering and Informatics and worked as a research fellow in the SATNEX V project, funded by European Space Agency. From 2018 to 2020, Naser served as a Marie Curie Research Fellow in the H2020-ITN-SECRET project. He has over 12 years' research experience in Antennas and Microwave Engineering. He is the author and co-author of several books/book chapters and more than 300 technical journals and conference papers. Naser is a member of the IEEE, MCAA, and EurAAP. He is also an active reviewer in various high-ranking journals and publishers such as IEEE Transactions, IET, Wiley, Springer, Elsevier, MDPI, etc. He was also appointed as Guest Editor of several special issues and on the Topic Board. He is also the recipient and co-recipient of various awards. Naser was included in the World's Top Scientists list in 2016, 2020, and 2021. His papers have received more than 5300 citations, corresponding to a h-index of 40, as reported by Google Scholar.

sensors MDPI

Editorial

Editorial: Special Issue "Antenna Design for 5G and Beyond"

Naser Ojaroudi Parchin [1,*], Chan Hwang See [1] and Raed A. Abd-Alhameed [2]

1 School of Engineering and the Built Environment, Edinburgh Napier University, Edinburgh EH10 5DT, UK; c.see@napier.ac.uk
2 Faculty of Engineering and Informatics, University of Bradford, Bradford BD7 1DP, UK; r.a.a.abd@bradford.ac.uk
* Correspondence: n.ojaroudiparchin@napier.ac.uk

Citation: Ojaroudi Parchin, N.; See, C.H.; Abd-Alhameed, R.A. Editorial: Special Issue "Antenna Design for 5G and Beyond". *Sensors* **2021**, *21*, 7745. https://doi.org/10.3390/s21227745

Received: 11 November 2021
Accepted: 17 November 2021
Published: 21 November 2021

The demand for high data rate transfer and large capacities of traffic is continuously growing as the world witnesses the development of the fifth generation (5G) of wireless communications with the fastest broadband speed yet and low latency [1]. Widespread coverage, adequate signal quality, and low latency are just a few of the benefits that have sparked interest in 5G networks. The commercialization of 5G communication has already started, as well as initial research into beyond technologies such as 6G [2]. The implementation of 5G technology is either in sub 6 GHz or in the millimetre-wave (MM-Wave) region. As a crucial part of future communication systems, breakthroughs in antenna systems will obviously improve the performance of the entire communication system [3]. Antenna design for 5G and beyond networks should apply careful considerations with some key requirements. Wide frequency coverage, the larger number of low-profile antenna elements, ease of fabrication, and conformity are some of the key parameters for the success of 5G antenna systems [4]. Substantial advances have been made in the design, optimization, and development of new antennas for 5G and beyond. In future antenna systems, two main antenna arrangements are considered, including multiple-input multiple-output (MIMO) and phased array with multiple elements. Multiple antenna elements are used to create increased multiple spatial channels, which leads to the improved data rate of the system [5].

This Special Issue covers various aspects of novel antenna designs for 5G and beyond applications. The featured topic articles in this issue highlight recent advances in designing antenna systems for smartphones, small cells, platform stations, massive MIMO, MM-Wave, front-end, metamaterials, and metasurface applications. In addition, other relevant subjects such as spectral efficiency improvement, K-user MIMO, and multi-mode band-pass filter design are discussed. This Special Issue is a collection of fifteen papers that are briefly explained in the following.

Seyyedesfahlan et al. [6] report on the design of a wideband and multi-port antenna for sub 6 GHz applications. The structure of the single element is composed of a circular radiation disk (with $\lambda/4$ diameter) with microstrip-line feeding and a modified triangular-shaped ground plane. It is designed to operate at the frequency range from 2 to 6 GHz. The antenna exhibits sufficient gain and efficiency results. In addition, two- and four-port MIMO configurations of the design with low mutual coupling are studied. The design antennas were fabricated, and their simulation and measurement results show good agreement. The multi-port antenna designs can be used in sub 6 GHz 5G wireless systems.

Ojaroudi Parchin et al. [7] propose a new ultra-wideband (UWB) MIMO antenna system design for future smartphones. It contains four pairs of compact microstrip-fed slot antennas placed at four different edge corners of the smartphone mainboard. The antenna elements provide radiation and polarization diversity, and they operate at a frequency of 2.5–10.2 GHz. Critical characteristics of the antenna elements and MIMO design, such as operation bandwidth, antenna gain and efficiency, and radiation patterns are investigated in detail. The authors also highlight the MIMO performance and channel capacity of the proposed smartphone antenna.

Ojaroudi Parchin et al. [8] demonstrate a new 5G smartphone antenna design with eight modified PIFA elements placed at four corners of the mainboard. It operates at three different frequencies of sub 6 GHz, namely 2.6, 3.6, and 5.8 GHz. In addition, in order to support the MM-Wave 5G spectrum, a compact phased array antenna with a wide bandwidth of 25–31 GHz is introduced for incorporation into a shared board. According to the presented results, quite good characteristics in terms of operation bandwidth, antenna gain, and radiation pattern are observed for the proposed antenna, both with and without the presence of user-head, user-hand, and smartphone components.

Sharaf et al. [9] developed a compact dual-band patch antenna to support 36 and 60 GHz for 5G mobile communications. The antenna configuration contains two electromagnetically coupled patches creating the first and second resonances. Due to its small profile, the proposed patch antenna design can be used either as a single antenna or an element to construct compact MIMO antenna systems. The antenna critical characteristics are assessed through numerical and experimental investigations and good results are observed. In addition, numerical comparisons show that the introduced patch antenna is superior to other published designs.

Dixit et al. [10] introduce a new antipodal Vivaldi array antenna with a 1×4 arrangement proposed for 5G MM-Wave applications. A corporate feeding network is applied on the top layer of the Vivaldi array design to feed the elements. It resonates in a wide frequency range of 24.19–40.47 GHz, covering n257~n261 bands of 5G. Using corrugations in the design of resonators, antenna characteristics such as gain level and front-to-back ratio are improved. The antenna exhibits high gain levels (varying from 8 to 13.2 dBi) versus its operation band. It is designed on a Rogers-5880 substrate material with overall dimensions of $24 \times 28.8 \times 0.254$ mm^3.

Song et al. [11] present a low-cost/low-profile metasurface transmit-array design, fed by a compact antenna array for 28 GHz 5G systems. The employed metasurface contains two metallic layers at different sides of the substrate along with four fixed vias connecting the layers. The transmit-array is composed of several unit cells with different dimensions. In order to optimize the phase distribution on the transmit-array and decrease the side-lobe level, the authors introduced particle swarm optimization. The optimized design achieves 27 dB gain with 11.8% 3 gain bandwidth, −30 dB side-lobe level, and aperture efficiency of 23%, which are improved from the unoptimized design.

Hamid et al. [12] focused on investigating the multibeam characteristics of a negative refractive index-shaped lens with high gain and narrow beamwidth to be operated at upper 5G frequency bands. Two types of negative refractive index lenses, including energy conservation and Abbe's sine lenses, were designed and their parameters are studied. They exhibit high gains and narrow beamwidth characteristics with 65~66% efficiency results. In addition, both designs provide optimum results for beam scanning up until 40°. The designed negative refractive index-shaped lens offers good characteristics in terms of fundamental properties and can be used in 5G mobile base stations.

Kozlov et al. [13] present a new 1-bit dual-polarized tiled transmit-array with beam-steering function for 5G communication systems. The configuration of a dual-polarized unit cell contains a pair of identical square-ring resonators via a pair of U-shaped feed loops. The proposed is highly flexible and adjustable for adding individual elements and using different feeding types. It offers 160 MHz impedance bandwidth from 5.67 to 5.83 GHz. An example of the proposed design with 10×10-element is presented, and its simulation and measurement performances are discussed in detail. The circuit model topology of the proposed beam-steerable transmit-array is discussed.

Ferreira-Gomes et al. [14] demonstrate a small-sized integrated metasurface antenna design with circular-polarized characteristics for 5G MM-Wave communications. Its configuration is composed of a chiral dielectric metasurface and a 2×2 arrangement of dielectric cylinders to reach optimized radiation patterns. The designed metasurface antenna is fed using the SIW-to-coax feeding technique. The proposed antenna operates from 25.3 to 31.6 GHz (22.6%) with 11.6% 3 dB axial ratio (AR). The investigated results show promising improvements in antenna gain, AR, and impedance bandwidth. The antenna has a planar structure and can be easily integrated into wireless systems.

Ramírez Arroyave et al. [15] developed a new design technique to minimize the nonlinear distortion and maximize power efficiency for an MM-Wave PIN diode in reconfigurable antennas. This is mainly achieved through the reduction of the mutual coupling between the internal switching and the external feeding ports obtained using a test-set with a nonlinear network analyser, on-wafer probe station, and a test fixture. The nonlinear models are extracted through X-parameter measurements and are validated by S-parameter in the low power signal regime and by harmonic measurements. A particular example of antenna design in 3.5 GHz is demonstrated to verify the minimum nonlinearities method.

Gaya et al. [16] proposed a novel feeding method for dielectric resonator antennas. In the proposed design, a ceramic-based dielectric resonator antenna is fed by a metallic patch structure via a cross-slot aperture on the back layer. Using the cross-slot aperture, the maximum power radiation along the broadside and the antenna gain are improved. In order to achieve the desired impedance match, the slot dimensions are optimized. The antenna is designed to operate at 26 GHz 5G band, and it offers high gain and efficiency characteristics over its operation band. These features make the design suitable to be used for indoor small cell applications.

Iqbal et al. [17] have developed a wideband half-mode substrate-integrated waveguide (HMSIW) filter for 5G front-end applications. In the proposed design, a semi-circular cavity resonator is applied to obtain a wide operation band of 3–7 GHz. In addition, in order to improve the out-of-band response, a U-shaped slot and a pair of L-shaped stubs are employed in the configuration of the design. Wide bandwidth, loss, planar structure, and ease of integration are some attractive features of the 5G filter design. In addition, it is better in terms of insertion loss when compared with other filters reported in the literature. The filter was fabricated on the RT/duroid 5880 substrate and exhibits good measurement results.

Luo et al. [18] present a new method to obtain continuous tuning without resonance blindness in the PIN diode. In this method, after achieving the equivalent impedance of the diodes, these nonlinear properties will be fitted with mathematical expressions. Using the mathematical equations, the PIN diode is modelled to be compatible with implementing co-simulation. Finally, in order to validate the usefulness of the proposed method, the co-simulation and experimental results are carried out at 5 GHz. In this approach, the active metamaterial structure contains two periodical cells with a PIN diode in each unit, inductance chips, and via holes between the top and bottom layers.

In their work, Dicandia and Genovesi [19] focused on improving the spectral efficiency of 5G massive MIMO systems using triangular lattice arrays. The authors investigated the beneficial effects of adopting a triangular lattice on phased arrays with regular and periodic grids. A particular study demonstrates the impact of the antenna array lattice at the system level. A planar array with 64 elements and an arrangement of 8×8 is analysed to better emphasize the advantages of the proposed method. The better performance achieved by the triangular lattice array makes it appealing for high-altitude platform station (HAPS) systems.

Yerrapragada and Kelley [20] propose a high-throughput wireless access framework (known as K-User MIMO) for beyond-5G or 6G networks. It drives the multi-user Shannon capacity formula to improve the spectral efficiency of the systems. The authors also define an OFDM frame structure to demonstrate the allocation of time-frequency resources for channel estimation. In addition, the performance of the proposed framework has been compared with related technologies. The introduced framework can cancel interference, demodulate, and maximize capacity through signal separation. It can also facilitate all-to-all communication between access points and mobile devices.

We would like to take this opportunity to appreciate and thank all authors for their excellent contributions and the reviewers for their fruitful comments and feedback. Special appreciation is also extended to the editorial board and editorial office of MDPI *Sensors* journal. We hope that readers will discover new and useful information on antenna design techniques for 5G and beyond applications.

Funding: This research received no external funding.

Conflicts of Interest: The authors declare no conflict of interest.

References

1. Wang, Y.; Li, J.; Huang, L.; Jing, Y.; Georgakopoulos, A.; Demestichas, P. 5G Mobile: Spectrum Broadening to Higher-Frequency Bands to Support High Data Rates. *IEEE Veh. Technol. Mag.* **2020**, *9*, 39–46. [CrossRef]
2. Nadeem, Q.; Kammoun, A.; Debbah, M.; Alouini, M. Design of 5G Full Dimension Massive MIMO Systems. *IEEE Trans. Commun.* **2018**, *66*, 726–740. [CrossRef]
3. Zhang, Z. *Antenna Design for Mobile Devices*; Wiley-IEEE Press: Hoboken, NJ, USA, 2011.
4. Zhang, J.; Ge, X.; Li, Q.; Guizani, M.; Zhang, Y. 5G Millimeter-Wave Antenna Array: Design and Challenges. *IEEE Wirel. Commun.* **2020**, *24*, 106–112. [CrossRef]
5. Jensen, M.; Wallace, J. A review of antennas and propagation for MIMO wireless communications. *IEEE Trans. Antennas Propag.* **2004**, *52*, 2810–2824. [CrossRef]
6. Seyyedesfahlan, M.; Uzun, A.K.; Skrivervik, A.; Tekin, I. Wideband Multiport Antennas. *Sensors* **2020**, *20*, 6960. [CrossRef] [PubMed]
7. Ojaroudi Parchin, N.; Jahanbakhsh Basherlou, H.; I.A. Al-Yasir, Y.; M. Abdulkhaleq, A.; A. Abd-Alhameed, R. Ultra-Wideband Diversity MIMO Antenna System for Future Mobile Handsets. *Sensors* **2020**, *20*, 2371. [CrossRef] [PubMed]
8. Ojaroudi Parchin, N.; Jahanbakhsh Basherlou, H.; Abd-Alhameed, R.A. Design of Multi-Mode Antenna Array for Use in Next-Generation Mobile Handsets. *Sensors* **2020**, *20*, 2447. [CrossRef] [PubMed]
9. Sharaf, M.H.; Zaki, A.I.; Hamad, R.K.; Omar, M.M.M. A Novel Dual-Band (38/60 GHz) Patch Antenna for 5G Mobile Handsets. *Sensors* **2020**, *20*, 2541. [CrossRef] [PubMed]
10. Dixit, A.S.; Kumar, S.; Urooj, S.; Malibari, A. A Highly Compact Antipodal Vivaldi Antenna Array for 5G Millimeter Wave Applications. *Sensors* **2021**, *21*, 2360. [CrossRef] [PubMed]
11. Song, C.; Pan, L.; Jiao, Y.; Jia, J. A High-Performance Transmitarray Antenna with Thin Metasurface for 5G Communication Based on PSO (Particle Swarm Optimization). *Sensors* **2020**, *20*, 4460. [CrossRef] [PubMed]
12. Hamid, S.A.; Rahman, N.H.A.; Yamada, Y.; Hung, P.V.; Quoc, D.N. Multibeam Characteristics of a Negative Refractive Index Shaped Lens. *Sensors* **2020**, *20*, 5703. [CrossRef] [PubMed]
13. Kozlov, D.; Munina, I.; Turalchuk, P.; Kirillov, V.; Shitvov, A.; Zelenchuk, D. Characterization of Tiled Architecture for C-Band 1-Bit Beam-Steering Transmitarray. *Sensors* **2021**, *21*, 1259. [CrossRef] [PubMed]
14. Ferreira-Gomes, B.; Oliveira, O.N., Jr.; Mejía-Salazar, J.R. Chiral Dielectric Metasurfaces for Highly Integrated, Broadband Circularly Polarized Antenna. *Sensors* **2021**, *21*, 2071. [CrossRef] [PubMed]
15. Ramírez Arroyave, G.A.; Barlabé, A.; Pradell, L.; Araque Quijano, J.L.; Cetiner, B.A.; Jofre-Roca, L. Design of Minimum Nonlinear Distortion Reconfigurable Antennas for Next-Generation Communication Systems. *Sensors* **2021**, *21*, 2557. [CrossRef] [PubMed]
16. Gaya, A.; Jamaluddin, M.H.; Ali, I.; Althuwayb, A.A. Circular Patch Fed Rectangular Dielectric Resonator Antenna with High Gain and High Efficiency for Millimeter Wave 5G Small Cell Applications. *Sensors* **2021**, *21*, 2694. [CrossRef] [PubMed]
17. Iqbal, A.; Jiat Tiang, J.; Kin Wong, S.; Alibakhshikenari, M.; Falcone, F.; Limiti, E. Multimode HMSIW-Based Bandpass Filter with Improved Selectivity for Fifth-Generation (5G) RF Front-Ends. *Sensors* **2020**, *20*, 7320. [CrossRef] [PubMed]
18. Luo, Y.; Liu, H.; He, Y.; Cui, H.; Yang, G. Continuous Resonance Tuning without Blindness by Applying Nonlinear Properties of PIN Diodes. *Sensors* **2021**, *21*, 2816. [CrossRef] [PubMed]
19. Dicandia, F.A.; Genovesi, S. Spectral Efficiency Improvement of 5G Massive MIMO Systems for High-Altitude Platform Stations by Using Triangular Lattice Arrays. *Sensors* **2021**, *21*, 3202. [CrossRef] [PubMed]
20. Yerrapragada, A.K.; Kelley, B. On the Application of K-User MIMO for 6G Enhanced Mobile Broadband. *Sensors* **2020**, *20*, 6252. [CrossRef] [PubMed]

sensors

MDPI

Article

Spectral Efficiency Improvement of 5G Massive MIMO Systems for High-Altitude Platform Stations by Using Triangular Lattice Arrays

Francesco Alessio Dicandia * and Simone Genovesi

Dipartimento di Ingegneria dell'Informazione, University of Pisa, 56122 Pisa, Italy; simone.genovesi@unipi.it
* Correspondence: alessio.dicandia@for.unipi.it

Abstract: The beneficial effects of adopting a triangular lattice on phased arrays with regular and periodic grids for high-altitude platform station (HAPS) systems are presented in the scenario of massive MIMO communications operating within the 5G NR n257 and n258 frequency bands. Assessment of a planar array with 64 elements (8×8) is provided for both a triangular lattice and a square one in terms of array gain, average sidelobe level (ASLL), and mutual coupling. Particular attention is devoted to illustrating the impact of the antenna array lattice at the system level by evaluating its significant merits, such as its spectral efficiency (SE) and signal-to-interference ratio (SIR). The better performance exhibited by the triangular lattice array in comparison to the square one makes it appealing for the 5G massive MIMO paradigm.

Keywords: phased array; massive MIMO; 5G; wideband array; triangular grid

Citation: Dicandia, F.A.; Genovesi, S. Spectral Efficiency Improvement of 5G Massive MIMO Systems for High-Altitude Platform Stations by Using Triangular Lattice Arrays. *Sensors* **2021**, *21*, 3202. https://doi.org/10.3390/s21093202

Academic Editors: Giuseppe Piro and Naser Ojaroudi Parchin

Received: 13 March 2021
Accepted: 3 May 2021
Published: 5 May 2021

Publisher's Note: MDPI stays neutral with regard to jurisdictional claims in published maps and institutional affiliations.

1. Introduction

The upcoming next generation (5G) of wireless communication networks is expected to drastically improve overall system performance, such as data throughput and energy efficiency [1,2]. Currently, in most wireless communication systems, the users inside a sector cell are served through a base station, whose radiation pattern consists of a fixed broad main beam. In this scenario, the wireless communications system turns out to be inefficient from the energy point of view, since most of the base station (BS)'s radiated signal propagates towards directions in which there are no users [3]. The exploitation of larger frequency bandwidths and the deployment of more BSs to reduce the cell area are adopted in order to tackle the ever-increasing data throughput. On the other hand, in upcoming 5G wireless technology, improvement of the data throughput is guaranteed mainly by massive multiple-input and multiple-output (MIMO) technology [4–6], which is capable of serving multiple users simultaneously within the same time–frequency resource, through a multibeam radiation pattern with a consequent increase in the spectral efficiency (SE) of the system [2]. The coverage of users inside a sector cell is achieved through the deployment of phased arrays with a massive number of antennas. These arrays are able to provide advanced beamforming and beam tracking to generate multiple concurrent beams that send the different streams of data allocated on the same time–frequency resource to separate users [7]. In addition to radiation pattern optimization, a large-frequency spectrum is pivotal for supporting communication links with increased data transmission rates between the base station and the users. For this reason, the limited available spectrum below 6 GHz no longer satisfies the system's needs; consequently, the millimeter-wave (mm-wave) band has recently drawn great attention for the next 5G wireless communications systems [8–11].

The design of massive MIMO phased array antennas represents one of the most challenging aspects to tackle in order to guarantee reliable performance. In fact, the array performance in terms of gain, beam steering, peak sidelobe level (PSLL), antenna mutual coupling, and inter-element distance directly affects the quality of the overall wireless communication. Some examples of antenna arrays for mm-wave communication are reported in Reference [12]. In Reference [13], a 28 GHz phased array composed of 64 elements (8×8) is described. Massive MIMO systems with 64 elements operating around 28 GHz and 40 GHz were designed with fully digital beamforming in References [14,15]. Dual-polarized phased array transceivers able to provide two concurrent independent beams, and hence double the channel capacity, have also been proposed [16,17].

However, some obstacles prevent the achievement of both seamless and ubiquitous wireless connectivity if only the terrestrial infrastructure is considered. In fact, terrestrial ground stations cannot be deployed in off-grid or inaccessible areas, such as rural zones, oceans, deserts, and generally harsh and remote environments. To this end, aerial wireless communication based on the employment of high-altitude platform stations (HAPSs) will play a paramount role in providing everywhere with access to the global network [18–20]. A HAPS consists of an unmanned aerial vehicle (UAV)—such as a gas-filled balloon, airship, or aircraft—operating in the stratosphere at an altitude of around 20 km [21] (Figure 1). HAPSs can be deployed in wireless communication networks with different topologies, within which they act mainly as aerial relays or aerial base stations to help improve the wireless communication [19]. In the former case, a HAPS collaborates with a ground BS by offering an alternative reliable link between it and a ground user by forwarding the data in case of a blockage between them. In the latter case, the HAPS acts as an aerial base station by offering wide wireless connectivity between ground users and the core network in the event of an inadequate terrestrial network or temporary ground station malfunction or maintenance. Moreover, thanks to their rapid deployment, HAPSs can assist in readily deploying communication networks after catastrophic events, such as earthquakes [22,23]. Furthermore, HAPSs can act as reliable relays between terrestrial users and CubeSats [24,25]—low Earth orbit (LEO) satellites—in order to form an airborne communication network (ACN) [26,27].

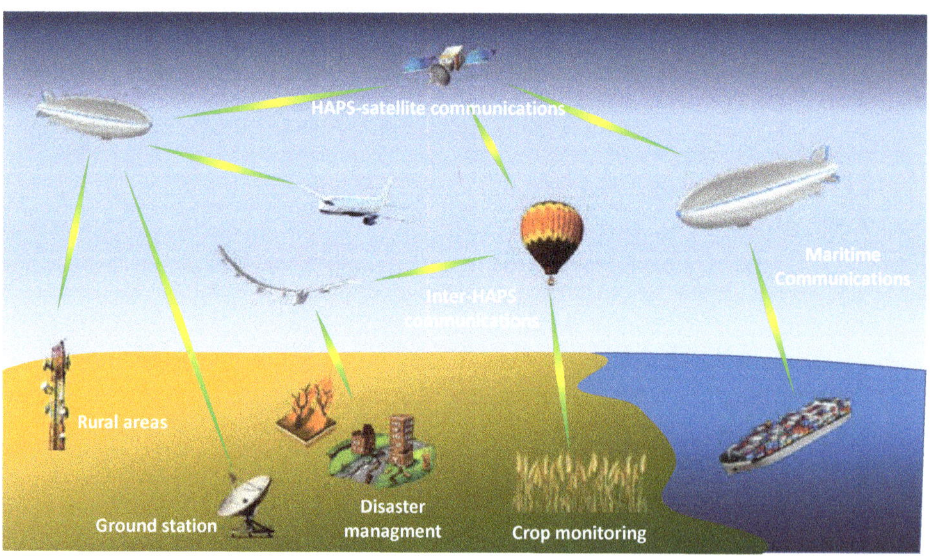

Figure 1. Examples of several scenarios where HAPSs can be exploited.

The antenna system certainly represents one of the most important factors for HAPSs where reliable performance is concerned [28]. In Reference [29] a multibeam lens antenna for a HAPS operating at L/S band is presented. A Ka-band phased array composed of

256 open-ended substrate-integrated square waveguides and a 4-channel beamformer circuit produced by Anokiwave was described in Reference [30].

In addition to the above-mentioned critical phased array aspects, another meaningful parameter that has to be accurately investigated for large phased arrays at mm-wave is the thermal aspect [31]. From the electromagnetic perspective, the most straightforward way of improving the cooling performance to dissipate heat is to increase the distance between the array elements. However, increasing the minimum distance between the single radiators too much can degrade the PSLL, with the appearance of grating lobes inside the visible region. This is detrimental to the signal-to-interference ratio (SIR) in massive MIMO systems with multiuser communication within the same time–frequency resource. Recently, a solution based on aperiodic antenna element arrangement capable of reducing the PSLL inside a defined sector cell for massive MIMO systems has been proposed in Reference [32]. However, the array aperiodicity considerably increases the design complexity of the feeding network, as well as complicating the array calibration.

Most of the designed phased arrays with uniform lattices rely on square or rectangular element grids, although a triangular lattice allows for an increase in the minimum antenna distance, avoiding the onset of grating lobes [33,34]. Few examples of triangular lattice 5G phased arrays are reported in the literature [35,36]. However, a comprehensive analysis addressing the overall performance of massive MIMO systems, including SE and SIR, has not been yet presented, although some recent studies have proven the advantages of triangular lattice arrays over square and rectangular ones for 5G massive MIMO system ground stations [37].

The previous mm-wave 5G phased antenna arrays were designed to operate in one specific band, even if there are different frequency bands allocated to 5G mm-wave worldwide. For this reason, it is advantageous to design phased arrays that can operate over a wide band in order to allow for multistandard operation or exploit interband carrier aggregation (CA) to enhance spectral efficiency [38].

In this paper, the beneficial effects of adopting a triangular lattice rather than a square one are described and tested on a phased array for HAPS communications within both the 5G New Radio (NR) n258 (24.25–27.5 GHz) and NR n257 (26.5–29.5 GHz) bands. A thorough analysis of the overall system performance—including array gain, average sidelobe level (ASLL), signal-to-interference ratio (SIR), and spectral efficiency (SSE)—is carried out in order to demonstrate the effects of the employed array lattice. Furthermore, the comparison to a planar array of 8×8 elements in terms of array gain, the minimum distance between elements, and ASLL is presented in Section 2. Section 3 is devoted to highlighting the superior robustness of triangular lattices for the antenna elements' impedance variation during beam steering inside the sector when compared to square lattices, via full-wave electromagnetic simulations. This represents a step forward with respect to Reference [37], since it better quantifies the effects of mutual coupling and matching efficiency (η_Γ). Massive MIMO metrics, such as SE and SIR, are evaluated in Section 4, whilst conclusions are summarized in Section 5.

2. Triangular vs. Square Lattice Planar Arrays

The radiation pattern (RP) generated by a planar array with $N \times M$ elements in the event of uniform amplitude excitation, and when neglecting the mutual coupling, is equal to [39]:

$$\begin{aligned} \mathrm{RP}(\theta,\phi) &= \sum_{n=0}^{N-1}\sum_{m=0}^{M-1} e^{j\varphi_{nm}} E_{nm}(\theta,\phi)\, e^{j\beta F} \\ F &= x_{nm}\sin(\theta)\cos(\phi) + y_{nm}\sin(\theta)\sin(\phi) \end{aligned} \tag{1}$$

where $\beta = 2\pi/\lambda_0$ is the phase constant; $E_{nm}(\theta,\phi)$ represents the elements' radiation pattern; and φ_{nm} represents the phase associated with the (n,m)-*th* element necessary to steer the main beam toward the desired direction (u_0, v_0), which depends on the element's position (x_{nm}, y_{nm}) within the employed lattice. The relation between the phase φ_{nm} and the element's position (x_{nm}, y_{nm}) is given by the following equation:

$$\begin{aligned} \varphi_{nm} &= \beta(x_{nm}u_0 + y_{nm}v_0) \\ u_0 &= \sin(\theta_0)\cos(\phi_0) \\ v_0 &= \sin(\theta_0)\sin(\phi_0) \end{aligned} \quad (2)$$

The minimum distance between elements (d_0) is a fundamental parameter for array radiation performance. Furthermore, d_0 should be selected in order to avoid the appearance of grating lobes within the visible region during the beam steering inside the angular sector selected to serve the users. In fact, high lateral lobes do not only provide a considerable reduction in the maximum array gain, but also produce interference for all other users served within the same time–frequency resource. In general, the maximum value of the minimum distance between the antenna elements depends on the maximum steering angle necessary to cover the area of interest.

The HAPS scenario differs from that of a 5G ground BS—whose coverage spans 30° in elevation and 120° in azimuth [37,40]—since a circular scan area turns out to be more appropriate [18]. Therefore, it is assumed to cover an angular sector with a maximum steering angle of 60° off broadside ($-60 \leq \theta_0 \leq 60°$) for all ϕ angles ($0 \leq \phi_0 \leq 180°$), as shown in Figure 2 in the u–v plane.

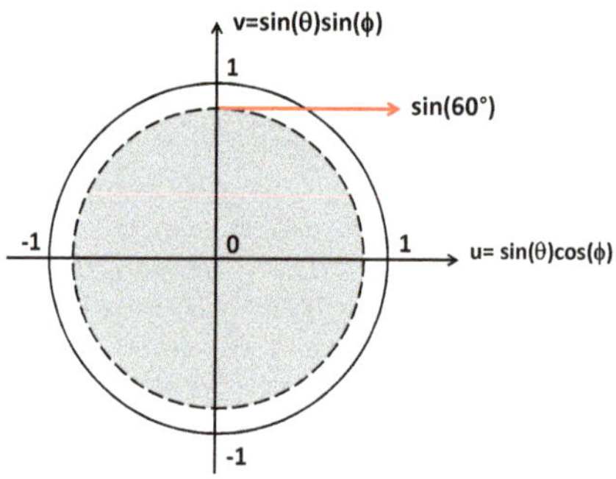

Figure 2. Circular sector cell in the u–v plane within which the HAPS antenna array has to steer the main beam to serve the 5G users.

As stated before, most of the designed phased arrays with uniform lattices rely upon a square arrangement of the antenna elements. In this case, the condition on d_0 for avoiding the appearance of grating lobes inside the visible region is given by:

$$d_0 \leq \frac{\lambda_{HF}}{R + \sin(\theta_{\max})} \quad (3)$$

where λ_{HF} represents the wavelength evaluated at the highest frequency (i.e., 29.5 GHz); $\theta_{\max}$ is the maximum antenna array steering angle, which is 60° for the considered circular angular sector shown in Figure 2; and R is a real number that represents the distance in the u–v plane between the closest grating lobes and the center of the visible region (u = 0, v = 0). Therefore, in order to guarantee the absence of undesirable high lobes inside the visible region, R has to be set to greater than 1.

In the case of an antenna array with uniform spacing but with a triangular grid, the spacing d_x and d_y—namely, the height and the base of the triangle (Figure 3)—are related to the γ angle by the following equations:

$$\begin{aligned} d_x &= D\sin(\gamma) \\ d_y &= 2D\cos(\gamma) \end{aligned} \quad (4)$$

where D is the distance between the (n,m)*-th* element and the $(n+1,m)$*-th*, as depicted in Figure 3.

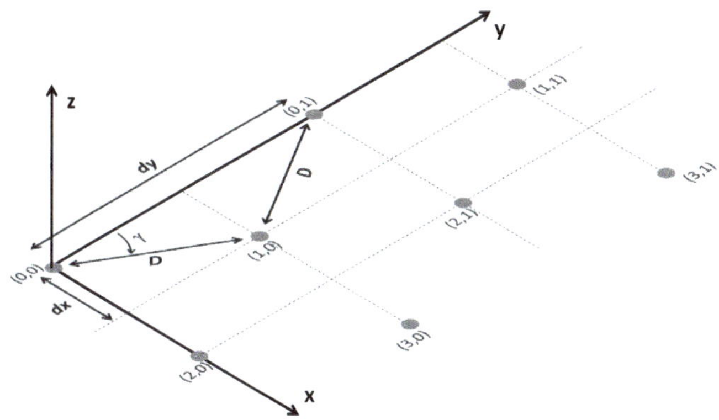

Figure 3. Planar antenna array with a triangular lattice.

Differently from a square lattice, there are different choices of grid spacing (d_x, d_y) in an array with a triangular grid, which are able to guarantee the absence of grating lobes. In this case the grating lobes are avoided if the following equations are satisfied:

$$\begin{aligned} \left(\frac{\lambda_{HE}}{2d_x} - u_0^*\right)^2 + \left(\frac{\lambda_{HE}}{d_y} - v_0^*\right)^2 \geq R^2 \\ \frac{2\lambda_{HE}}{d_y} - \sin(\theta_{\max}) \geq R \\ \frac{\lambda_{HE}}{d_x} - \sin(\theta_{\max}) \geq R \\ u_0^* = \sin(\theta_{\max})\cos(\gamma) \\ v_0^* = \sin(\theta_{\max})\sin(\gamma) \end{aligned} \quad (5)$$

where u_0^* and v_0^* represent the u–v plane coordinates where the grating lobes can appear. It is worthwhile to note that the previous grating lobes absence equations are only valid for equilateral or isosceles triangular lattices. Scalene lattices have not been considered since they provide a lower value for d_0. Equation (5) was used to understand the effect of γ on the element spacing in a circular scan sector. The elements' spacing (d_x, d_y), along with the minimum distance between them, were evaluated for γ within the interval [20°,80°] in the event of R = 1.1. The plot summarizing the results is shown in Figure 4.

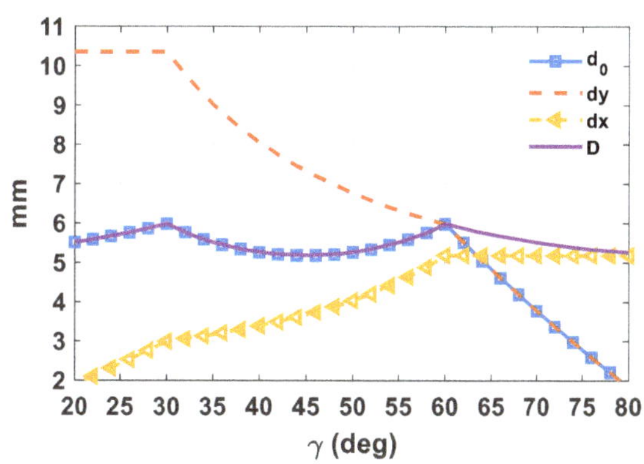

Figure 4. Elements' spacing (d_x, d_y), the antenna elements' distance (D), and its minimum (d_0) as a function of the γ value for a triangular lattice array in the event of R = 1.1.

It is apparent that the γ value has a considerable effect on d_0. Specifically, γ = 30° and γ = 60° provide the largest minimum distance between elements, which is equal to 5.97 mm (0.587 λ_{HF}) in the addressed scenario. Moreover, Figure 4 emphasizes how the triangular lattice provides the designer more degrees of freedom due to different possible combinations of element spacing (d_x, d_y). For example, let us consider the *RP* of an 8 × 8 array of isotropic elements (i.e., $E_{nm}(\theta,\phi) = 1$) when the main beam is steered along one of the ϕ directions where there are grating lobes and when θ = 60°. Figure 5 illustrates the outcomes for two values of γ, namely, 30° and 60°. It is apparent that the different antenna element spacing (Figure 4) determines a different pattern shape, although the minimum distance remains the same. In particular, in the event of γ = 30° (Figure 5a), the antenna array presents a superior spatial resolution along the v plane, whereas the equilateral triangular lattice (γ = 60°) provides an almost circular footprint in the u–v plane. Moreover, in both cases the grating lobe is evidently outside the visible region (red circle), located at a distance of R = 1.1 from the center (u = 0, v = 0), as expected. Therefore, according to the desired spatial resolution in the u–v plane, it is possible to select the proper γ value.

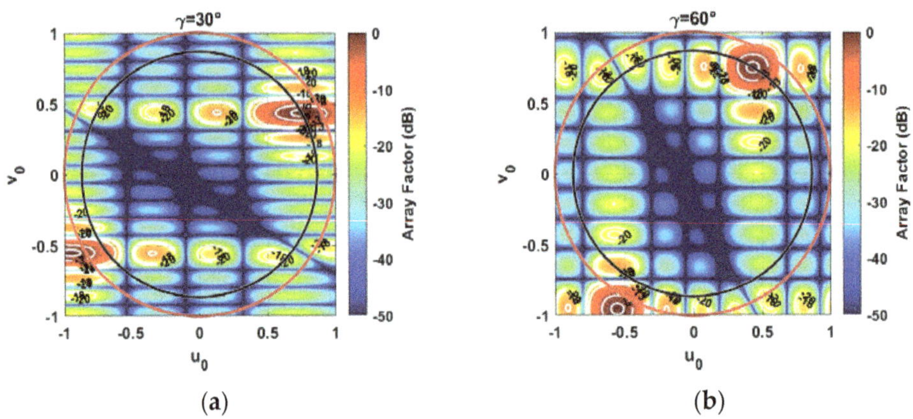

Figure 5. RP in the case of an antenna array with 8 × 8 isotropic elements arranged in a triangular lattice in the event of (**a**) γ = 30°, and (**b**) γ = 60°. The red circle represents the visible region, whereas the black circle highlights the circular sector cell.

In order to find the most suitable triangular lattice grid—and hence, the γ value—the angular average gain (η_{Gain}), namely, the mean value of the gain attained during the beam steering inside the all-circular sector, in the case of a planar array composed of 8 × 8 elements, was evaluated as a function of the γ value within the addressed bandwidth (24.25–29.5 GHz), and by considering the beam scan within the circular sector cell, as shown in Figure 2. An element pattern shape equal to $E(\theta,\phi) = cos(\theta)$, with a half-power beamwidth (HPBW) of 90°, was assumed as the element factor during the average gain evaluation.

From the color map of Figure 6 it can be concluded that η_{Gain} always increases with frequency, although it exhibits higher value fluctuations with respect to the γ value, especially in the lower band. Furthermore, η_{Gain} grows almost linearly up to γ = 30°, then gradually decreases with the increase of the γ value up to around γ = 45°, where a reversal of the trend occurs. Subsequently, the angular average gain reaches the second peak at γ = 60°, after which it quickly drops again. Therefore, by considering the minimum element spacing (Figure 4) and the angular average gain (Figure 6), it is possible to infer that with a circular sector cell there are two optimal γ values in cases of triangular grids, namely, γ = 30° and γ = 60°. In fact, these values ensure the largest minimum antenna element distance and, hence, the lowest performance degradation due to the elements' mutual coupling [41,42], in addition to having the highest angular average gain as a function of the frequency.

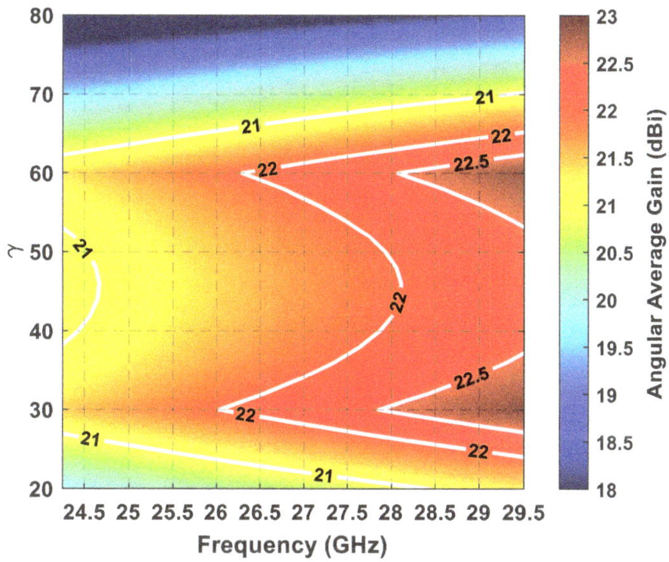

Figure 6. Angular average gain (η_{Gain}) as a function of the γ value within the working frequencies (24.25 –29.5 GHz) in the case of a triangular lattice planar array of 8 × 8 elements.

In view of highlighting the impact of the antenna array lattice, two 8 × 8 planar arrays were considered—the former with elements arranged on a square lattice, and the latter on an equilateral triangular ($\gamma = 60°$) grid. It is worth noting that, as stated before, in a triangular lattice placement of the elements there are two most suitable values, namely, $\gamma = 30°$ and $\gamma = 60°$. However, in the following performance comparison, it was decided to use an equilateral triangular lattice due to its more uniform footprint in the u–v plane, as shown in Figure 5b. By considering the Equations (3)–(5), with $R = 1.1$, for a square lattice d_0 turns out to be 5.17 mm (0.508 λ_{HF}), whereas the equilateral triangular grid exhibits a minimum element distance of 5.97 mm (0.587 λ_{HF}), hence offering a 13.5% improvement. It is worth observing that values of R greater than 1 assure the absence of grating lobes during the beam steering inside the circular sector [43]. The same R value for both lattices provides the same distance in the u–v plane between the closest grating lobes and the center of the visible region ($u = 0$, $v = 0$), and hence, the same PSLL value as a function of the beam steering inside the circular sector cell. Furthermore, PSLL is around −13.3 dB at the broadside, whereas with the increase in the θ steering angle, PSLL degrades up to around −9 dB for $\theta = 60°$ due to the cosine-shaped elements' radiation pattern. The larger minimum distance obtained via a triangular lattice arrangement allows us to achieve a lower level of mutual coupling among antenna elements by ensuring greater robustness of the active impedance of the array elements along the beam steering, a superior linearity of the employed power amplifiers (PAs), and a general improvement of the massive MIMO performance [41]. The wider element distance offered by the triangular lattice can be also considered advantageous for the thermal aspect, by helping the cooling system of the transceiver, which is critical in large mm-wave phased arrays [31]. The overlapped array geometry of triangular and square lattices is illustrated in Figure 7.

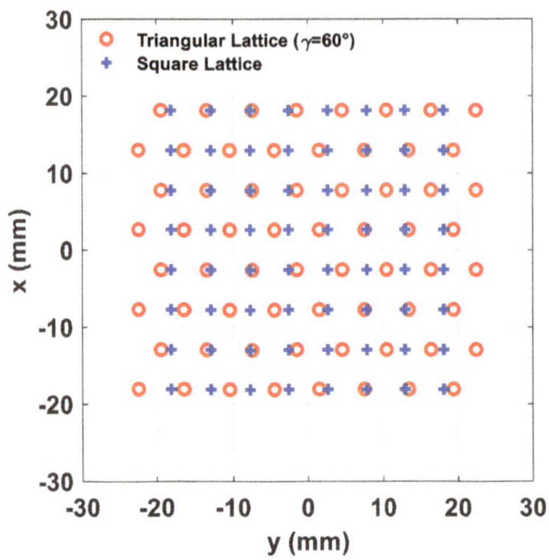

Figure 7. Array geometry of a planar array of 64 elements (8 × 8) in a triangular lattice (γ = 60°) and a square lattice.

The larger array area of the triangular element arrangement also guarantees a better angular resolution and, in a massive MIMO scenario, a reduced angular interval within which users cannot be spatially resolved [2].

The array gain as a function of the main beam direction (θ_0, ϕ_0) for both arrays of Figure 7 is illustrated in Figure 8 by considering an element pattern equal to $E(\theta,\phi) = cos(\theta)$. It can be noted that, although the employed array lattices present similar trends during beam steering, the triangular lattice outperforms the square one due to a higher array gain. Furthermore, the gain value presents the highest value along the broadside ($\theta_0 = 0°$), then decreases during the main beam steering due to beam widening and the approaching of the grating lobes inside the visible region [39]. For a more comprehensive analysis, the impact of the employed lattice on the array gain was also examined from a statistical point of view, by calculating the mean value (η_{Gain}), the variance ($\sigma^2{}_{Gain}$), the minimum value (min_{Gain}), and the maximum value (max_{Gain}). The results reported in Table 1 emphasize that the triangular lattice enables the attainment of an average linear array gain improvement of around 13 %, when compared to a square lattice. Moreover, as visible from the color maps of Figure 8, the use of a triangular lattice provides both a superior minimum value (min_{Gain}) and maximum value (max_{Gain}) to those of a square grid.

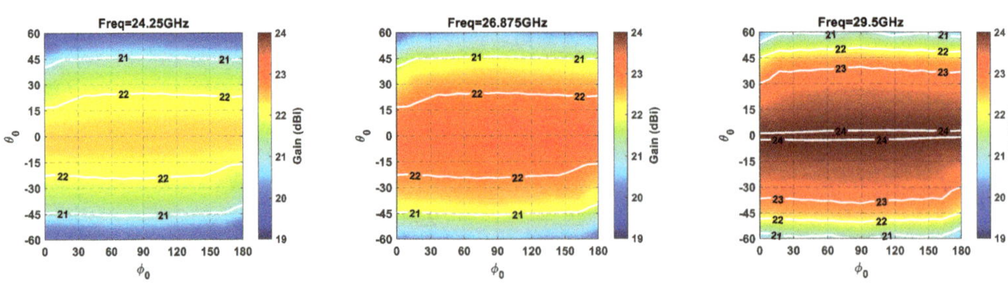

(a) (b) (c)

Figure 8. *Cont.*

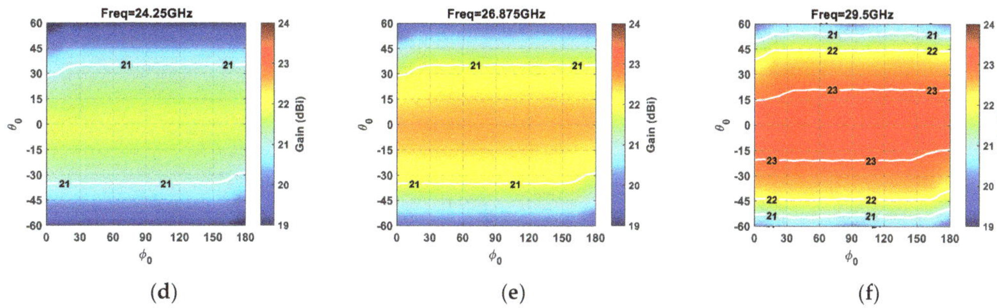

Figure 8. Gain in dBi of a planar array composed of 64 elements (8 × 8 array) as a function of the main beam steering inside the circular angular sector ($-60° \leq \theta_0 \leq 60°$, $0° \leq \phi_0 \leq 180°$). Equilateral triangular lattice at (**a**) 24.25 GHz, (**b**) 26.875 GHz, and (**c**) 29.5 GHz; square lattice at (**d**) 24.25 GHz, (**e**) 26.875 GHz, and (**f**) 29.5 GHz.

Table 1. Statistical comparison between the gain (dBi) of a square lattice and an equilateral triangular lattice for an 8 × 8 array evaluated within the circular scan sector.

	24.25 GHz		26.875 GHz		29.5 GHz	
	Tri	Squ	Tri	Squ	Tri	Squ
η_{Gain}	21.38	20.77	22.18	21.57	22.87	22.28
σ^2_{Gain}	0.89	0.83	0.97	0.9	1.12	0.99
min_{Gain}	19.17	18.73	19.81	19.38	20.24	19.83
max_{Gain}	22.42	21.78	23.26	22.61	24.03	23.37

Since the user's interference plays a key role in massive MIMO systems, the impact of the employed array lattice on the average sidelobe level (ASLL) was evaluated from a statistical point of view. For the ASLL evaluation, the addressed region was selected for each main beam pointing at a user by the following ellipse equation:

$$\begin{array}{c} \frac{(u-u_0)}{r_u^2} + \frac{(v-v_0)}{r_v^2} > 1 \\ r_u = 1.2\frac{\lambda}{L_x},\ r_v = 1.2\frac{\lambda}{L_y} \end{array} \quad (6)$$

where L_x and L_y represent the array side length along the x and y directions; λ is the wavelength; (u_0,v_0) the desired direction in the u–v plane in which to steer the main beam; and r_u and r_v identify the main beam in the u and v planes, respectively. Once the sidelobe region has been selected for a desired (u_0,v_0) direction, the ASLL is calculated by averaging all of the array's radiation pattern values inside the sidelobe region. Moreover, since the array can steer the main beam inside a circular area (Figure 2), we decided to evaluate the ASLL mean (η_{ASLL}), minimum (min_{ASLL}), and maximum (max_{ASLL}) values by considering different scan angles (N_θ = 15, N_ϕ = 91) uniformly distributed inside the circular sector through the following equations:

$$\begin{array}{c} \eta_{ASLL} = \frac{1}{N_\theta N_\phi} \sum_{i=1}^{N_\theta} \sum_{j=1}^{N_\phi} ASLL(i,j) \\ \min_{ASLL} = \min\{ASLL(i,j)\},\ i = 1,2,3,\ldots,N_\theta, j = 1,2,3,\ldots,N_\phi \\ \max_{ASLL} = \max\{ASLL(i,j)\},\ i = 1,2,3,\ldots,N_\theta, j = 1,2,3,\ldots,N_\phi \end{array} \quad (7)$$

where ASLL(*i*,*j*) represents the ASLL when the array main beam direction is toward the direction identified by (*i*,*j*). The calculated ASLL mean (η_{ASLL}), minimum (min_{ASLL}), and maximum (max_{ASLL}) values are illustrated in Figures 9 and 10. Specifically, Figure 9 highlights the ASLL statistical comparison in the event that the sidelobe region is represented by the intersection between Equation (6) and the u–v points inside the unit radio's circle, whereas the ASLL assessment of Figure 10 takes into account a sidelobe region represented by the intersection between Equation (6) and the u–v points inside the circular sector cell shown in Figure 2.

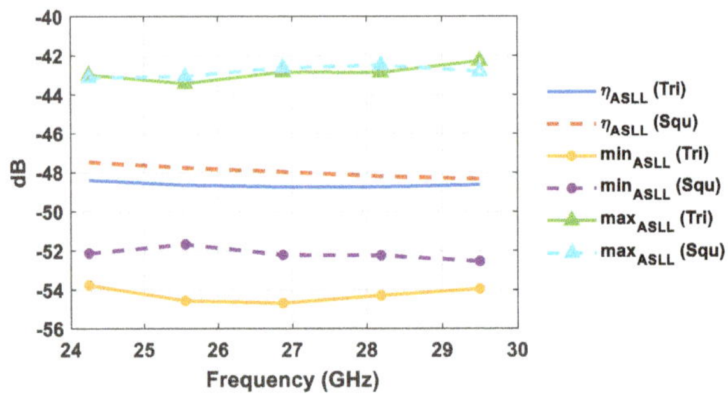

Figure 9. ASLL statistical comparison of square and equilateral triangular lattices by considering the whole visible region.

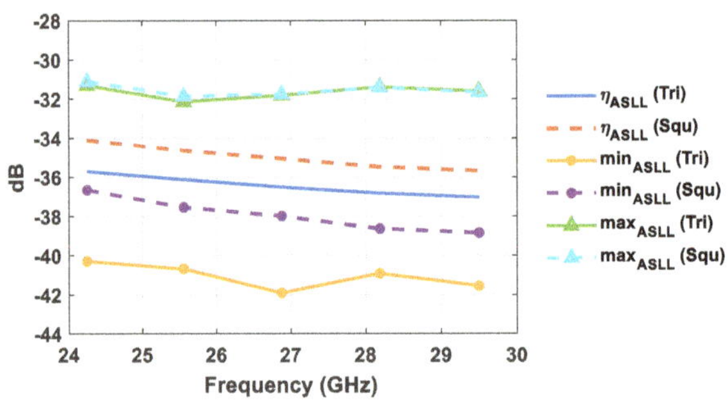

Figure 10. ASLL statistical comparison of square and equilateral triangular lattices by considering the circular sector cell.

By looking at Figures 9 and 10 it is evident that the use of triangular lattices in planar antenna arrays outperforms that of square ones from a statistical point of view, in terms of ASLL. The better performance of triangular lattices is highlighted by a lower ASLL mean value, a reduced minimum value, and a similar maximum value. Specifically, triangular lattices allow for a percentage reduction of the ASLL mean value (η_{ASLL}) of 1–2% within the desired bandwidth, compared to square lattices, when considering the whole visible region. The ASLL superiority of triangular lattices is further confirmed by taking into account only the investigated circular sector cell (Figure 10). Indeed, the ASLL mean value percentage reduction in triangular lattices compared to square lattices turns out to be between 3.5% and 4.5%. This feature is attractive for 5G massive MIMO systems, since it allows for a reduction in both the intra-cell and inter-cell interference. Since lateral lobes related to the main beam pointing at one user generate interference for all other users served within the same time–frequency resource, the lower ASLL guaranteed by a triangular lattice is appealing for use in a HAPS 5G massive MIMO system.

3. Phased Array Comparison of Triangular and Square Lattices

In this section, a planar array composed of 64 elements (8 × 8) is analyzed to better emphasize the advantages of using a triangular lattice. The full-wave electromagnetic simulations were carried out using Ansys HFSS [44]. Each single element consists of a square patch antenna printed on a 0.7 mm thick grounded dielectric layer (RO5880), fed through a coaxial cable. As mentioned in the previous section, the minimum antenna element spacing for a square lattice is 5.17 mm (0.508 λ_{HF}), whereas the equilateral triangular grid provides a minimum element distance of 5.97 mm (0.587 λ_{HF}).

Since antenna array design is beyond the scope of this paper, the lattice comparison as a function of frequency was pursued by tuning the antenna array elements to be matched in one narrowband frequency band at time. The finite array with 64 elements, with the antenna elements tuned to 29.5 GHz, is illustrated in Figure 11. Specifically, the x–z plane represents the E-plane of the array, whereas y–z represents the H-plane. At the beginning, the mutual coupling (S_{ij}) among antenna elements was addressed. The simulated mutual coupling average value (η_{Sij}) among antenna elements, and the maximum mutual coupling as a function of the frequency, are illustrated in Table 2. It can be seen that a triangular lattice enables us to considerably reduce both the mutual coupling average value and the maximum mutual coupling. For instance, the peak mutual coupling achieved at the highest frequency (29.5 GHz) with a square lattice (−14.4 dB) turns out to be worse than the peak mutual coupling achieved at the lowest frequency (24.25 GHz) for the triangular lattice planar array (−14.91 dB). In general, the mutual coupling difference between triangular and square lattices is more pronounced at the lowest frequencies, and then the coupling difference tends to reduce with the increase in frequency.

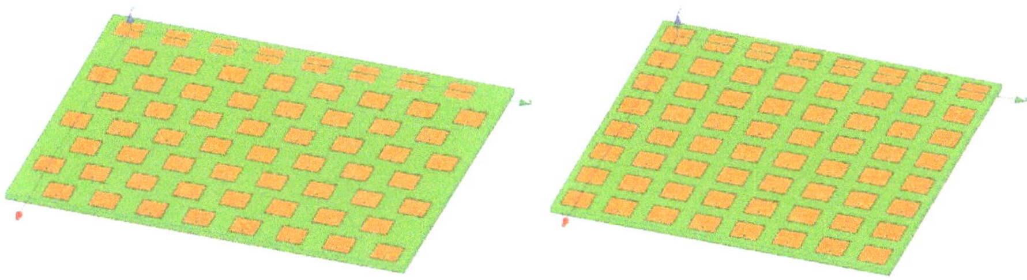

(**a**) (**b**)

Figure 11. Planar array with 64 elements arranged with (**a**) an equilateral triangular lattice; and (**b**) a square lattice.

Table 2. Simulated mutual coupling average values (η_{Sij}) among antenna elements and the maximum value comparison in dB between triangular and square lattice planar arrays composed of 64 elements (8 × 8).

	24.25 GHz		**26.875 GHz**		**29.5 GHz**	
	Tri	Squ	Tri	Squ	Tri	Squ
η_{Sij}	−48.6	−42.52	−49.35	−44.8	−50.74	−46.2
Max(S_{ij})	−14.91	−11.84	−15.62	−12.55	−16.9	−14.4

One of the detrimental effects of mutual coupling is the variation of the antenna elements' input impedance during beam steering. The impact of the employed array lattice on active S_{ii} parameters (i = 1, . . . ,64) was evaluated from a statistical point of view through the cumulative distribution function (CDF) of the active S_{ii} of all of the antenna array elements during the coverage. From Figure 12, a better statistical behavior of active S_{ii} parameters in triangular lattice than in square one is clearly recognizable. Indeed, the triangular lattice, thanks to a lower mutual coupling (Table 2), provides an average value of the active S_{ii} of around 13.8 dB within the investigated frequencies, whereas the square one is characterized by a mean value between −11.8 dB and −12.5 dB.

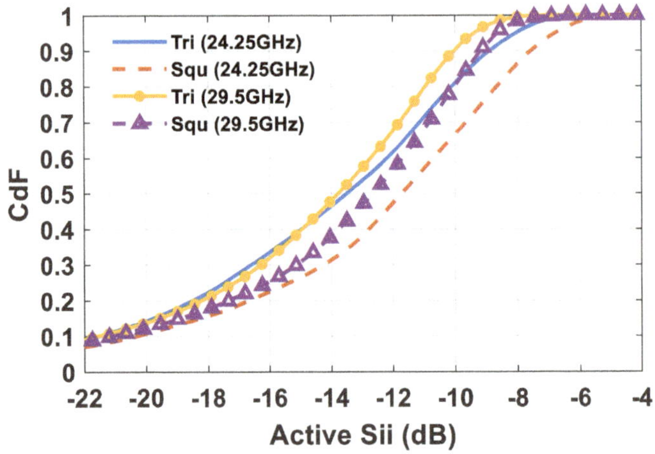

Figure 12. Active S_{ii} CDF for square and equilateral triangular lattice planar arrays composed of 64 elements at 24.25 and 29.5 GHz.

To further emphasize the superior robustness of the antenna elements' impedance variation when using a triangular lattice, the simulated array matching efficiency (η_Γ) was evaluated for all of the investigated steering angles by, using the following equation:

$$\eta_\Gamma = \frac{\sum_{i=1}^{64}\left(1 - |S_{ii}|^2\right)}{64} \tag{8}$$

The color maps of Figure 13 confirm the advantages of adopting triangular lattices in planar arrays, rather than square ones, in that they provide higher efficiency values for all investigated frequencies (24.25–29.5 GHz). Specifically, the triangular lattice provides a higher matching efficiency value than the square lattice for all of the investigated steering angles, except around ϕ = 90° plane (H-plane of the array) for θ angles greater than 40°. In particular, the triangular lattice shows a better matching efficiency in approximately 90% of the covered area within the addressed frequency band.

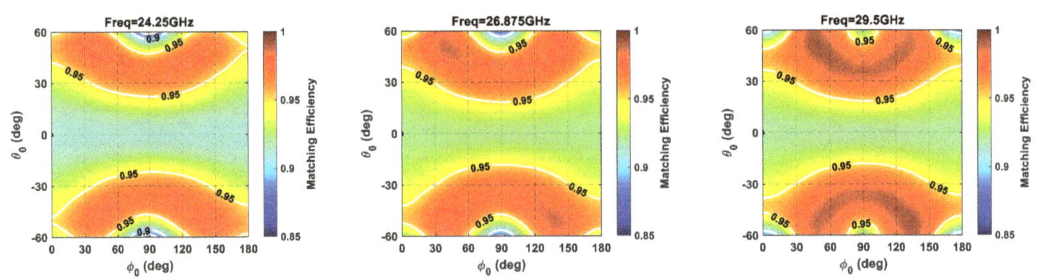

(a) (b) (c)

Figure 13. *Cont.*

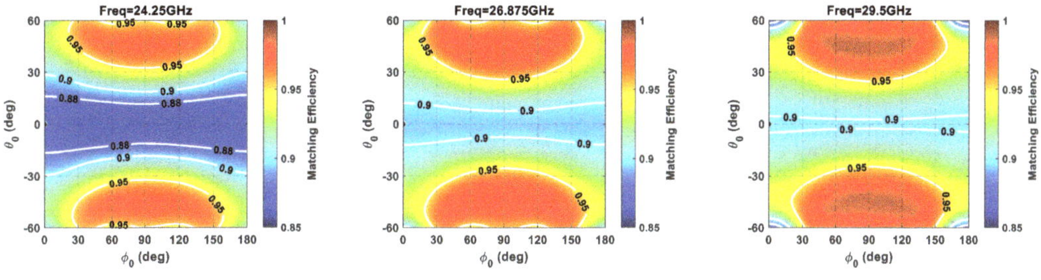

(d) (e) (f)

Figure 13. Matching efficiency (η_Γ) of a planar array composed of 64 elements (8 × 8) as a function of the main beam steering inside the circular angular sector ($-60^\circ \leq \theta_0 \leq 60^\circ, 0^\circ \leq \phi_0 \leq 180^\circ$). Equilateral triangular lattice at (**a**) 24.25 GHz, (**b**) 26.875 GHz, and (**c**) 29.5 GHz; square lattice at (**d**) 24.25 GHz, (**e**) 26.875 GHz, and (**f**) 29.5 GHz.

The main reason the triangular lattice's matching efficiency undergoes a value reduction in the principal planes for θ angles close to 60° at the highest frequency is due to scan blindness onset. This phenomenon can be observed through the active S_{11} parameter related to the array's central element at the highest frequency (29.5 GHz), as a function of the beam steering shown in Figure 14. To emphasize the scan blindness onset, the active S_{11} parameter was considered during beam steering within the whole visible region. More in detail, the color maps highlight that the triangular lattice does not avoid the scan blindness onset—characterized by a strong antenna element mismatch—but only reduces it by pushing the involved angular sector a bit further [43]. In fact, by considering the simulated phased array in the event of beam steering, the square lattice provides an active S_{11} higher than –10dB within the angular ranges $|\theta_0| > 60^\circ$, $\phi_0 < 60^\circ$, and $\phi_0 > 120^\circ$ (Figure 14b). On the other hand, the triangular lattice presents a milder mismatch of the central element for $|\theta_0| > 60^\circ$, $\phi_0 < 30^\circ$, $\phi_0 > 150^\circ$, and $80^\circ < \phi_0 < 100^\circ$ (Figure 14a). Moreover, by considering the angular sector with $|\theta_0| > 60^\circ$—hence, outside the desired scan angle area (highlighted by dashed red lines)—the triangular lattice presents an S_{11} central array element higher than −10dB in 40 % of cases, whereas the square lattice does so in 50 % of cases.

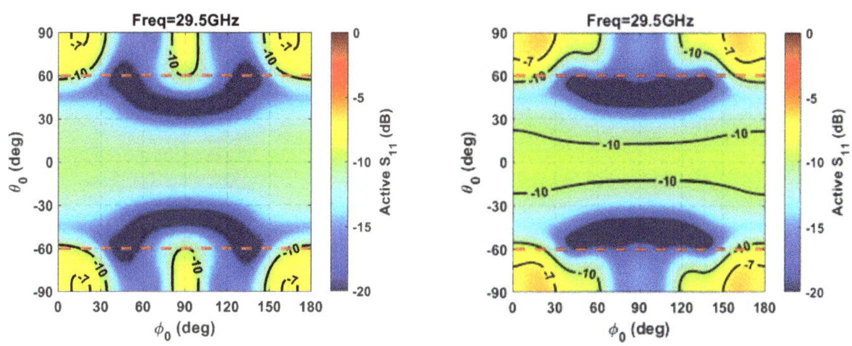

(a) (b)

Figure 14. Active S_{11} parameter related to the central array element as a function of beam steering for a planar array composed of 64 elements with (**a**) a triangular lattice; and (**b**) a square lattice.

With the aim of evaluating the impact of the antenna array lattice on the radiation pattern shape, the realized gain as a function of the θ angle at a broadside direction is reported in Figure 15.

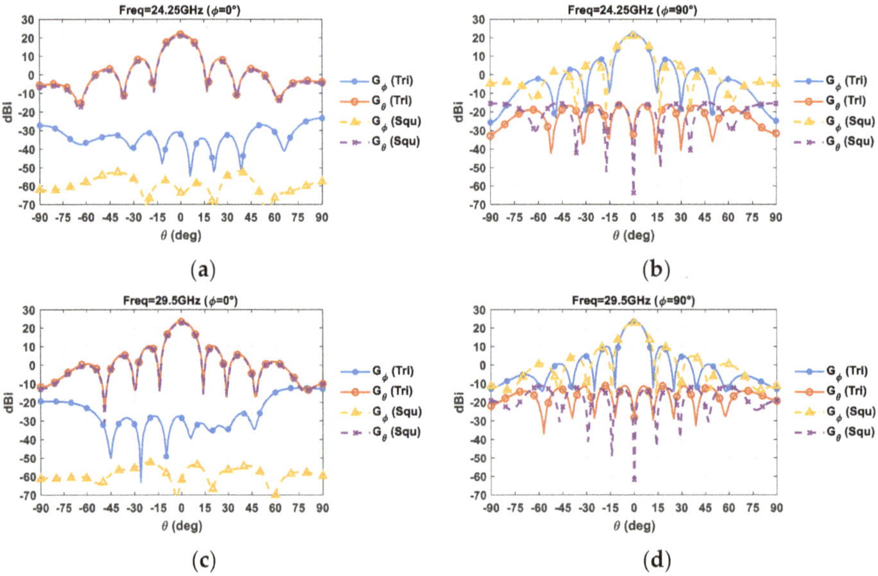

Figure 15. Simulated realized gain for a planar array composed of 64 elements at a broadside direction: (**a**) 24.25 GHz $\phi = 0°$; (**b**) 24.25 GHz $\phi = 90°$; (**c**) 29.5 GHz $\phi = 0°$; and (**d**) 29.5 GHz $\phi = 90°$.

Figure 15 confirms that a triangular lattice provides a narrower main beam in the $\phi = 90°$ plane, thus providing a higher maximum gain and better angular resolution. Moreover, a triangular lattice leads to comparable cross-polar levels to square one in the $\phi = 90°$ plane, whereas it introduces a slight degradation in the E-plane of the array (Figure 15a,c)—although a cross-polar level of about 50 dB around the mean beam is still guaranteed.

4. Massive MIMO Performance Evaluation

The comparison of triangular and square lattices was carried out in terms of array gain, ASLL, active S_{ii} parameters, and matching efficiency, by considering a circular sector in which to serve users. However, one of the key factors that will allow us to drastically improve the data throughput of the upcoming 5G wireless technology will be the use of massive MIMO technology capable of serving different users within the same time–frequency resource through multibeam radiation patterns. In general, various beamforming methods are available for the achievement of a multibeam system [10,45,46].

Let us consider a HAPS equipped with an array of $M = 64$ (8×8) antenna elements that serves K concurrent ground users, equipped with a single isotropic antenna, located inside the above-mentioned circular angular sector (Figure 2), through a multibeam radiation pattern in a line-of-sight (LOS) scenario. The assumption of a LOS channel is consistent since, at mm-wave, the LOS ray represents the predominant mode of propagation between the base station (BS) and the users, due to large path loss as well as the use of high-gain antennas [47,48]. Furthermore, the LOS channel condition turns out to be further emphasized in the case of HAPS wireless communication [49].

The received signal-to-interference-plus-noise ratio (SINR) for the nth user in case of a multi-user scenario can be written as [40,50,51]:

$$SINR_n = \frac{\delta \eta_\Gamma(\theta_n, \phi_n) G(\theta_n, \phi_n)}{\delta \sum_{i=1}^{K} G(\theta_i, \phi_i) |h_i * W_n|^2 + 1} \tag{9}$$

where $G(\theta_n,\phi_n)$ represents the HAPS array's gain toward the nth user located at (θ_n,ϕ_n); δ represents the ratio between the transmitted power (P_t) and the noise power (σ_0); $h_n \in 1\times M$ (n = 1,2,..,K) corresponds to the channel propagation between the nth user and the HAPS antenna array; and $w_n \in M\times 1$ consists of the precoding vector that, in general, depends on the selected beamforming method. It is worth noting that, differently from Reference [37], the simulated matching efficiency (η_Γ) was included in (9) in order to achieve a more accurate SINR estimate. To evaluate the system quality of service, a maximum ratio (MR) precoding technique [2] with a perfect channel state information (CSI) was assumed in the following analysis. However, it is worth noting that more efficient precoding and combining algorithms able to reduce interference could be exploited [52] but, in general, they require more complexity and turn out to be more sensitive to channel estimation error. Additionally, since array elements' mutual coupling (MC) undermines the MIMO performance [53] due to the unwanted correlation among elements, the channel matrix H is modelled as [54]:

$$H = H_0\left(\mathrm{I}_{mxm} - SS^H\right) \tag{10}$$

where $H_0 \in K \times M$ denotes the complex channel gain among the users and the HAPS antenna elements in the absence of antenna mutual coupling; $I_{m\times m}$ is an identity matrix; and $S \in M\times M$ corresponds to the scattering parameters matrix of the HAPS planar array. Unlike [37], the channel matrix was evaluated by taking into account in (10) the full-wave simulation of the planar array, and not just the antenna element distance. Once the SINR is known, the maximum achievable bitrate over 1 Hz of bandwidth for the nth user, which is the spectral efficiency (SE) per user, is:

$$SE_n = \log_2(1 + SINR_n) \tag{11}$$

For the SE assessment both the simulated array gain and the S-matrix of the planar arrays shown in Figure 11 were used. Furthermore, 10,000 sets of K-concurrent users randomly distributed inside the circular sector were adopted. It is worth observing that a smart user's selection inside the sector can improve the massive MIMO performance and the energy efficiency of the wireless communication [55]. However, this aspect was neglected, since the main goal of the paper is to highlight the performance differences between square and triangular lattices. The achievable SE as a function of the δ value, which is the ratio between the transmitted power (P_t) and the noise power (σ_0), is reported in Figure 16 for both square and triangular lattices in the event of eight simultaneous users inside the investigated circular sector.

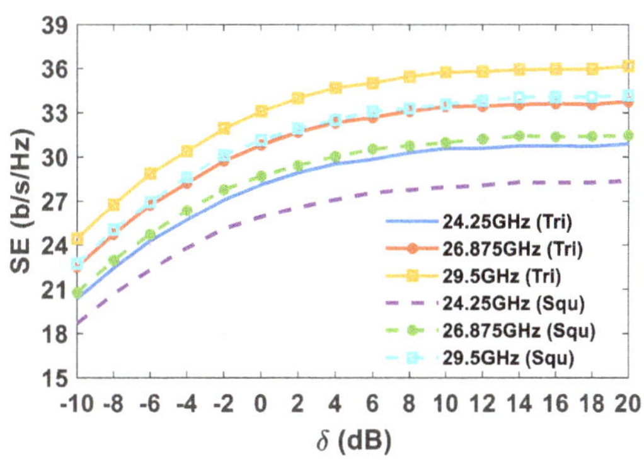

Figure 16. SE comparison for a planar array composed of 8 × 8 elements, in cases of square and triangular lattices, as a function of the δ ratio in dB in the event of 8 concurrent users.

It can be seen that the triangular lattice deployment of the array antenna elements allows for the achievement of a considerable improvement in the SE. The SE starts to grow almost linearly with the increase in the δ value, after which it reaches a superior limit due to the SIR, and the system becomes interference limited. Therefore, once the upper SE boundary is reached, further increasing the transmitted power does not represent an efficient way of improving the SE. The improved SE in the triangular lattice array is apparent from looking at the statistical behavior of the CDF of the SIR (Figure 17), and both the average SIR (η_{SIR}) and 90% of the SIR occurrence ($SIR_{90\%}$), as highlighted in Table 3.

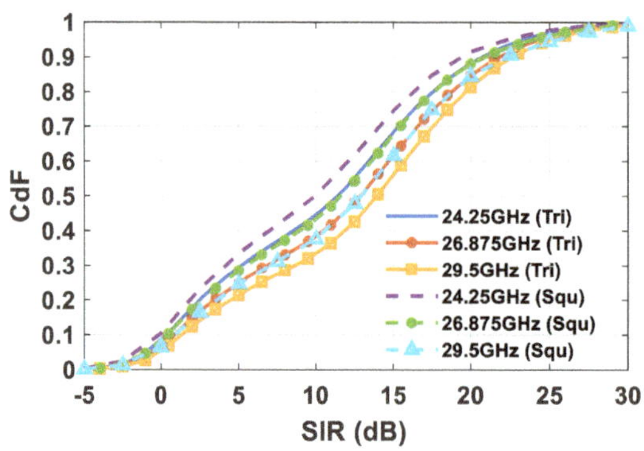

Figure 17. SIR CDF comparison for a planar array composed of 8x8 elements, in cases of square and triangular lattices, in the event of 8 concurrent users.

Table 3. Simulated average SIR (η_{SIR}) in dB and $SIR_{90\%}$ across triangular and square lattice planar arrays composed of 64 elements (8 × 8), in the event of 8 served users inside the circular sector.

	24.25 GHz		26.875 GHz		29.5 GHz	
	Tri	Squ	Tri	Squ	Tri	Squ
η_{SIR}	11.1	10	12	11.25	13.3	12.2
$SIR_{90\%}$	0.4	0	0.9	0.5	1.4	1

Since to improve the SE it is better to serve more users inside the predefined sector cell rather than increasing the transmitted power, the SE is plotted as a function of the number of users (K) for both triangular and square lattices in the event of a δ = 20 dB (Figure 18).

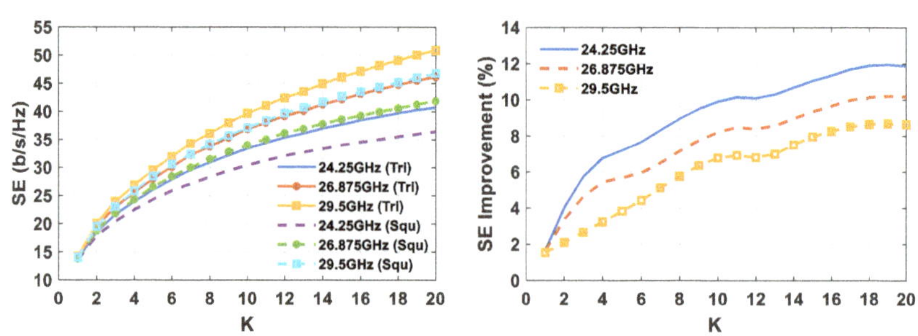

(a) (b)

Figure 18. (**a**) SE comparison as a function of the number of users (K) between square and triangular lattice planar arrays composed of 64 elements; and (**b**) percentage SE improvement in the event of δ = 20 dB.

The remarkable advantage of using a triangular lattice arrangement of the array elements instead of a square one is confirmed by observing the SE in Figure 18a. To emphasize the SE comparison, the percentage SE improvement is calculated and reported in Figure 18b. As can be seen, the achievable SE between the two lattices turns out to be comparable for few users, even if the triangular lattice's SE is somewhat higher due to a superior array gain value (Figure 8). Conversely, with the increase in the number of concurrent users, the SE improvement guaranteed by the triangular lattice array grows continuously, and reaches a value between 8.5% and 12 % in the event of 20 simultaneous users inside the sector.

In view of highlighting the advantageous impact of the triangular lattice on planar arrays in massive MIMO systems, the percentage SE improvement of the most suitable triangular lattice, compared to rectangular or square lattices, is reported in Table 4 for a case of 20 concurrent users deployed inside the different sector cells. Table 4 confirms that, although with some performance differences, the triangular lattice turns out to be the most efficient array grid regardless of the sector cell or the adopted frequency within the considered range. It is worth observing that, in practice, a minimum user distance within the same time–frequency resource is required in order to avoid high SIR and, hence, increase the SE. However, SE improvement between triangular and square lattices remains substantially unchanged. A further analysis aimed at understanding the role of array gain and the elements' mutual coupling (MC) was performed. The SE improvement between triangular and square lattices, as shown in Figure 18b, was recalculated under the hypothesis of having the same array gain (or the same received signal) for both of the employed lattices (20 dBi for all the frequencies), and in the case of an ideal array without MC (w/o MC).

Table 4. Percentage SE improvement offered by the most suitable triangular lattice, compared to rectangular/square lattices, in the case of planar arrays composed of 64 elements (8 × 8) for 20 served users inside the different sector cells.

	Sector	Lattice	SE Improvement	
			24.25 GHz	29.5 GHz
[37]	Rectangular	$\gamma = 36.6°$ vs Rect	14 %	10 %
This Paper	Circular	$\gamma = 60°$ vs Squ	12 %	8.5 %

Figure 19a highlights how the higher array gain values, as a function of the steering angles obtained in the case of the triangular lattice planar array (Figure 8), allow for an increase in the SE for just a few users (up to three). In fact, the SE improvement with the same array gain in the case of just a single user is equal to zero at all frequencies, whereas by increasing the number of concurrent users the curves are overlapped to the case of different gain values. Therefore, the higher SE provided by the triangular lattice in the event of many users is due to the better interference robustness guaranteed by the superior angular resolution. The SE improvement comparison of Figure 19b emphasizes how the MC effects contribute to a further increase in the percentage SE improvement of the triangular lattice compared to the square lattice, mainly at the lowest frequency (24.25 GHz). Conversely, the MC effect on percentage SE improvement gradually vanishes as the frequency increases, and becomes irrelevant at the highest frequency (29.5 GHz).

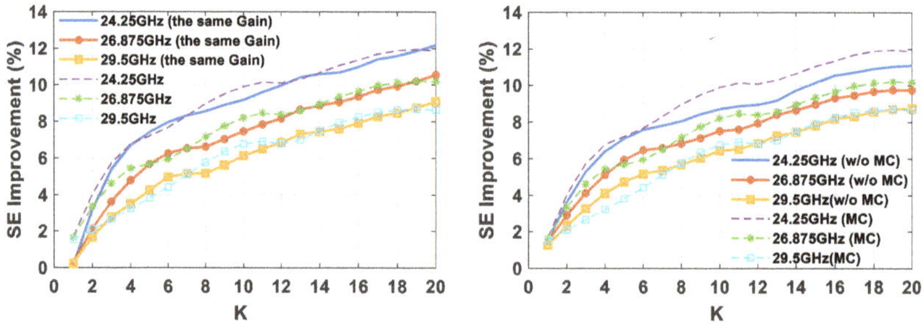

(a) (b)

Figure 19. Percentage SE improvement as a function of the number of users (*K*) between square and triangular lattice planar arrays composed of 64 elements in the event of (**a**) the same HAPS array gain for all of the frequencies for both lattices; and (**b**) an ideal planar array equipped with 64 elements without MC (w/o MC).

5. Conclusions

An extensive analysis of HAPS massive MIMO systems employing triangular lattices has been presented within the 5G NR n257 and n258 frequency bands (24.25–29.5 GHz). The noteworthy massive MIMO performance improvement when adopting a triangular lattice arrangement of the array elements instead of a square one in a planar array with a regular lattice has been highlighted. Specifically, a triangular lattice allows for the achievement of a superior array gain and ASLL reduction, as well as greater robustness of the antenna elements' impedance variation during beam steering by exploiting lower MC levels. Moreover, the larger minimum distance between elements in the case of a triangular grid guarantees a better angular resolution that, in a massive MIMO scenario, provides superior interference robustness and, hence, higher SE. The advantages offered by the regular and periodic triangular lattice arrangement of antenna elements make it appealing for 5G massive MIMO applications.

Author Contributions: Conceptualization, F.A.D.; validation, F.A.D. and S.G.; investigation, F.A.D. and S.G.; writing—original draft preparation, F.A.D.; writing—review and editing, F.A.D. and S.G. All authors have read and agreed to the published version of the manuscript.

Funding: Work partially supported by the Italian Ministry of Education and Research (MIUR) in the framework of the CrossLab project (Departments of Excellence).

Institutional Review Board Statement: Not applicable.

Informed Consent Statement: Not applicable.

Data Availability Statement: Not applicable.

Conflicts of Interest: The authors declare no conflict of interest.

References

1. Larsson, E.G.; Edfors, O.; Tufvesson, F.; Marzetta, T.L. Massive MIMO for next Generation Wireless Systems. *IEEE Commun. Mag.* **2014**, *52*, 186–195. [CrossRef]
2. Björnson, E.; Hoydis, J.; Sanguinetti, L. Massive MIMO Networks: Spectral, Energy, and Hardware Efficiency. *Found. Trends®Signal Process.* **2017**, *11*, 154–655. [CrossRef]
3. *Millimeter-Wave Circuits for 5G and Radar*, 1st ed.; Hueber, G.; Niknejad, A.M. (Eds.) Cambridge University Press: Cambridge, UK, 2019; ISBN 978-1-108-68639-6.
4. Lu, L.; Li, G.Y.; Swindlehurst, A.L.; Ashikhmin, A.; Zhang, R. An Overview of Massive MIMO: Benefits and Challenges. *IEEE J. Sel. Top. Signal Process.* **2014**, *8*, 742–758. [CrossRef]
5. Dicandia, F.A.; Genovesi, S.; Monorchio, A. Analysis of the Performance Enhancement of MIMO Systems Employing Circular Polarization. *IEEE Trans. Antennas Propag.* **2017**, *65*, 4824–4835. [CrossRef]

6. Dicandia, F.A.; Genovesi, S.; Monorchio, A. Circularly Polarized MIMO Antennas for Wireless LAN Applications. In Proceedings of the 2016 IEEE International Symposium on Antennas and Propagation (APSURSI), Fajardo, PR, USA, 26 June–1 July 2016; pp. 59–60.
7. Hong, W.; Jiang, Z.H.; Yu, C.; Zhou, J.; Chen, P.; Yu, Z.; Zhang, H.; Yang, B.; Pang, X.; Jiang, M.; et al. Multibeam Antenna Technologies for 5G Wireless Communications. *IEEE Trans. Antennas Propag.* **2017**, *65*, 6231–6249. [CrossRef]
8. Helander, J.; Zhao, K.; Ying, Z.; Sjöberg, D. Performance Analysis of Millimeter-Wave Phased Array Antennas in Cellular Handsets. *IEEE Antennas Wirel. Propag. Lett.* **2016**, *15*, 504–507. [CrossRef]
9. Rappaport, T.S.; Sun, S.; Mayzus, R.; Zhao, H.; Azar, Y.; Wang, K.; Wong, G.N.; Schulz, J.K.; Samimi, M.; Gutierrez, F. Millimeter Wave Mobile Communications for 5G Cellular: It Will Work! *IEEE Access* **2013**, *1*, 335–349. [CrossRef]
10. Roh, W.; Seol, J.-Y.; Park, J.; Lee, B.; Lee, J.; Kim, Y.; Cho, J.; Cheun, K.; Aryanfar, F. Millimeter-Wave Beamforming as an Enabling Technology for 5G Cellular Communications: Theoretical Feasibility and Prototype Results. *Commun. Mag. IEEE* **2014**, *52*, 106–113. [CrossRef]
11. Zhang, J.; Ge, X.; Li, Q.; Guizani, M.; Zhang, Y. 5G Millimeter-Wave Antenna Array: Design and Challenges. *IEEE Wirel. Commun.* **2017**, *24*, 106–112. [CrossRef]
12. Naqvi, A.H.; Lim, S. Review of Recent Phased Arrays for Millimeter-Wave Wireless Communication. *Sensors* **2018**, *18*, 3194. [CrossRef] [PubMed]
13. Kibaroglu, K.; Sayginer, M.; Phelps, T.; Rebeiz, G.M. A 64-Element 28-GHz Phased-Array Transceiver With 52-DBm EIRP and 8–12-Gb/s 5G Link at 300 Meters Without Any Calibration. *IEEE Trans. Microw. Theory Tech.* **2018**, *66*, 5796–5811. [CrossRef]
14. Yang, B.; Yu, Z.; Lan, J.; Zhang, R.; Zhou, J.; Hong, W. Digital Beamforming-Based Massive MIMO Transceiver for 5G Millimeter-Wave Communications. *IEEE Trans. Microw. Theory Tech.* **2018**, *66*, 3403–3418. [CrossRef]
15. Kuai, L.; Chen, J.; Jiang, Z.H.; Yu, C.; Guo, C.; Yu, Y.; Zhou, H.; Hong, W. A N260 Band 64 Channel Millimeter Wave Full-Digital Multi-Beam Array for 5G Massive MIMO Applications. *IEEE Access* **2020**, *8*, 47640–47653. [CrossRef]
16. Gu, X.; Liu, D.; Baks, C.; Tageman, O.; Sadhu, B.; Hallin, J.; Rexberg, L.; Parida, P.; Kwark, Y.; Valdes-Garcia, A. Development, Implementation, and Characterization of a 64-Element Dual-Polarized Phased-Array Antenna Module for 28-GHz High-Speed Data Communications. *IEEE Trans. Microw. Theory Tech.* **2019**, *67*, 2975–2984. [CrossRef]
17. Sadhu, B.; Tousi, Y.; Hallin, J.; Sahl, S.; Reynolds, S.K.; Renstrom, O.; Sjogren, K.; Haapalahti, O.; Mazor, N.; Bokinge, B.; et al. A 28-GHz 32-Element TRX Phased-Array IC With Concurrent Dual-Polarized Operation and Orthogonal Phase and Gain Control for 5G Communications. *IEEE J. Solid-State Circuits* **2017**, *52*, 3373–3391. [CrossRef]
18. Shibata, Y.; Kanazawa, N.; Konishi, M.; Hoshino, K.; Ohta, Y.; Nagate, A. System Design of Gigabit HAPS Mobile Communications. *IEEE Access* **2020**, *8*, 157995–158007. [CrossRef]
19. Arum, S.C.; Grace, D.; Mitchell, P.D. A Review of Wireless Communication Using High-Altitude Platforms for Extended Coverage and Capacity. *Comput. Commun.* **2020**, *157*, 232–256. [CrossRef]
20. Qiu, J.; Grace, D.; Ding, G.; Zakaria, M.D.; Wu, Q. Air-Ground Heterogeneous Networks for 5G and Beyond via Integrating High and Low Altitude Platforms. *IEEE Wirel. Commun.* **2019**, *26*, 140–148. [CrossRef]
21. Widiawan, A.K.; Tafazolli, R. High Altitude Platform Station (HAPS): A Review of New Infrastructure Development for Future Wireless Communications. *Wirel. Pers. Commun.* **2007**, *42*, 387–404. [CrossRef]
22. Alam, M.S.; Kurt, G.K.; Yanikomeroglu, H.; Zhu, P.; Đào, N.D. High Altitude Platform Station Based Super Macro Base Station Constellations. *IEEE Commun. Mag.* **2021**, *59*, 103–109. [CrossRef]
23. Deaton, J. High Altitude Platforms for Disaster Recovery: Capabilities, Strategies, and Techniques for Emergency Telecommunications. *EURASIP J. Wirel. Commun. Netw.* **2008**, *2008*. [CrossRef]
24. Dicandia, F.A.; Genovesi, S. A Compact CubeSat Antenna With Beamsteering Capability and Polarization Agility: Characteristic Modes Theory for Breakthrough Antenna Design. *IEEE Antennas Propag. Mag.* **2020**, *62*, 82–93. [CrossRef]
25. Dicandia, F.A.; Genovesi, S. Characteristic Modes Analysis of Non-Uniform Metasurface Superstrate for Nanosatellite Antenna Design. *IEEE Access* **2020**, *8*, 176050–176061. [CrossRef]
26. Cao, X.; Yang, P.; Alzenad, M.; Xi, X.; Wu, D.; Yanikomeroglu, H. Airborne Communication Networks: A Survey. *IEEE J. Sel. Areas Commun.* **2018**, *36*, 1907–1926. [CrossRef]
27. Curzi, G.; Modenini, D.; Tortora, P. Large Constellations of Small Satellites: A Survey of Near Future Challenges and Missions. *Aerospace* **2020**, *7*, 133. [CrossRef]
28. Karapantazis, S.; Pavlidou, F. Broadband Communications via High-Altitude Platforms: A Survey. *IEEE Commun. Surv. Tutor.* **2005**, *7*, 2–31. [CrossRef]
29. Cai, R.; Yang, M.; Zhang, X.; Li, M.; Liu, X. A Novel Multi-Beam Lens Antenna for High Altitude Platform Communications. In Proceedings of the 2012 IEEE 75th Vehicular Technology Conference (VTC Spring), Yokohama, Japan, 6–9 May 2012; pp. 1–5.
30. Stoneback, M.; Madsen, K. A Planar All-Silicon 256-Element Ka-Band Phased Array for High-Altitude Platforms (HAPs) Application. In Proceedings of the 2018 IEEE/MTT-S International Microwave Symposium—IMS, Philadelphia, PA, USA, 10–15 June 2018; pp. 783–786.
31. Aslan, Y.; Puskely, J.; Janssen, J.H.J.; Geurts, M.; Roederer, A.; Yarovoy, A. Thermal-Aware Synthesis of 5G Base Station Antenna Arrays: An Overview and a Sparsity-Based Approach. *IEEE Access* **2018**, *6*, 58868–58882. [CrossRef]
32. Aslan, Y.; Roederer, A.; Yarovoy, A. System Advantages of Using Large-Scale Aperiodic Array Topologies in Future Mm-Wave 5G/6G Base Stations: An Interdisciplinary Look. *IEEE Syst. J.* **2021**, 1–10. [CrossRef]

33. Haupt, R.L. *Antenna Arrays: A Computational Approach*; Wiley-IEEE Press: Hoboken, NJ, USA, 2010; ISBN 978-0-470-40775-2.
34. Sharp, E. A Triangular Arrangement of Planar-Array Elements That Reduces the Number Needed. *IRE Trans. Antennas Propag.* **1961**, *9*, 126–129. [CrossRef]
35. Sadhu, B.; Gu, X.; Valdes-Garcia, A. Compact 28-GHz Phased Array Antenna for 5G Access. *IEEE Microw. Mag.* **2019**, *20*, 32–50. [CrossRef]
36. Theunissen, W.; Jain, V.; Menon, G. Development of a Receive Phased Array Antenna for High Altitude Platform Stations Using Integrated Beamformer Modules. In Proceedings of the 2018 IEEE/MTT-S International Microwave Symposium—IMS, Philadelphia, PA, USA, 10–15 June 2018; pp. 779–782.
37. Dicandia, F.A.; Genovesi, S. Exploitation of Triangular Lattice Arrays for Improved Spectral Efficiency in Massive MIMO 5G Systems. *IEEE Access* **2021**, *9*, 17530–17543. [CrossRef]
38. Yin, Y.; Ustundag, B.; Kibaroglu, K.; Sayginer, M.; Rebeiz, G.M. Wideband 23.5-29.5-GHz Phased Arrays for Multistandard 5G Applications and Carrier Aggregation. *IEEE Trans. Microw. Theory Tech.* **2020**, 1. [CrossRef]
39. Mailloux, R.J. *Phased Array Antenna Handbook*, 2nd ed.; Artech House antennas and propagation library; Artech House: Boston, MA, USA, 2005; ISBN 978-1-58053-689-9.
40. Aslan, Y.; Puskely, J.; Roederer, A.; Yarovoy, A. Trade-Offs between the Quality of Service, Computational Cost and Cooling Complexity in Interference-Dominated Multi-User SDMA Systems. *IET Commun.* **2020**, *14*, 144–151. [CrossRef]
41. Ge, X.; Zi, R.; Wang, H.; Zhang, J.; Jo, M. Multi-User Massive MIMO Communication Systems Based on Irregular Antenna Arrays. *IEEE Trans. Wirel. Commun.* **2016**, *15*, 5287–5301. [CrossRef]
42. Zhang, S.; Chen, X.; Pedersen, G.F. Mutual Coupling Suppression With Decoupling Ground for Massive MIMO Antenna Arrays. *IEEE Trans. Veh. Technol.* **2019**, *68*, 7273–7282. [CrossRef]
43. Pozar, D.; Schaubert, D. Scan Blindness in Infinite Phased Arrays of Printed Dipoles. *IEEE Trans. Antennas Propag.* **1984**, *32*, 602–610. [CrossRef]
44. Ansys | Engineering Simulation Software. Available online: https://www.ansys.com/ (accessed on 4 May 2021).
45. Ahmed, I.; Khammari, H.; Shahid, A.; Musa, A.; Kim, K.S.; De Poorter, E.; Moerman, I. A Survey on Hybrid Beamforming Techniques in 5G: Architecture and System Model Perspectives. *IEEE Commun. Surv. Tutor.* **2018**, *20*, 3060–3097. [CrossRef]
46. Hefnawi, M. Hybrid Beamforming for Millimeter-Wave Heterogeneous Networks. *Electronics* **2019**, *8*, 133. [CrossRef]
47. Orikumhi, I.; Kang, J.; Jwa, H.; Na, J.-H.; Kim, S. SINR Maximization Beam Selection for MmWave Beamspace MIMO Systems. *IEEE Access* **2020**, *8*, 185688–185697. [CrossRef]
48. Rappaport, T.S.; Xing, Y.; MacCartney, G.R.; Molisch, A.F.; Mellios, E.; Zhang, J. Overview of Millimeter Wave Communications for Fifth-Generation (5G) Wireless Networks—With a Focus on Propagation Models. *IEEE Trans. Antennas Propag.* **2017**, *65*, 6213–6230. [CrossRef]
49. AL-Hourani, A.; Chandrasekharan, S.; Kaandorp, G.; Glenn, W.; Jamalipour, A.; Kandeepan, S. Coverage and Rate Analysis of Aerial Base Stations [Letter]. *IEEE Trans. Aerosp. Electron. Syst.* **2016**, *52*, 3077–3081. [CrossRef]
50. Salman, S.; Aslan, Y.; Puskely, J.; Roederer, A.; Yarovoy, A. System Modeling and Simulation in 5G: A Hybrid Beamforming Approach With Power Flux Equalization in the Elevation Plane. In Proceedings of the 2019 49th European Microwave Conference (EuMC), Paris, France, 1–3 October 2019; pp. 746–749.
51. Abdul Haleem, M. On the Capacity and Transmission Techniques of Massive MIMO Systems. *Wirel. Commun. Mob. Comput.* **2018**, *2018*, 1–9. [CrossRef]
52. Björnson, E.; Sanguinetti, L.; Hoydis, J.; Debbah, M. Optimal Design of Energy-Efficient Multi-User MIMO Systems: Is Massive MIMO the Answer? *IEEE Trans. Wirel. Commun.* **2015**, *14*, 3059–3075. [CrossRef]
53. Kildal, P.-; Rosengren, K. Correlation and Capacity of MIMO Systems and Mutual Coupling, Radiation Efficiency, and Diversity Gain of Their Antennas: Simulations and Measurements in a Reverberation Chamber. *IEEE Commun. Mag.* **2004**, *42*, 104–112. [CrossRef]
54. Artiga, X.; Devillers, B.; Perruisseau-Carrier, J. Mutual Coupling Effects in Multi-User Massive MIMO Base Stations. In Proceedings of the 2012 IEEE International Symposium on Antennas and Propagation, Chicago, IL, USA, 8–14 July 2012; pp. 1–2.
55. Ge, J.; Zhu, B.; Sun, C.; Li, J. Low Complexity User Scheduling Algorithm for Energy-Efficient Multiuser Multiple-Input Multiple-Output Systems. *IET Commun.* **2014**, *8*, 343–350. [CrossRef]

sensors

MDPI

Article

Continuous Resonance Tuning without Blindness by Applying Nonlinear Properties of PIN Diodes

Yong Luo [1], Hongtao Liu [1], Yiming He [1], Hengrong Cui [2,*] and Guangli Yang [1]

[1] School of Communication and Information Engineering, Shanghai University, Shanghai 200444, China; y_luo@foxmail.com (Y.L.); LHT_shu@163.com (H.L.); yiming_he@foxmail.com (Y.H.); guangli.yang@shu.edu.cn (G.Y.)
[2] College of Information Sciences and Technology, Donghua University, Shanghai 201620, China
* Correspondence: hengrong@dhu.edu.cn

Abstract: Metamaterial antennas consisting of periodical units are suitable for achieving tunable properties by employing active elements to each unit. However, for compact metamaterials with a very limited number of periodical units, resonance blindness exists. In this paper, we introduce a method to achieve continuous tuning without resonance blindness by exploring hence, taking advantage of nonlinear properties of PIN diodes. First, we obtain the equivalent impedance of the PIN diode through measurements, then fit these nonlinear curves with mathematical expressions. Afterwards, we build the PIN diode model with these mathematical equations, making it compatible with implementing co-simulation between the passive electromagnetic model and the active element of PIN diodes and, particularly, the nonlinear effects can be considered. Next, we design a compact two-unit metamaterial antenna as an example to illustrate the electromagnetic co-simulation. Finally, we implement the experiments with a micro-control unit to validate this method. In addition, the nonlinear stability and the supplying voltage tolerance of nonlinear states for both two kinds of PIN diodes are investigated as well. This method of obtaining smooth tuning with nonlinear properties of PIN diodes can be applied to other active devices, if only PIN diodes are utilized.

Keywords: active metamaterial antenna; continuous tuning; resonance blindness; EM co-simulation; nonlinear property

Citation: Luo, Y.; Liu, H.; He, Y.; Cui, H.; Yang, G. Continuous Resonance Tuning without Blindness by Applying Nonlinear Properties of PIN Diodes. *Sensors* **2021**, *21*, 2816. https://doi.org/10.3390/s21082816

Academic Editor: Naser Ojaroudi Parchin

Received: 12 March 2021
Accepted: 14 April 2021
Published: 16 April 2021

1. Introduction

Electromagnetic metamaterials (EM MTMs) [1] employ periodical units, that are derived from split-ring resonators (SRRs) [2], composite right–left-handed (CRLH) structures [3], and high-impedance structures (HISs) [4], to obtain a negative refractive index, negative phase constant, and high surface impedance, thereby achieving the unique properties of super-lens [5], back–forward radiation [6], and field enhancement [7]. Thanks to EM MTMs that are characterized by periodical configuration, it is possible to realize multiple tunable states either in spectrum resonances [8] or spatial radiation patterns [9] by applying active components to each periodical unit. This kind of tuning mechanism benefits from a periodical array with n unit cells, while each unit can be tuned individually with m states utilizing active elements such as PIN diodes [8,10–12], varactors [13–20], or MEMS [21], thus, ideally speaking, we can possess in total as many as m^n tunable states. This means that extremely large MTMs with an infinite ($n \to \infty$) number of units have an infinite number of tunable states, leading to continuous tuning. Conversely, compact MTMs with a very limited number of units have only several discrete tunable states. The absence of continuous tunability in an active MTM design is called tuning blindness, and it is has two causes: the MTM design has very limited periodical units, such as two, three, or five cells; RF switches as the active component in the MTM only have two tunable states (ON/OFF). For instance, as demonstrated in Figure 1a, we simulate an MTM antenna containing two-unit (n = 2, m = 2) HIS structures, and it indeed demonstrates several tunable resonances, but they are discrete with unavoidably induced resonance blindness (as shown in the shadow area). Similarly, in [9], where programmable radiations are realized with PIN diodes, and in [21], where programmable spectrum resonances are achieved with MEMS, there exists tuning blindness as well. More specifically, in [9], though scanning beams from roughly −60° to +60° are obtained, as the shadow area demonstrates in Figure 1b, scanning blindness occurs from −15° to +15°. In this paper, we explore another method to achieve continuous tuning with PIN diodes: investigating the equivalent impedance in the transition zone between completed ON and OFF, and exploring the nonlinear zone in between. Thanks to the PIN diodes possessing this nonlinear zone, we can achieve a continuous spectrum tuning without blindness and, meanwhile, with low actuated voltages less than 1.5 V, which is suitable for NB-IoT scenery that requires many tunable but narrow-band spectrum channels and with low power consumption.

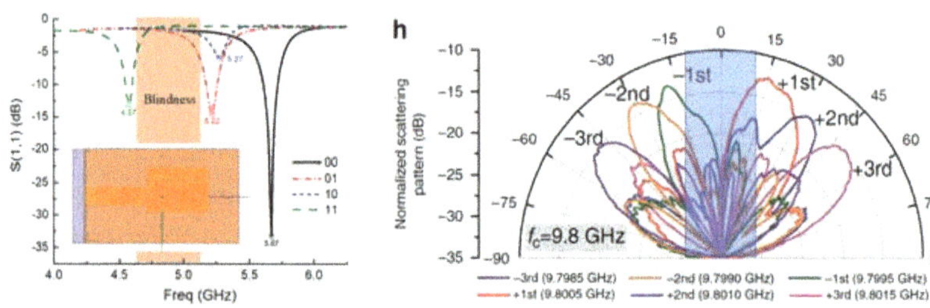

(**a**) (**b**)

Figure 1. Tuning blindness exists in (**a**) a compact MTM antenna with two periodical units and in (**b**) programmable spatial radiation patterns [9].

In practice, MTMs with a limited number of periodical units are quite common, and in some cases, they are even preferred due to their compact size. In addition to the method proposed in this paper, another common method for avoiding blindness is to increase the tunable states m possessed by the individual cell, while unit number n is kept to a small value for a compact size. For instance, [13–20] introduce the tunable antenna using varactor diodes, or variable capacitors, to obtain multiple tunable states m = 9, 7, 6, respectively. These good works with multi-state tuning indeed increase the tuning continuity with discrete structures, but usually require variable voltage to even as high as 20 V, which might not be compatible with low-power applications such as narrow bandwidth Internet of Things (NB-IoT) [22–24].

This paper is arranged as follows: Section 2 investigates the nonlinear property of PIN diodes; Section 3 introduces the electromagnetic (EM) co-simulation; Section 4 presents experiments; Sections 5 and 6 provide the discussion and conclusion.

2. Nonlinear Properties

PIN diodes are conventionally utilized as RF switches with two states (ON/OFF). However, there exists a transition zone in between. In order to investigate this nonlinear property, we study the relationship between the equivalent impedance and the actuated voltage using the PIN diode A (MACOM MA4AGBLP912, MACOM, Lowell, MA, USA). First, we measure the PIN diode by employing a microstrip line in a 5 GHz band. As shown in Figure 2a, we make a slot in the middle of the standard 50 Ω microstrip line and integrate the surface-mounted PIN diode A there, then apply two inductors (Murata LQW18AN22NG00, Mutrata, Nagaokakyo, Kyoto, Japan) with a large value (22 nH) to block the interference from the DC supplier. Second, we apply transmission line (TL) theory to analyze this equivalent circuit model, as shown in Figure 2b. The equivalent model includes Z_c = 50 Ω that represents the characteristic impedance of the standard transmission line with length l_0, the equivalent impedance Z_{pin} of the PIN diodes, and the port impedance Z_{port} = 50 Ω. According to TL theory, the equivalent impedance of PIN diodes Z_{pin} can be retrieved from input impedance Z_{in} as [25]

$$Z_{pin} = \frac{Z_C \cdot (Z_{in} - Z_L) + j \cdot \tan(\beta l_0) \cdot (Z_{in} \cdot Z_L - Z_C^2)}{Z_C - j \cdot Z_{in} \cdot \tan(\beta l_0)} \tag{1}$$

where β is the phase constant, and the input impedance Z_{in} is measured in experiments.

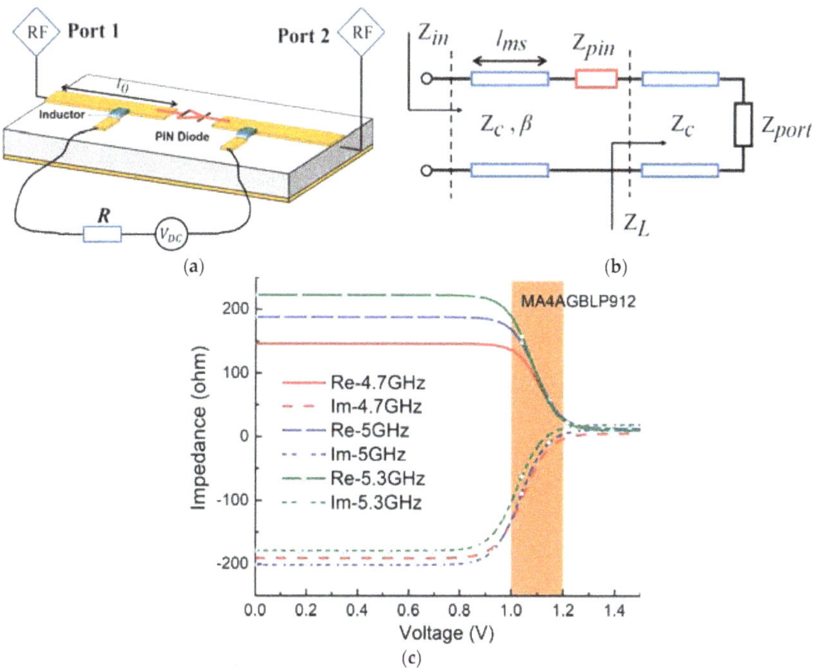

Figure 2. (**a**) Measurement setup and (**b**) its equivalent circuit model for investigating the nonlinear zone of PIN diode A (MACOM MA4AGBLP912); (**c**) extracted equivalent impedance including resistance and reactance according to different voltages actuated to the PIN diode.

As shown in Figure 2c, since actuated voltages are varied from 0 V to 1.5 V, the equivalent impedance of the PIN diodes Z_{pin} is changed accordingly; the resistance ranges from 225 Ω to a very small value close to 0 Ω, and the reactance varies from −200 Ω to a very small value as well. Particularly, we can observe that there exists a transition zone (as marked by the shadow area in Figure 2c) between the PIN diodes' OFF zone where impedance is around 200–200 j, and the ON zone where the impedance is a very small value close to zero. In this transition zone, the actuated voltage is around 1–1.2 V and, accordingly, the impedance varies nonlinearly and smoothly from the OFF state to the ON state.

In order to accommodate the EM co-simulation including passive EM models and nonlinear active components, we build a PIN diode model with respect to the nonlinear properties and considering parameters of actuated voltages and frequencies. Referring to impedance curves as shown in Figure 2c, curves in the transition zone are nonlinear in an S shape, which is close to the Boltzmann function [26]. Thus, we select the Boltzmann function to fit them. Based on Boltzmann's mathematical model, the real part Z_{Re} and imaginary part Z_{Im} are

$$Z_{Re} = Z_{Re\_on} + \frac{Z_{Re\_off} - Z_{Re\_on}}{1 + e^{\frac{V - V_{Re\_0}}{d_{Re}}}} \tag{2}$$

$$Z_{Im} = Z_{Im\_on} + \frac{Z_{Im\_off} - Z_{Im\_on}}{1 + e^{\frac{V - V_{Im\_0}}{d_{Im}}}} \tag{3}$$

in which V is the actuated voltage for PIN diode A, Z_{Re_off} and Z_{Re_on} are the measured Z_{Re} when the diode is in the OFF state with V = 0 V and the ON state with V = 1.5 V. Similarly, Z_{Im_off} and Z_{Im_on} are Z_{Im} when V = 0 V (OFF state) and 1.5 V (ON state). V_{Re_0} is defined as the voltage when Z_{Re} equals the mean of Z_{Re_off} and Z_{Re_on}, while V_{Im_0} is the voltage when Z_{Im} equals the mean of Z_{Im_off} and Z_{Im_on}. Parameters d_{Re} and d_{Im} are the slope of curves Z_{Re} and Z_{Im} when $V = V_{Re\_0}$ and $V = V_{Im\_0}$.

Until now, the above equations have concerned only one frequency point, but we need to consider the whole frequency band. This means all the parameters in (2) and (3), Z_{Re_off}, Z_{Re_on}, V_{Re_0}, d_{Re} and Z_{Im_off}, Z_{Im_on}, V_{Im_0}, d_{Im}, need to be related to frequencies. We select several frequency points located at the relatively low, moderate, and high frequency sections of the band, and fit them to the equations, thereby involving the whole frequency band when describing the nonlinear properties. Particularly, according to these curves' shapes concerning frequencies, the mean function and Gaussian function are applied to fit the real part Z_{Re} and the imaginary part Z_{Im}, respectively. For Z_{Re}, the relative parameters in respect to frequencies can be described as

$$Z_{Re\_on} = \frac{1}{3} \sum_{f = f_0, f_1, f_2} Z_{Re}(f) \tag{4}$$

$$Z_{Re\_off} = Z_{Re\_off}(f_2) + d_1 \cdot (f - f_2)/10^9 \tag{5}$$

$$V_{Re\_0} = V_{Re\_0}(f_2) + d_2 \cdot (f - f_2)/10^9 \tag{6}$$

$$d_{Re} = d_{Re}(f_2) + d_3 \cdot (f - f_2)/10^9 \tag{7}$$

where three typical frequency points are f_0 = 4.7 GHz, f_1 = 5 GHz, and f_2 = 5.3 GHz. Other parameters are Z_{Re_on} = 11.18 Ω, Z_{Re_off} (f_2) = 223.8 Ω, V_{Re_0} (f_2) = 1.08 V, d_{Re} (f_2) = 0.04751, d_1 = 126.57, d_2 = −0.05024, and $d_3 = 7.45 \times 10^{-3}$. Similarly, we use (8)–(11) for the imaginary part Z_{Im}:

$$Z_{Im\_on} = Z_{Im\_on}(f_2) + d_4 \cdot (f - f_2)/10^9 \tag{8}$$

$$Z_{Im\_off} = Z_{Im\_off}(f_2) + d_5 \cdot e^{-0.5 \cdot \left(\frac{f/10^9 - f_3}{d_6}\right)^2} \tag{9}$$

$$V_{Im\_0} = V_{Im\_0}(f_2) + d_7 \cdot (f - f_2)/10^9 \tag{10}$$

$$d_{Im} = d_{Im}(f_2) + d_8 \cdot (f - f_2)/10^9 \tag{11}$$

where the relative parameters Z_{Im_on} (f_2) = 18.63 Ω, Z_{Im_off} (f_2) = −171.38 Ω, V_{Im_0} (f_2) = 1.023 V, and d_{Im} (f_2) = 0.04902. Other parameters d_4 = 22.59, d_5 = −27.24, d_6 = 0.23727, d_7 = −0.03697, and $d_8 = 7.966 \cdot 10^{-4}$. Especially, the parameter f_3 = 4.922 is derived from the peak position of the Gaussian function. Finally, we achieve the completed equations to express the nonlinear property of the PIN diode as follows

$$Z_{Re} = \frac{1}{3} \sum_{f = f_0, f_1, f_2} Z_{Re}(f) + \frac{Z_{Re\_off}(f_2) + d_1 \cdot (f - f_2)/10^9 - \frac{1}{3} \sum_{f = f_0, f_1, f_2} Z_{Re}(f)}{1 + e^{\frac{V - (V_{Re\_0}(f_2) + d_2 \cdot (f - f_2)/10^9)}{d_{Re}(f_2) + d_3 \cdot (f - f_2)/10^9}}} \tag{12}$$

$$Z_{Im} = Z_{Im\_on}(f_2) + d_4 \cdot (f - f_2)/10^9 + \frac{Z_{Im\_off}(f_2) + d_5 \cdot e^{-0.5 \cdot \left(\frac{f/10^9 - f_3}{d_6}\right)^2} - \left(Z_{Im\_on}(f_2) + d_4 \cdot (f - f_2)/10^9\right)}{1 + e^{\frac{V - (V_{Im\_0}(f_2) + d_7 \cdot (f - f_2)/10^9)}{d_{Im}(f_2) + d_8 \cdot (f - f_2)/10^9}}} \quad (13)$$

Note that we fit the measured impedance curves of the PIN diode with these above-mentioned equations through several typical frequency points f_0, f_1, and f_2, thus, we need to double check if they can represent the whole frequency band. We randomly select the frequencies 4.75 GHz, 4.9 GHz, and 5.13 GHz in the band, and compare the fitting curves with the measured results. As shown in Figure 3a, Z_{Re} and Z_{Im} match well with the measured ones, implying the equivalent effectiveness of the nonlinear property in the whole frequency band. In this way, we obtain the mathematical expressions to describe the nonlinear properties of the PIN diode, and accordingly model this PIN diode in ANSYS Electronics Desktop, ensuring the nonlinear property is considered in the EM co-simulation.

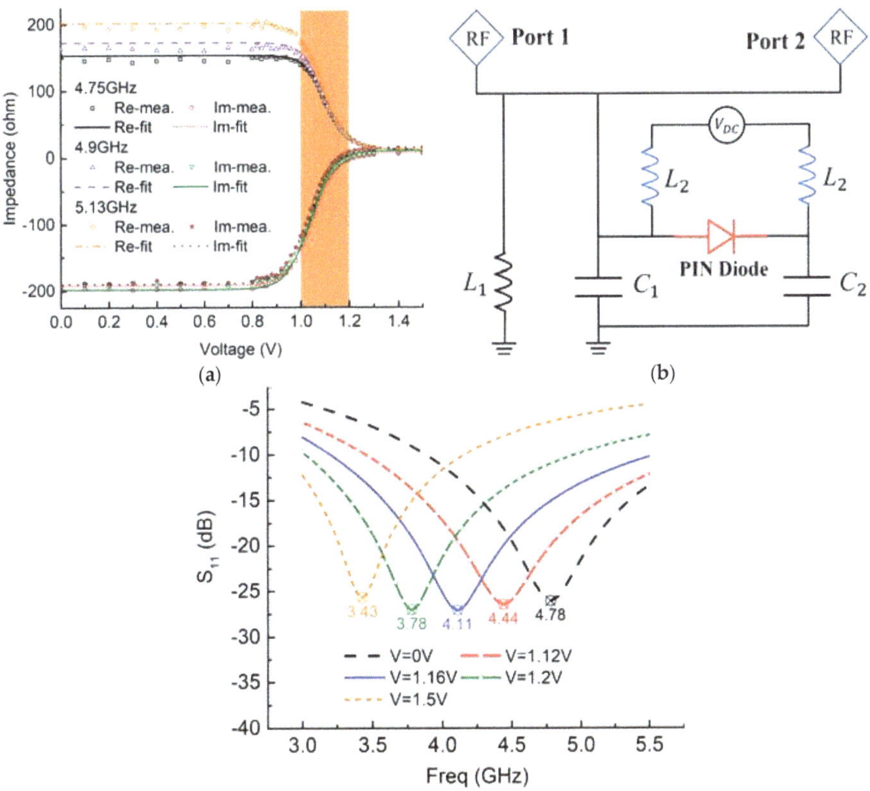

Figure 3. (**a**) Comparisons between fitting results by using the Boltzmann function and measured results; (**b**) the circuit contains parallel LC and PIN diode A (MACOM MA4AGBLP912) model, thereby implementing the proof-of-concept simulation to prove (**c**) continuous and smooth resonance tuning without blindness.

To demonstrate that the nonlinear properties can be exploited for achieving smooth and uniform resonance tuning, we implement a proof-of-concept level simulation with the PIN diode A (MACOM MA4AGBLP912). As shown in Figure 3b, it is a parallel L_1C_1 circuit model with the parameters L_1 = 1 nH and C_1 = 1 pF. IN particular, we put another capacitance C_2 in the shunt direction with the same value C_2 = 1 pF, but it can be connected or disconnected parallel to the L_1C_1 circuit via the PIN diode, which can be controlled by the supplying voltages from the OFF state, nonlinear states to ON state. An inductance of L_2 = 1 H is used to block the interference from the DC suppliers. Theoretically speaking, there should be a continuous resonance tuning between the resonance $1/\left(2\pi\sqrt{L_1C_1}\right) = 5.03$ GHz when the PIN diode is ideally open, and $1/\left(2\pi\sqrt{2L_1C_1}\right) = 3.56$ GHz when the diode is ideally short, through middle states while actuating the diode in the nonlinear zone. As shown in Figure 3c, with controlling the actuated voltages to make the PIN diode work in OFF, ON, and transition states, the resonances are tuned from 3.43 GHz to 4.78 GHz via nonlinear states 3.78 GHz, 4.11 GHz, and 4.44 GHz, respectively. This continuous and smooth resonance tuning verifies the concept of eliminating the resonance blindness with nonlinear properties by PIN diodes.

In a brief summary, PIN diodes have the advantages of nonlinear properties while the actuated voltages fall in the transition zone, providing the potential capability of continuous tuning in MTM antenna even with a very limited number of units. That is either different from varactors that rely on a large dynamic voltage range, or different from MEMS that have a noncontinuous equivalent capacitance value variation due to the beam membrane pulled in the 1/3 length position [27].

3. Layout, Design, and EM Co-Simulation

We design a compact MTM antenna using PIN diodes to introduce the EM co-simulation, and take advantage of its nonlinear properties to realize the smooth tuning and eliminate the resonance blindness. As in Figure 4a, the active MTM structure comprises two periodical cells which produce a compact size, a PIN diode that plays the role of the active element in each unit, and inductance chips for blocking the interference from DC suppliers. Via holes are made between the top and bottom layers (Figure 4b) to connect the micro-control unit (MCU) for DC supply. As seen in the side view in Figure 4c,d, PIN diodes placed in the two slots of each unit electrically connect/disconnect these slots, thus manipulating zeroth-order resonances (ZORs) of the MTM antenna. Thanks to the MTM configuration separating units from each other, voltages for actuating each PIN diode can be controlled independently. FR4 material with permittivity ε_r = 4.3, tanδ = 0.02, and thickness h = 2.5 mm is used as the substrate. The unit cell is designed according to CRLH-TL theory [11], in which the equivalent circuit model has inductances and capacitances in both series and shunt directions, thereby producing the zeroth-order resonance (ZOR) resonating at the frequency β = 0. The mechanism can be qualitatively demonstrated by the equivalent circuit model, as shown in Figure 4e. The left-handed capacitance is equivalently considered as $C_L = 2C_{L1} + C_L$, where capacitance C_{L1} is formed by the gaps between adjacent units and C_{L2} is produced by the two symmetric J-shaped patches. Left-handed inductance L_L is generated by a strip patch in the x-direction, and is regarded as being connected to the ground through another capacitance C_g induced between the edge patch and the ground. Series inductance L_R and shunt capacitance C_R are formed from the conventional microstrip line. According to CRLH theory, the ZOR ω_{zor} is related to shunt-directed resonances ω_{sh} [11]:

$$\omega_{zor} = \omega_{sh} = \sqrt{\frac{1}{L_L C_g} + \frac{1}{L_L C_R}} \tag{14}$$

which implies what the active element PIN diodes are particularly utilized to tune: shorting/opening PIN diodes alter the effective area of the edge patch, leading to equivalently varying the capacitance C_g, hence, tuning the ZORs. Moreover, we design the unit operating in the ZOR mode, because at this resonance, the phase constant β = 0, and its guided wavelength, is infinite, leading to the favorable characteristic that its resonance is independent of the physical length [11]. Therefore, we have the freedom to employ arbitrary numbers of periodical units. For a compact MTM antenna to demonstrate ZOR tuning without blindness, we utilize two periodical units as an example.

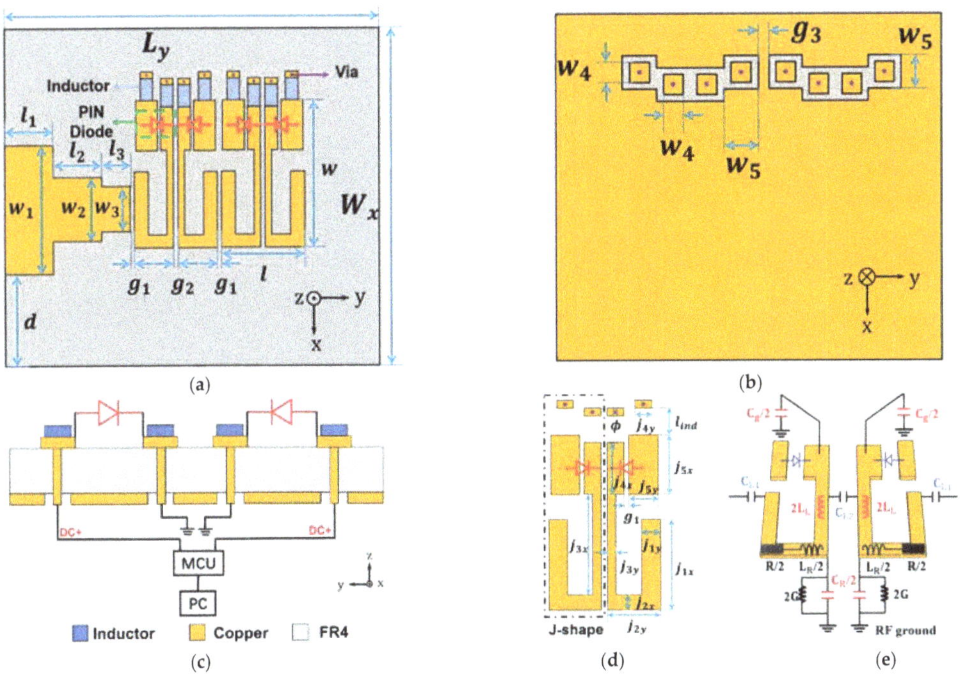

Figure 4. Design model of the proposed MTM antenna: (**a**) top view, (**b**) bottom view, (**c**) side view, and (**d**) unit design with dimensions (mm): $W_x = 26$, $L_y = 34.5$, $l_1 = 3.5$, $w_1 = 9$, $l_2 = 6.5$, $w_2 = 4$, $l_3 = 2$, $w_3 = 2.7$, $w = 9.1$, $l = 5.2$, $g_1 = 0.2$, $g_2 = 0.3$, $d = 7.85$, $j_{1x} = 4.7$, $j_{1y} = 0.9$, $j_{2x} = 0.8$, $j_{2y} = 2.5$, $j_{3x} = 5.2$, $j_{3y} = 0.4$, $j_{4x} = 2.7$, $j_{4y} = 0.8$, $j_{5x} = 3.1$, $j_{5y} = 1.4$, $g_1 = 0.4$, $l_{ind} = 1.4$, $\Phi = 0.2$, $w_4 = 0.8$, $w_5 = 1.4$; (**e**) the equivalent CRLH circuit model of unit cell.

We use ANSYS Electronics Desktop to simulate the whole design including the passive EM model and the active element PIN diodes, as illustrated in Figure 5a. First, we design the antenna with passive simulation in HFSS without any diodes. That means in the passive full-wave simulation, with/without rectangular patches are utilized to imitate ON/OFF states of PIN diodes, thus considering a preliminary simulation with electric field distributions and radiation patterns. Afterward, in the EM model, lumped ports are set up where the active elements are placed, so we have chances to insert the active element model there. Then, we build the SPICE model for PIN diode A MACOM MA4AGBLP912 with the mathematical equations shown above, thereby involving its nonlinear property. In addition, the S2P file of the inductor (Murata LQW18AN22NG00, Mutrata, Nagaokakyo, Kyoto, Japan) is employed in the EM co-simulation as well. As shown in Figure 5b, a simulation is conducted with the S2P file of the inductor, and it exhibits good isolation of less than −20 dB between the DC supplier and RF signals. As the active elements are ready, finally, we can implement the EM co-simulation by considering the S2P file of the inductor and the SPICE model of PIN diode A for the lumped ports. Particularly, four DC voltage sources are connected to the lumped ports as well to supply PIN diodes accordingly. In such a method, we can achieve the results of co-simulation easily within a few minutes.

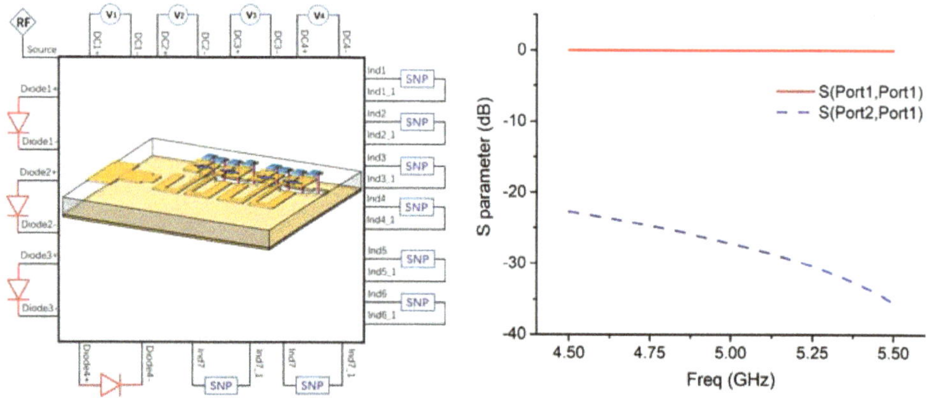

(a) (b)

Figure 5. (**a**) EM co-simulation model including passive EM model, active element PIN diodes, and inductance chip; (**b**) simulated S-parameters from the SNP file of inductor (Murata LQW18AN22NG00).

With the co-simulation method, we obtain the active MTM antenna simulations as plotted in Figure 6, in which both the linear and nonlinear cases are illustrated. In the linear case, as shown in Figure 6a, there are OFF and ON states, and we code PIN diodes in the OFF state as state 0 when 0V is applied, and code the ON state as state 1.5 when 1.5 V is applied. For example, 0-1.5-0-1.5 means the second and fourth PIN diodes are actuated to the ON state while other two diodes are in the OFF state. In a nonlinear case, as shown in Figure 6b, however, more nonlinear states where actuated voltages fall in the transition zone are shown. We code these nonlinear states exactly as the voltages actuated to PIN diodes. For instance, 0-1.1-1.23-1.09 indicates the four PIN diodes are actuated with 0 V, 1.1 V, 1.23 V, and 1.09 V, respectively. In Figure 6b, the ZORs of the nonlinear case are tuned from 4.71 GHz to 5.31 GHz via 4.82 GHz, 4.96 GHz, 5.11 GHz, and 5.19 GHz, while in the linear case, as shown in Figure 6a, the resonances are tuned from 4.71 GHz to 5.31 GHz but the tuning is not smooth and continuous, and there is blindness in the band from 4.83 GHz to 5.07 GHz. Since each unit can provide four coding sequences, 0-0, 0-1.5, 1.5-0, and 1.5-1.5, an MTM antenna consisting of two units has all 16 coding states to cover 4.71 GHz to 5.31 GHz, while for the nonlinear case, it has more middle states. As shown in Figure 6c, by comparing the 16 states of the linear case and 30 selected states of the nonlinear case, we find that nonlinear advantages allow the ZOR tuning to be smooth, continuous, and uniform, without resonance blindness.

In summary, we apply the S2P file of the inductor, the SPICE model of PIN diode A, and the DC voltage model to the EM co-simulation. These kinds of two-port models have the advantages of not needing to consider the complicated equivalent circuit model with all detailed parameters of R, L, and C, because all of these circuit parameters are included in the S2P model or SPICE model. Thanks to the EM co-simulation considering nonlinear properties of PIN diodes, we can simulate an active MTM antenna with continuous and uniform resonance tuning, and eliminate the resonance blindness. The nonlinearity of PIN diodes not only prevents frequency tuning blindness due to the compact MTM design with limited discrete states, but also makes frequency tuning uniform.

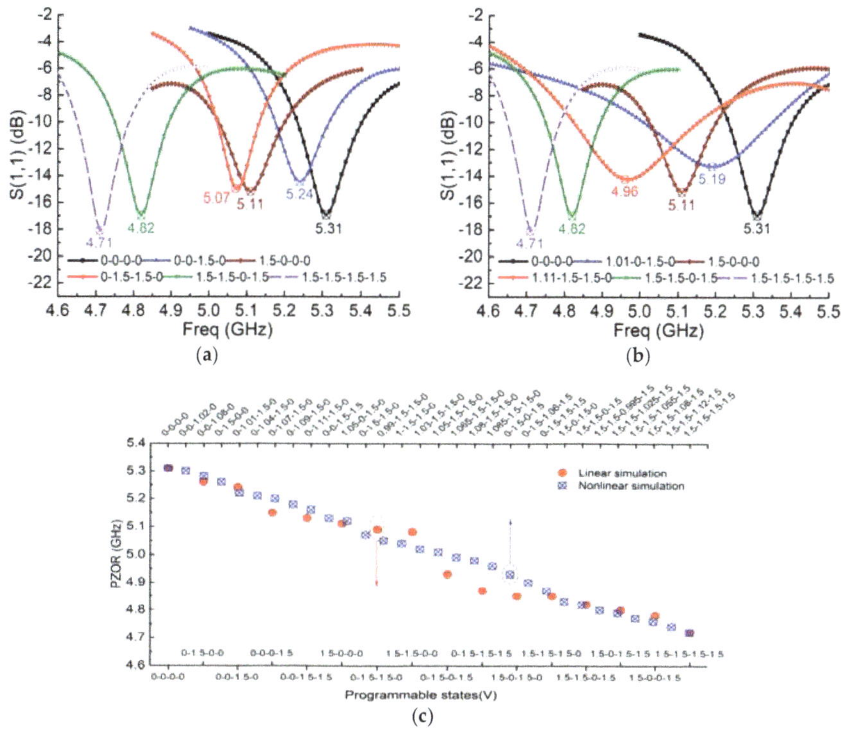

Figure 6. Simulation results of S11 with co-simulation method using PIN diode model as RF switch: (**a**) 6 linear states; (**b**) 6 nonlinear states with uniform tuning; and (**c**) ZOR comparisons of all 16 linear states with 30 nonlinear states.

4. Experimental Implementation

According to the previous design, the compact MTM antenna consisting of two cells is fabricated as shown in Figure 7a. The configuration and layout are exactly that in Figure 4: the FR4 substrate has the parameters of $\varepsilon_r = 4.3$, $\tan\delta = 0.02$, and PIN diode A and the inductance chip are MACOM MA4AGBLP912 and Murata LQW18AN22NG00, respectively. As demonstrated in Figure 7b, via holes go through the substrate to connect four pairs of wires, so as to supply these PIN diodes through the micro-control unit (MCU). In this design, four channels of the DC supply can be manipulated independently because of the isolated and periodical configuration of the MTM. Figure 7c shows the setup for anechoic chamber measurements, in which a laptop is utilized to output C language for controlling the MCU for voltage manipulations.

We measure both the linear case, which includes OFF (actuated voltage 0 V, indicated as 0) and ON states (actuated voltage 1.5 V, indicated as 1.5), and the nonlinear case (state coded as the actuated voltage) which considers applying voltages in the transition zone. In Figure 8a,b, several ZORs of the linear and nonlinear cases are demonstrated, and it is seen that as resonances are tuned from 4.7 GHz to 5.3 GHz via many tuning states, the resonance tuning of the linear case is not uniform, while that of the nonlinear case is uniform and smooth. More specifically, as shown in Figure 8c, more tunable states are compared. For the linear case, which considers all the completed $m^n = 4^2 = 16$ tuning states ($m = 4$ represents the four PIN diodes, and $n = 2$ indicates PIN diodes' ON/OFF states) for the compact two-cell MTM antenna, we can clearly observe that its tuning is nonuniform and blindness clearly exists in the frequency band of 4.9~5.1 GHz and 5.1~5.25 GHz. For instance, states 0-0-1.5-1.5 and 1.5-0-0-0 have almost the same resonant point and overlap at 5.12 GHz, while states 1.5-1.5-0-0 and 0-1.5-0-1.5 are separated by roughly 0.2 GHz and are recognized as tuning blindness. For the nonlinear case with manipulated supply voltages in the transition zone of 1 V to 1.2 V, however, the tuning is very uniform, leading to continuous resonance tuning without blindness. In this case, we code the supplied voltage of the PIN diode working in the nonlinear zone. For example, state 1.05-1.5-1.5-0 means the four PIN diodes from left to right are actuated with 1.05 V, 1.5 V, 1.5 V, and 0 V, respectively. Thanks to the PIN diode possessing the nonlinear property, we can obtain many tunable states. Thirty tunable states are illustrated in Figure 8c, and it is seen that ZORs are tuned uniformly with a step around 0.02 GHz in the range from 4.7 GHz to 5.3 GHz, eliminating the resonance blindness and indicating the nonlinear advantages of PIN diodes. In addition, as shown in Figure 8c, in both linear and nonlinear cases, simulated ZORs agree well with measured ones, validating the effectiveness of the nonlinear model and EM co-simulation.

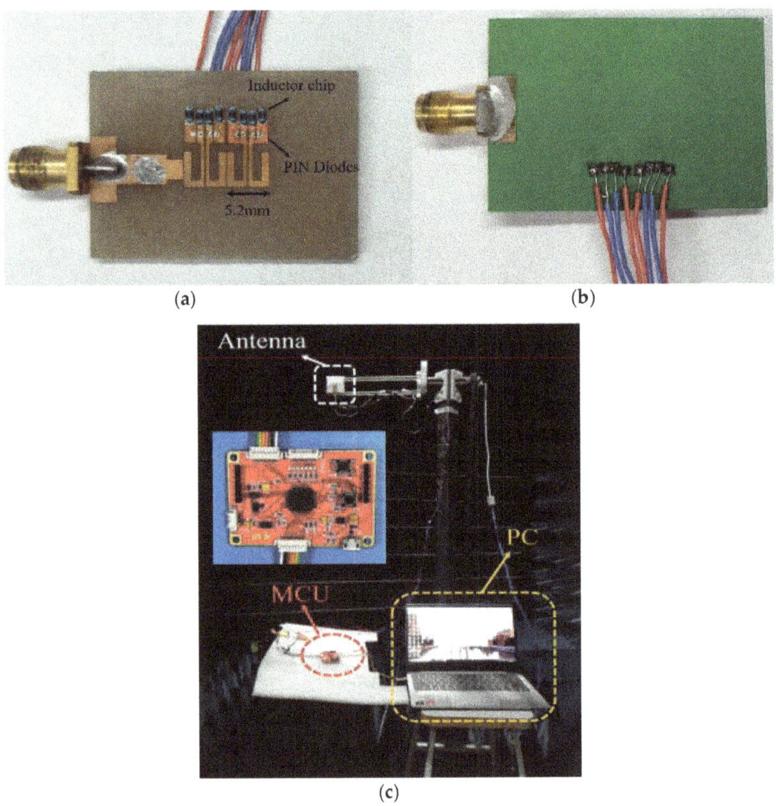

Figure 7. Fabricated MTM antenna includes (**a**) top view of MTM units with PIN diodes and inductor chips; (**b**) bottom view of four pairs of wires for DC voltage supply; (**c**) anechoic chamber measurements in which a PC controls the MCU for outputting DC voltages to PIN diodes.

Note that, as shown in Figure 8b, the bandwidth is varied when resonances are tuned in different states. This can be explained by the fact that when the active element PIN diodes are used to tune the effective area of the edge patch, they vary circuit parameter C_g, as shown in Figure 4e. Meanwhile, the PIN diode itself induces resistance as well, which varies the conductance G. Hence, the Q factor and bandwidth are changed. More specifically, according to the CRLH theory, the resonance ω_{zor} is dominated by the shunt-directed resonances ω_{sh}, as indicated in Equation (14). Thus, the Q factor and bandwidth are investigated and discussed in terms of the shunt-directed circuit part. As shown in

Figure 4e, which illustrates the equivalent circuit model, the shunt admittance can be written as

$$Y = j\omega C_R - \frac{j\omega C_g}{\omega^2 L_L C_g - 1} + G \tag{15}$$

The quality factor Q is

$$Q = \frac{1}{2}\frac{\omega}{G}\left(\frac{C_R^2 + \omega^2 C_R^2 L C_g + C_R C_g}{C_g}\right) \tag{16}$$

Consequently, the bandwidth can be expressed as

$$BW = \frac{\omega}{2\pi Q} = \frac{G C_g}{2\pi C_R (C_g + C_R)} \tag{17}$$

This equation can qualitatively explain the relationship between the bandwidth and different tunable states. Employing PIN diodes in an active MTM antenna electrically opens/shorts the gaps in the edge patch, resulting in varying the parameter C_g. On the other hand, the resistance variations in the PIN diodes in the shunt direction affect the conductance G. That indicates that tunable states vary both the C_g and G. According to Equation (17), these two variables change the bandwidth. Therefore, as seen in Figure 8b, the bandwidth is changed according to different tunable states.

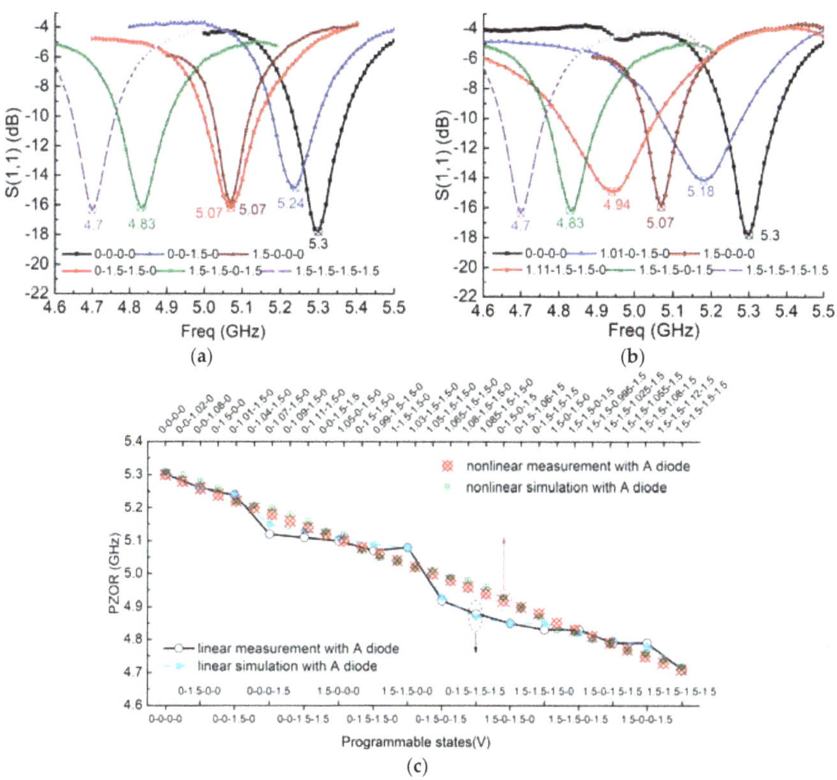

Figure 8. (**a**) Measured ZOR tuning in linear case including states: 0-0-0-0, 0-0-1.5-0, 1.5-0-0-0, 0-1.5-1.5-0, 1.5-1.5-0-1.5, 1.5-1.5-1.5-1.5 and (**b**) in nonlinear case including states: 0-0-0-0, 1.01-0-1.5-0, 1.5-0-0-0, 1.11-1.5-1.5-0, 1.5-1.5-0-1.5, and 1.5-1.5-1.5-1.5; (**c**) ZOR comparisons between linear case of all 16 linear states, and nonlinear case of 30 states.

We study the gains, efficiency, and radiation patterns of the active MTM antenna with PIN diode A (MACOM MA4AGBLP912). In particular, two extreme states of completed ON and OFF states and four nonlinear states are investigated, while in other states, gains and the radiation efficiency are on the same level, and radiation patterns are quite similar. As shown in Figure 9, two extreme states, 0-0-0-0 and 1.5-1.5-1.5-1.5, that indicate PIN diodes are completely OFF/ON, have gains of 3.73 dBi and 2.27 dBi, respectively. For another four nonlinear states, 0-0-1.02-0, 0-1.01-1.5-0, 1.05-1.5-1.5-0, and 1.5-1.5-1.09-1.5, the measured gains are 3.41 dBi, 2.77 dBi, 2.51 dBi, and 2.4 dBi, which are between the gains of the two extreme cases. The corresponding radiation efficiencies of the nonlinear states are 49%, 43.5%, 37.7%, and 36.4%, which are between the two extreme states of 54% (OFF state) and 36% (ON state). In terms of radiation patterns, as illustrated in Figure 10a–f, all the states, including completed ON/OFF states and four nonlinear states, demonstrate similar radiation patterns, and the measured radiation patterns agree well with the simulated patterns.

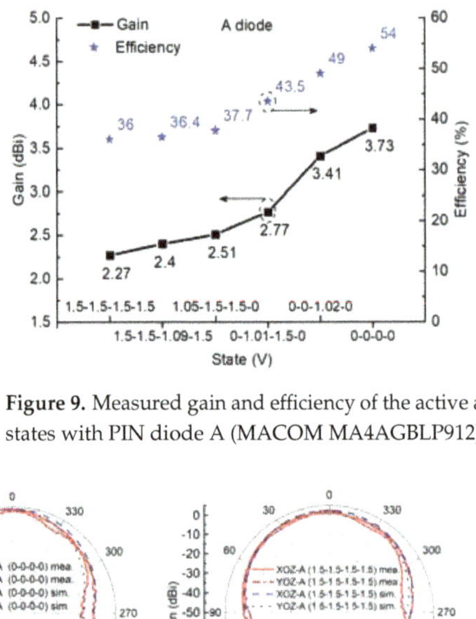

Figure 9. Measured gain and efficiency of the active antenna in two extreme states and four nonlinear states with PIN diode A (MACOM MA4AGBLP912).

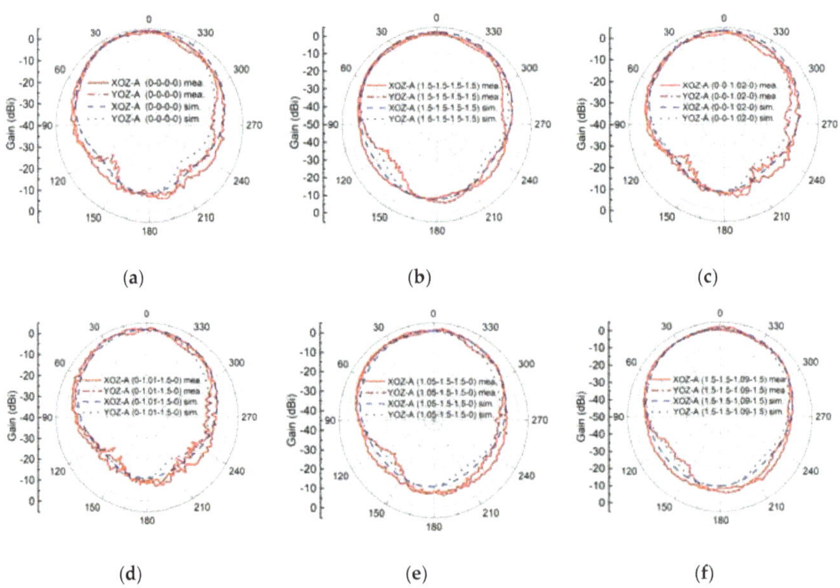

Figure 10. Comparisons between simulated and measured radiation patterns of the two extreme states: (**a**) 0-0-0-0 and (**b**) 1.5-1.5-1.5-15, and four nonlinear states: (**c**) 0-0-1.02-0, (**d**) 0-1.01-1.5-0, (**e**) 1.05-1.5-1.5-0, and (**f**) 1.5-1.5-1.09-1.5.

In this part, based on the EM co-simulation, we implement experiments with PIN diodes, which demonstrate nonlinear advantages over the linear case in eliminating resonance blindness, and in realizing uniform and continuous ZOR tuning.

5. Discussion

In this section, several interesting items associated with the nonlinearity of PIN diodes are discussed. First, we keep the same MTM antenna design but change it to employ PIN diode B (MACOM MA4FCP300), to study the generality of this kind of nonlinear property. As shown in Figure 11a, it demonstrates similar nonlinear properties: there exists a nonlinear zone where the actuated voltages fall in the transition zone 0.6–0.7 V, and by taking advantage of the nonlinear property, we can achieve similar nonlinear advantages over the linear case in achieving uniform and continuous ZOR tuning without blindness in the range 4.7 GHz to 5.3 GHz. Relative radiation patterns are quite similar to that of PIN diode A, and gains and the radiation efficiency are illustrated in Figure 11b; gains and radiation efficiency for the two extreme cases are 1.34 dBi and 3.46 dBi, and 38.41% (ON state) and 51.4% (OFF state), respectively, while for other nonlinear states of 0-0-0.66-0, 0-0.69-1-0, 0.64-1-1-0, and 1-1-0.61-1, the relative values are on the same level but between that of the completed ON and OFF states. That means, whether for PIN diode A or B, the nonlinear property is not a special case and can exist similarly and generally in other kinds of PIN diodes.

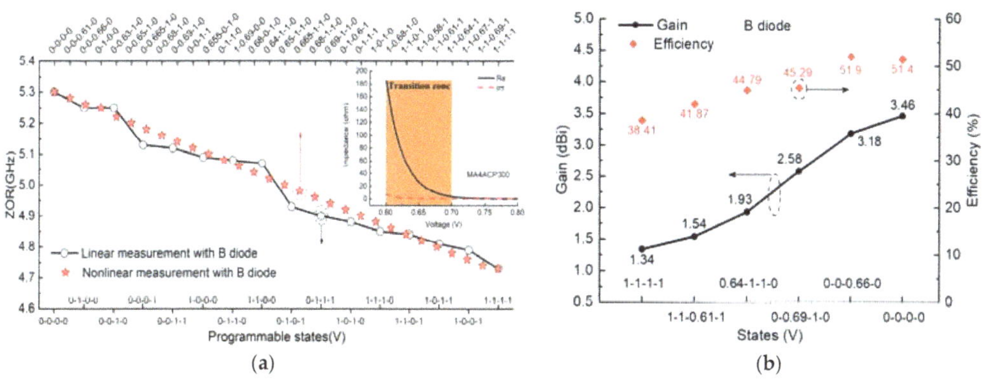

Figure 11. (**a**) Similar nonlinear advantages over linear case in achieving uniform tuning with diode B working in the transition zone; (**b**) measured gains and radiation efficiency of the two extreme states and four nonlinear states.

Second, we study the stability of the nonlinear property, namely, how stable the PIN diodes are while they work in the nonlinear zone. As shown in Figure 12a, we measure the four nonlinear states of 1.01-0-1.5-0, 1.05-1.5-1.5-0, 1.11-1.5-1.5-0, and 1.5-1.5-1.09-1.5 when using PIN diode A (MACOM MA4AGBLP912) and another four nonlinear states of 0-0-0.66-0, 0.61-0-1-0, 1-1-0.64-0, and 1-1-0.61-1 when using PIN diode B (MACOM MA4FCP300, MACOM, Lowell, MA, USA), four times on different dates. In the measurements for the two different PIN diodes, the ZORs are kept the same with a slight variation, indicating good stability of the nonlinear property. For example, for PIN diode A, state 1.01-0-1.5-0 provides the same resonance at 5.18 GHz at different measurement times, and other nonlinear states have variations less than 0.005 GH.

Third, voltage tolerance needs to be investigated because ZORs seem very sensitive to voltage variation when PIN diodes operate in the nonlinear transition zone. Figure 13a shows the supplying voltage has good tolerance and gets rid of the risk of excessive sensitive voltage variations regardless of the type of PIN diode. For instance, considering state 1.01-0-1.5-0 for PIN diode A, we can achieve a stable resonant frequency at 5.18 GHz despite varying the supplying voltage from 1.005 V to 1.015 V, meaning that we have a voltage tolerance of 0.01V. In terms of PIN diode B with the state 0.61-0-1-0, as shown in Figure 13b, similarly, we achieve a stable resonant frequency of 5.18 GHz despite varying the supplying voltage from 0.605 V to 0.621 V, which indicates that we have a voltage tolerance of 0.016 V.

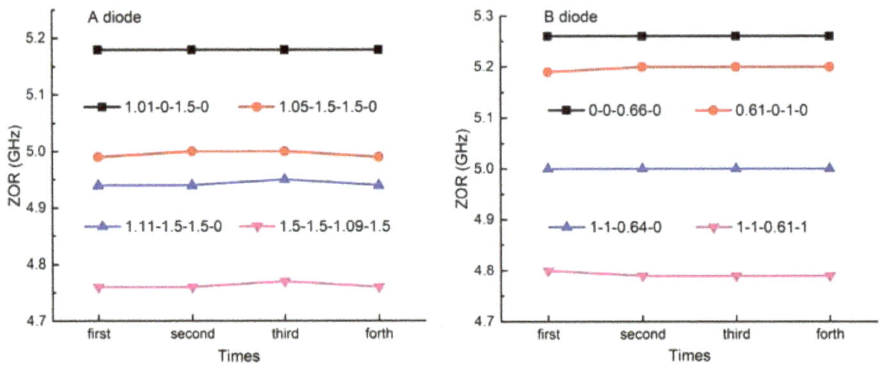

(a) (b)

Figure 12. The stability of the nonlinear states with (**a**) PIN diode A and (**b**) PIN diode B.

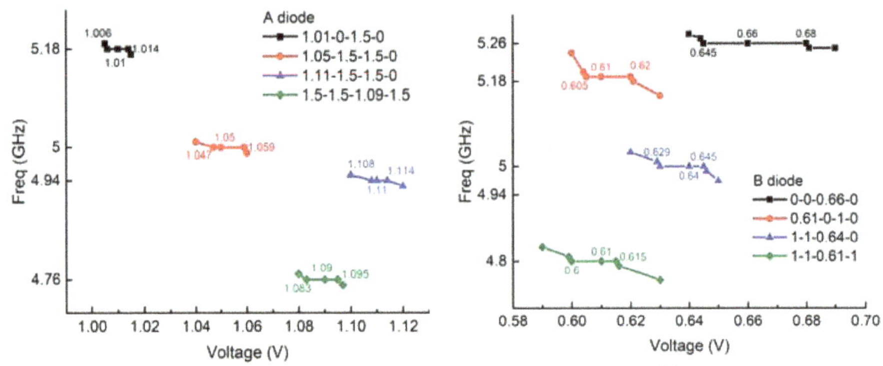

(a) (b)

Figure 13. The supplying voltage tolerance of nonlinear states of (**a**) PIN diode A (**b**) PIN diode B.

Finally, for the proposed active MTM antenna, we investigate the influence of the active components, including inductance, MCU, and PIN diodes on radiation gains and the efficiency. Looking at Figure 14a, several states for both active and passive cases are shown, and it is seen that the gains with active components decrease by 1 or 2 dBi compared to those without active components, while the radiation efficiency of the active case, as shown in Figure 14b, is lower than that of the passive case but no more than 10%.

Employing active components, as compared in Table 1, indeed shows the nonlinear advantages in eliminating resonance blindness over the passive case or the case only applying the RF switches with only OFF/ON states. Meanwhile, this proposed active MTM antenna requires actuated voltages lower than 1.5 V, which can be applied to 5G narrow bandwidth Internet of Things (NB-IoT) with low power capacities.

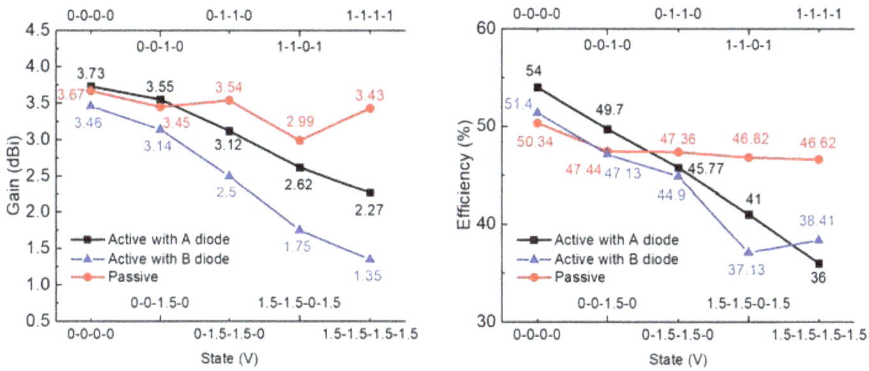

(a) (b)

Figure 14. Comparisons between the active antenna and passive antenna with both PIN diodes A and B for (**a**) measured gains and (**b**) the measured efficiency.

Table 1. Comparison of different types of tuning.

References	Active Devices	Number	Tuning Style	Tuning Stated	Bias Voltage (V)	Resonance Blindness
[8]	PIN diode	15	discrete	1.97 GHz and 2.37 GHz	0–0.7	yes
[10]	PIN diode and varactor diode	2 and 2	continuous	3.04 GHz to 5.89 GHz	0–30	no
[12]	PIN diode	6	discrete	5.95 GHz and 7.2 GHz	/	yes
[14]	Varactor diode	1	continuous	1.6 GHz to 2.23 GHz	2–20	no
[17]	Varactor diode	2	continuous	1.94 GHz to 2.44 GHz	0–20	no
This work	PIN diode	4	continuous	4.7 GHz to 5.3 GHz	0–1.5	no

6. Conclusions

In this paper, we study the nonlinear property of PIN diodes, fit it to an EM co-simulation, and, particularly, apply it to an active MTM antenna to eliminate resonance tuning blindness. We conclude that the nonlinear property indeed possesses the advantages to help achieve smooth resonance tuning with low actuated voltages, and it can be generally extended to other PIN diodes with good stability and voltage tolerance. The active MTM antenna with uniform and smooth frequency tuning slices the frequency spectrum into many narrow-band channels, which can be applied to 5G narrow bandwidth Internet of Things (NB-IoT), which requires spectrum channels of narrow bandwidth and low power capacities.

Author Contributions: Conceptualization, Y.L. and H.C.; validation, H.L. and Y.L.; writing—original draft preparation, H.L.; writing—review and editing, Y.H., and Y.L.; funding acquisition, G.Y. and H.C. All authors have read and agreed to the published version of the manuscript.

Funding: This research was funded by the National Natural Science Foundation of China (Grant No. 61801284).

Institutional Review Board Statement: Not applicable.

Informed Consent Statement: Not applicable.

Data Availability Statement: The data presented in this study are openly available.

Conflicts of Interest: The authors declare no conflict of interest.

References

1. Veselago, V.G. The electro dynamics of substances with simultaneously negative values of ε and μ. *Sov. Phys. Usp.* **1968**, *10*, 509–514. [CrossRef]
2. Smith, D.R.; Padilla, W.J.; Vier, D.C.; Nemat-Nasser, S.C.; Schultz, S. Composite Medium with Simultaneously Negative Permeability and Permittivity. *Phys. Rev. Lett.* **2000**, *84*, 4184–4187. [CrossRef]
3. Lai, A.; Caloz, C.; Itoh, T. Composite right/left-handed transmission line metamaterials. *IEEE Microw. Mag.* **2004**, *5*, 34–50. [CrossRef]
4. Sievenpiper, D.; Zhang, L.; Broas, R.; Alexopolous, N.; Yablonovitch, E. High-impedance electromagnetic surfaces with a forbidden frequency band. *IEEE Trans. Microw. Theory Tech.* **1999**, *47*, 2059–2074. [CrossRef]
5. Giovampaola, C.D.; Engheta, N. Digital metamaterials. *Nat. Mater.* **2014**, *13*, 1115–1121. [CrossRef] [PubMed]
6. Sievenpiper, D. Forward and backward leaky wave radiation with large effective aperture from an electronically tunable textured surface. *IEEE Trans. Antennas Propag.* **2005**, *53*, 236–247. [CrossRef]
7. Cai, J.; Zhou, Y.; Zhang, Y.; Li, Q. Gain-assisted ultra-high-Q spoof plasmonic resonator for the sensing of polar liquids. *Opt. Express* **2018**, *26*, 25460–25470. [CrossRef] [PubMed]
8. Khan, M.S.; Capobianco, A.-D.; Iftikhar, A.; Asif, S.; Ijaz, B.; Braaten, B.D. A Frequency-Reconfigurable Series-Fed Microstrip Patch Array with Interconnecting CRLH Transmission Lines. *IEEE Antennas Wirel. Propag. Lett.* **2016**, *15*, 242–245. [CrossRef]
9. Zhang, L.; Chen, X.; Liu, S.; Zhang, Q.; Zhao, J.; Dai, J.; Bai, G.; Wan, X.; Cheng, Q.; Castaldi, G.; et al. Space-time-coding digital metasurfaces. *Nature Commun.* **2018**, *9*, 4334. [CrossRef]
10. Li, T.; Zhai, H.; Li, L.; Liang, C. Frequency-Reconfigurable Bow-Tie Antenna with a Wide Tuning Range. *IEEE Antennas Wirel. Propag. Lett.* **2014**, *13*, 1549–1552. [CrossRef]
11. Caloz, C.; Itoh, T. *Electromagnetic Metamaterials: Transmission Line Theory and Microwave Applications*; John Wiley & Sons, Inc.: New York, NY, USA, 2005; Volume 7, pp. 316–322. [CrossRef]
12. Qin, P.-Y.; Weily, A.R.; Guo, Y.J.; Bird, T.S.; Liang, C.-H. Frequency Reconfigurable Quasi-Yagi Folded Dipole Antenna. *IEEE Trans. Antennas Propag.* **2010**, *58*, 2742–2747. [CrossRef]
13. Kim, J.; Kim, G.; Seong, W.; Choi, J. A Tunable Internal Antenna with an Epsilon Negative Zeroth Order Resonator for DVB-H Service. *IEEE Trans. Antennas Propag.* **2009**, *57*, 4014–4017. [CrossRef]
14. Mirzaei, H.; Eleftheriades, G.V. A Compact Frequency-Reconfigurable Metamaterial-Inspired Antenna. *IEEE Antennas Wirel. Propag. Lett.* **2011**, *10*, 1154–1157. [CrossRef]
15. Boukarkar, A.; Lin, X.Q.; Jiang, Y. A Dual-Band Frequency-Tunable Magnetic Dipole Antenna for WiMAX/WLAN Applications. *IEEE Antennas Wirel. Propag. Lett.* **2015**, *15*, 492–495. [CrossRef]
16. Yu, Y.; Xiong, J.; Li, H.; He, S. An Electrically Small Frequency Reconfigurable Antenna with a Wide Tuning Range. *IEEE Antennas Wirel. Propag. Lett.* **2011**, *10*, 103–106. [CrossRef]
17. Huang, H.-J.; Tsai, C.-H.; Lai, C.-P.; Chen, S.-Y. Frequency-Tunable Miniaturized Strip Loop Antenna Fed by a Coplanar Strip. *IEEE Antennas Wirel. Propag. Lett.* **2016**, *15*, 1000–1003. [CrossRef]
18. Chi, P.-L.; Waterhouse, R.; Itoh, T. Compact and Tunable Slot-Loop Antenna. *IEEE Trans. Antennas Propag.* **2011**, *59*, 1394–1397. [CrossRef]
19. Ko, J.; Kim, D. A Wideband Frequency-Tunable Dipole Antenna Based on Antiresonance Characteristics. *IEEE Antenna Wirel. Propag. Lett.* **2017**, *16*, 3067–3070. [CrossRef]
20. Takemura, N. Tunable Inverted-L Antenna with Split-Ring Resonator Structure for Mobile Phones. *IEEE Trans. Antennas Propag.* **2012**, *61*, 1891–1897. [CrossRef]
21. Chaabane, G.; Madrangeas, V.; Chatras, M.; Arnaud, E.; Huitema, L.; Blondy, P. High Linearity 3-Bit Frequency Tunable Planar Inverted F-Antenna for RF Applications. *IEEE Antennas Wirel. Propag. Lett.* **2017**, *16*, 983–986. [CrossRef]
22. Narayanan, S.; Tsolkas, D.; Passas, N.; Merakos, L. NB-IoT: A Candidate Technology for Massive IoT in the 5G Era. In Proceedings of the 2018 IEEE 23rd International Workshop on Computer Aided Modeling and Design of Communication Links and Networks (CAMAD), Barcelona, Spain, 17–19 September 2018; pp. 1–6.
23. Mahjoubi, A.E.; Mazri, T.; Hmina, N. NB-IoT and eMTC: Engineering Results Towards 5G/IoT Mobile Technologies. In Proceedings of the 2018 International Symposium on Advanced Electrical and Communication Technologies (ISAECT), Rabat, Morocco, 21–23 November 2018.
24. Yang, W.; Wang, M.; Zhang, J.; Zou, J.; Hua, M.; Xia, T.; You, X. Narrowband Wireless Access for Low-Power Massive Internet of Things: A Bandwidth Perspective. *IEEE Wirel. Commun.* **2017**, *24*, 138–145. [CrossRef]
25. Pozar, D.M. Transmission Line Theory. In *Microwave Engineering*, 4th ed.; John Wiley & Sons, Inc.: Hoboken, NJ, USA, 2007; Chapter 2; pp. 56–58.
26. Dubois, J.-M.; Ouanounou, G.; Rouzaire-Dubois, B. The Boltzmann equation in molecular biology. *Prog. Biophys. Mol. Biol.* **2009**, *99*, 87–93. [CrossRef] [PubMed]
27. Luo, Y.; Xu, J.; Yang, G.; Toshiyoshi, H. EM radiation from electrostatic nonlinear pull-in instability of MEMS. *Electron. Lett.* **2018**, *54*, 68–70. [CrossRef]

sensors

MDPI

Article

Circular Patch Fed Rectangular Dielectric Resonator Antenna with High Gain and High Efficiency for Millimeter Wave 5G Small Cell Applications

Abinash Gaya [1], Mohd Haizal Jamaluddin [1,*], Irfan Ali [1] and Ayman A. Althuwayb [2]

1 Wireless Communication Centre, School of Electrical Engineering, Universiti Teknologi Malaysia, Skudai 81310, Malaysia; abinashgaya@gmail.com (A.G.); irfan_lrk_15@yahoo.com (I.A.)
2 Electrical Engineering Department, Jouf University, Sakaka 72388, Aljouf, Saudi Arabia; aaalthuwayb@ju.edu.sa
* Correspondence: haizal@fke.utm.my; Tel.: +60-7-553-6107 (ext. 36107)

Abstract: A novel method of feeding a dielectric resonator using a metallic circular patch antenna at millimeter wave frequency band is proposed here. A ceramic material based rectangular dielectric resonator antenna with permittivity 10 is placed over a rogers RT-Duroid based substrate with permittivity 2.2 and fed by a metallic circular patch via a cross slot aperture on the ground plane. The evolution study and analysis has been done using a rectangular slot and a cross slot aperture. The cross-slot aperture has enhanced the gain of the single element non-metallic dielectric resonator antenna from 6.38 dB from 8.04 dB. The Dielectric Resonator antenna (DRA) which is designed here has achieved gain of 8.04 dB with bandwidth 1.12 GHz (24.82–25.94 GHz) and radiation efficiency of 96% centered at 26 GHz as resonating frequency. The cross-slot which is done on the ground plane enhances the coupling to the Dielectric Resonator Antenna and achieves maximum power radiation along the broadside direction. The slot dimensions are further optimized to achieve the desired impedance match and is also compared with that of a single rectangular slot. The designed antenna can be used for the higher frequency bands of 5G from 24.25 GHz to 27.5 GHz. The mode excited here is characteristics mode of TE_{1Y1}. The antenna designed here can be used for indoor small cell applications at millimeter wave frequency band of 5G. High gain and high efficiency make the DRA designed here more suitable for 5G indoor small cells. The results of return loss, input impedance match, gain, radiation pattern, and efficiency are shown in this paper.

Keywords: 5G; dielectric resonator antenna; aperture coupled; millimeter wave; 26 GHz; small cell

Citation: Gaya, A.; Jamaluddin, M.H.; Ali, I.; Althuwayb, A.A. Circular Patch Fed Rectangular Dielectric Resonator Antenna with High Gain and High Efficiency for Millimeter Wave 5G Small Cell Applications. *Sensors* **2021**, *21*, 2694. https://doi.org/10.3390/s21082694

Academic Editor: Naser Ojaroudi Parchin

Received: 13 March 2021
Accepted: 8 April 2021
Published: 11 April 2021

1. Introduction

To address the diversified requirements from the envisioned 5G usage scenarios, 5G needs access to "high", "medium", and "low" level of frequencies. The sub 6 GHz and millimeter wave 30 GHz band (e.g., 24.25 GHz–29.5 GHz and 37 GHz–43.5 GHz) will be most populated frequency bands for 5G. The base station antennas to be used for millimeter wave frequency bands must support high data rate transmission and high efficiency. 5G millimeter wave transmission upgrades to low latency transmissions with high data rate. Microstrip patch, dielectric resonator, and many such antennas have been used in millimeter wave frequency bands, but dielectric resonator antenna (DRA) has gained more popularity because of light weight, small size, zero surface wave loss, and metallic losses. It can achieve wider impedance bandwidth and gain compared to metallic antennas like as micro strip patch antenna. Metallic antennas like microstrip patch antenna has maximum metallic losses but ceramic based dielectric resonator antennas has minimum metallic losses at millimeter wave frequencies [1]. Dielectric resonator antennas (DRAs) are the most suitable candidates to replace the conventional radiating elements at millimeter wave frequencies and especially for indoor applications of millimeter wave frequency bands [2]. DRAs do not have conduction losses and are importantly characterized by high radiation efficiency when get excited with desired radiating mode [3,4]. In aperture coupled technique the slot dimensions made on the ground plane can have the effects of impedance and capacitance as the DRA is placed over the metallic ground plane [5]. The physical dimensions of a Dielectric Resonator is the function of its dielectric permittivity and loss tangent of the material used. So, the actual dimension of a DRA can be controlled to its minimal with larger Permittivity value range from 10 to 100. The resonant mode used depends on the geometry of the resonator and the required radiation pattern [6]. The Rectangular Dielectric Resonators have practical advantages over other shapes. Further, for a given resonant frequency, two aspect ratios of a rectangular DRA (height/length and width/length) can be chosen independently. Maximum impedance bandwidth can be achieved using a impedance match between the connector and the DRA [7]. The major advantage of using the rectangular DRA is characterized by three independent geometrical dimensions along three different coordinate axis, and the height of the DRA is d, which enhances maximum flexibility in the physical dimensions of rectangular DRA while compared to the cylindrical DRA [8,9]. A Dielectric Resonator can be excited through a strip line, Aperture, Coaxial, or substrate integrated waveguide techniques [10,11]. Using a proper excitation or feed technique, a dielectric resonator structure can act as a radiator at desired resonating frequencies. It has to notice that, for any given or desired resonant frequency, the actual size of a dielectric resonator is inversely proportional to its relative permittivity of the constitutive material. An average permittivity of 10 has better impedance bandwidth in DRAs [12]. The essential principle of operation of dielectric resonator is comparable to that of the cavity resonators. Based on such conventional design approaches the most popular radiating dielectric resonators are the cylindrical and the rectangular ones [3,13]. The aspect ratio and Q factor of DRA can be compared with Figure 1. An indoor small cell base station requires highly efficient antennas with lowest impedance mismatch. The complexity of the antenna design at millimeter wave frequency band is high because of impedance mismatch and small size. Dielectric resonator antennas offer wide bandwidth and high efficiency at millimeter wave frequency and which enhances the signal strength to overcome reflection losses and throughput in channel [14]. The use of circular patch as feed has been most popular as shown in previous work [15,16] but the measured gain at low frequencies is quite low. Moreover, a circular patch has been most convenient way to feed the antenna. As metallic antennas has losses associated with its metallic properties. An analytical study comparing a microstrip patch and dielectric resonator antenna has been shown by Guha and Kumar [8]. Using all basic feed mechanism. A dielectric resonator can exhibit as a magnetic dipole under basic feeding mechanism. The impedance variation in different feeding techniques helps to excite the DRA with better reflection coefficient and with desired mode of excitation. Circular patch antennas have been used as more convenient antennas because of low profile characteristics. In this paper an aperture coupled technique has been used fed by a microstrip circular patch antenna placed on the other side of the substrate. The impedance bandwidth of a Dielectric Resonator is the function of materials permittivity and Length to Height ratio. Because of Advantages like low loss, small size, wide bandwidth, and easy of excitation the dielectric resonators are used at mm wave transmissions. The most conventional feed technique in DRAs is using a slot over the ground plane. The impedance match allows the antenna to deliver maximum radiating power along the desired direction and the slot apertures made on the ground plane excites the required resonant modes of the DRA [17,18]. The antenna designed here has used a cross type slot over the ground plane to enhance the gain of the DRA. This gain enhancement can be considered as a better feed mechanism compared to other feeding schemes. Small cells in 5G needs high gain and wideband antenna system for indoor cellular services. The throughput and efficiency of the channel can be enhanced using the high gain and efficient antenna designs. The interference and internal reflections reduce the signal strength, so for a high gain antenna system can be a better system. Indoor small cells need high gain antenna performances as the reflections across the walls will generate maximum attenuation.

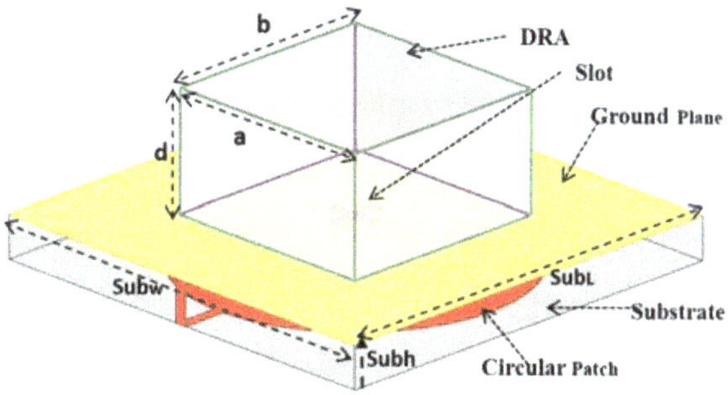

Figure 1. Dielectric Resonator Antenna fed by a Circular Patch.

The design proposed in this paper which can be used for indoor 5G applications in the frequency band of 24.25 GHz–27.5 GHz. This novel design method can also distinguish between the Dielectric Resonator Antennas to other conventional antennas as microstrip patch. The simulation design and study are carried out using High frequency structure simulator (HFSS). The optimization results of slot dimensions, input impedance to the antenna and DRA dimensions are also presented here. Here the simulation work is carried out using HFSS and DRA dimensions are calculated using Mat Lab. In Section 2 the antenna design and basic calculations are expressed with study of feed dimensions, aperture coupled mechanism and gain enhancement. Section 3 explains about the optimization study and analysis of DRA dimensions and radiated field effects and its comparison of rectangular slot with the cross slot. The limitations of such antenna design are also discussed in the Section 4 of this article. A manual fabrication and glue technique used in fixing the DRA over the substrate need high accuracy and proficiency.

2. Antenna Design and Analysis

A ceramic ECCOS-TOCK Hik material type Dielectric Resonator antenna is used with permittivity $\varepsilon_r = 10$ and loss tangent 0.002 over a substrate of Roger RT/Duroid 5880 with permittivity 2.2 and loss tangent 0.009. The ground plane is above the substrate and a micro strip patch is used as feed and is placed below the substrate. The slot is created via the ground plane over which the DRA is placed. The slot dimensions are calculated with respect to the resonating wavelength 11.52 mm and are further optimized to match with required input impedance of 50 Ohm. The calculated dimensions of the DRA are a = 2.9 mm, b = 2.6 mm and d = 1.4 mm which is shown in Figure 1. The calculated dimensions of the substrate are $\mathrm{Sub_L}$ = 5.76 mm, $\mathrm{Sub_W}$ = 5.76 mm, $\mathrm{Sub_h}$ = 0.254 mm. The feed line dimensions calculated are $\mathrm{L_1}$ = 0.63 mm, $\mathrm{L_{1W}}$ = 0.15 mm. The position of the DRA can be moved either along x or y direction to achieve an efficient coupling. Here a high-quality factor (Q) vale of 14 has been achieved theoretically considering the permittivity of DRA as 10. The resonant frequency of the DRA is proportional to $\epsilon_r^{-0.5}$, so for a wide range of permittivity values can be used to resonate the antenna at required frequency bands. Figure 1 represents the DRA design method. The theoretical calculations for resonating frequency of a rectangular dielectric resonator antenna are shown in Equations (1) and (2).

$$k_x \times \tan(k_x d/2) = \sqrt{(\varepsilon_r - 1)k_0^2 - k_x^2}, \tag{1}$$

where

$$k_0 = \frac{2\pi}{\lambda_0} = \frac{2\pi f_0}{c}$$

$$k_y = \frac{m\pi}{b}$$

and

$$k_z = \frac{n\pi}{d},$$

and

$$k_x^2 + k_y^2 + k_z^2 = \varepsilon_r \times k_0^2, \quad (2)$$

where c is velocity of light, ε_r is relative permittivity of DRA, k_0 is free space wave number, m and n are called as half-wave field variations along the y and z directions, respectively. The symbols k_x, k_y, and k_z represent the wave numbers in the x, y, and z-directions, respectively, and a, b, and d indicates the dimensions of DRA which are proportional to the square root of dielectric constant values of the ceramic based DRA.

The measured dimensions of the DRA are calculated using Mat lab and the simulated and optimized details of DRA dimensions are shown in Table 1. Figure 2 shows the top view and bottom view of the antenna design. The resonant frequency of the rectangular DRA can be found from Equation (1). The rectangular dielectric resonator offers three degree of freedom as it has three coordinate axes with respect to length, width, and height of the DRA. k_X, k_Y, and k_Z are the three coordinate axis wave numbers along x, y, and z direction of the DRA. A rectangular DRA can have three different characteristic's modes which are called as the fundamental modes of DRA TE_{X11}, TE_{1Y1}, and TE_{11Z} [19].

Table 1. Antenna Design Specifications. (DRA: Dielectric Resonator Antenna).

Antenna Parameters	Parameter Details	Values in mm
a	DRA Width	2.9
b	DRA Length	2.9
d	DRA Height	1.4
S_W	Slot Width	0.2
S_L	Slot Length	0.8
X	Ground Plane Width	5.76
Y	Ground Plane Length	5.76
Subh	Substrate Height	0.254
SubL	Substrate Length	5.76
SubW	Substrate Width	5.76
r	Radius of Patch	2.25
L1	Patch Feed Length	0.63
L1W	Patch feed Width	0.15

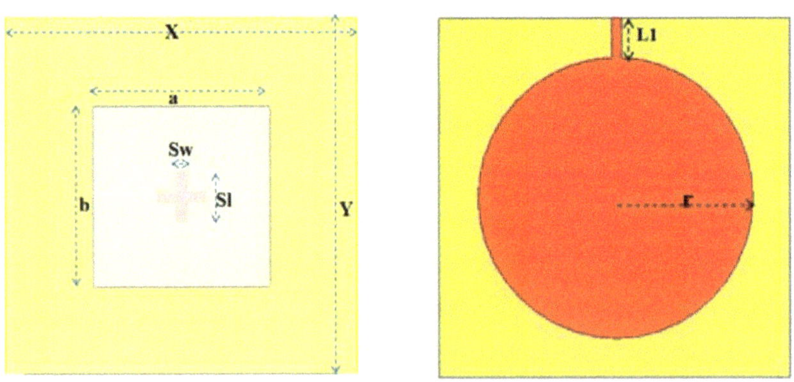

(a) (b)

Figure 2. Aperture Coupled Dielectric Resonator Antenna: (**a**) Top View showing the Cross Slot; (**b**) Bottom View showing the circular patch.

Ansys HFSS is used here for design and simulation of the antenna. All the boundary conditions are satisfied to create a perfect electric field environment. The unit cell design is placed at the center of the coordinating axes which is $x = 0$ and $y = 0$. The DRA is placed above the substrate so the height along the z axis is the substrate height. The perfect electric field ground plane is generated to match with the potential differences of the antenna. The DRA dimensions that are optimized further to match with the required impedance bandwidth and the calculation for the quality factor of the DRA are shown in the Table 2. Here both the length and the width of the DRA are kept similar and the height of the DRA is optimized to excite the DRA under the characteristic's mode. The aspect ratio of the DRA can further be optimized with the varying dimensions height to length ratio. The mode of excitation either can be TE (Transverse Electric) or TM (Transverse Magnetic) depending upon the physical dimension and electric field distribution of the DRA. The coupling of radiated field between the patch and the DRA depends on the physical dimensions of the cross slot made on the ground plane. Table 2 has the details of aspect ratio, quality factor (Q) and impedance bandwidth of the antenna. The radiation Q-factor is then found by determining the radiated power and stored energy. Further the quality factor can be analyzed with the dimension ratio of the DRA and the aspect ratio requirements for wider bandwidth. The substrate and the ground plane dimensions are (5.76 mm × 5.76 mm). The slot is made at the center of the ground plane which is placed over the substrate.

Table 2. Antenna Design Calculations for aspect ratio and bandwidth.

ϵ_r	a (mm)	a/b (mm)	d/b (mm)	b (mm)	d (mm)	Q	Bandwidth (%)
10	2.9	1	0.48	2.9	1.4	14.13	5 (Simulated) 4 (Measured)

2.1. The Metallic Circular Patch as the Feed to DRA

The circular patch element is used to feed the DRA across the cross slot made over the ground plane. The coupling of electric fields across the slot depends on the power radiated by the patch. Figure 3a shows the fields radiated across the patch and Figure 3b shows the fields radiated over the DRA. The radius of the patch is further optimized to check with the input impedance of the DRA. Both the DRA and the patch is centered at the same point and the electric field radiated by the patch element is easily coupled to the DRA.

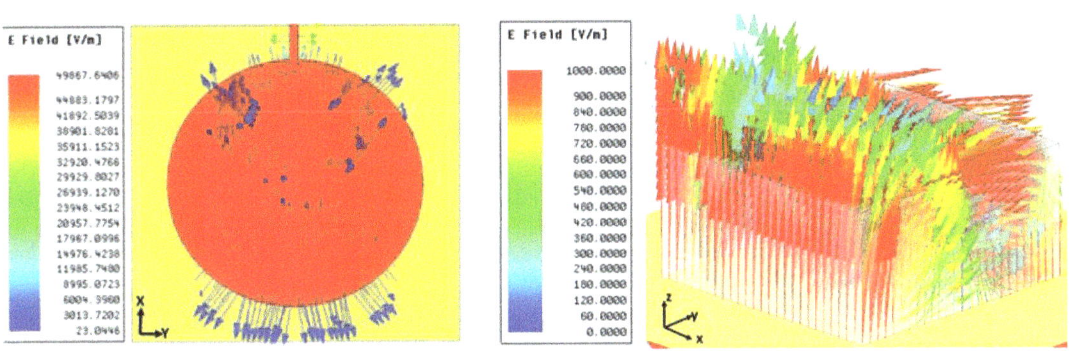

(**a**) (**b**)

Figure 3. Electric field distribution over (**a**) metallic circular patch (**b**) DRA.

The power radiated by the circular patch depends upon the slot impedance as well as the load impedance at the connector. The difference between the radiated power and the input power will measure the radiation efficiency of the DRA. Here, the slot impedance and the load impedance (impedance offer at the connector to the microstrip line is maintained to minimum such that maximum power can be delivered to the radiated circular patch. Figure 4 shows the magnetic dipole moment distribution over the patch. The power radiated by the circular patch can be used to measure the conductance and the directivity of the antenna. As the circular patch is placed at the center of the substrate and at position z = 0, maximum power will be radiated to the DRA through the slots made on the ground plane.

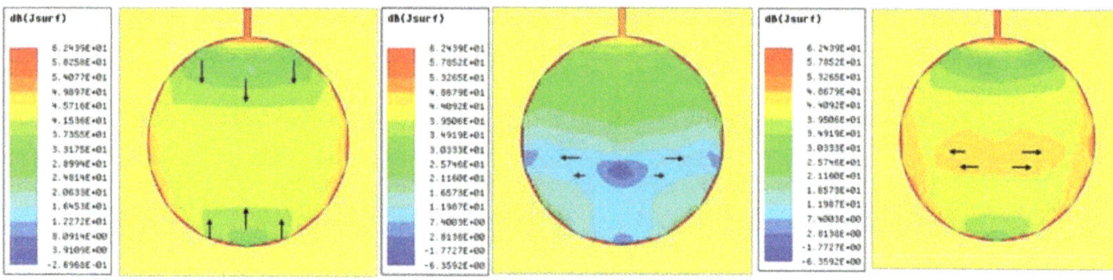

Figure 4. Electric field distribution over DRA radiating like a dipole moment.

2.2. *Aperture Coupling Mechanism and Calculations*

The DRA is placed over the ground plane where slot is been made. The slot dimensions as slot length and slot width can be calculated as the desired resonant frequency of the antenna. The DRA is placed exactly over the slot apertures above to a height of 0.254 mm which is the thickness of the substrate. The metallic circular patch is placed on the other side of the substrate. The dimensions of both the ground plane is maintained same as the substrate dimensions and calculated in terms of wavelength. The optimization study of the partial ground plane effect in done in the next part of the paper. Equations (3) and (4) represents the equations for calculating the slot dimensions.

The slot dimensions can be calculated as

Slot Length,

$$S_L = \frac{0.4\lambda_0}{\sqrt{\varepsilon_{eff}}} \tag{3}$$

where ε_{eff} is called as effective permittivity which is calculated as

$$\varepsilon_{eff} = \frac{\varepsilon_r + \varepsilon_s}{2}$$

where ε_r and ε_s are the dielectric constants of the DRA and the substrate, respectively.

Slot Width,

$$S_w = 0.2 \times S_L \tag{4}$$

The calculated vales slot length and slot width are at the resonating frequency of 26 GHz as shown in Table 3. The electric fields radiated into the DRA depends on the calculated aperture dimensions and the amount of coupling depends on electric field distributed over the DRA. Equation (5) represents the coupling factor from feed line to DRA through a slot on the defined ground plane.

Table 3. Slot dimension Calculations.

Resonating Frequency f_r (GHz)	DRA Permittivity ε_r	Substrate Permittivity ε_s	Effective Permittivity ε_{eff}	Slot Length S_L (mm)	Slot Width S_w (mm)
26	10	2.2	6.1	1.86 (Theoretical) 0.8 (Optimized)	0.37 (Theoretical) 0.2 (Optimized)

The coupling factor (C) of the DRA can be expressed as

$$C = \int_v [(E_{DRA} \cdot Js)dV], \quad (5)$$

Here E_{DRA} is the electric field vector distribution over the transmission line and Js is a uniform current source. The distributed surface current need to be controlled over the thickness of the dielectric substrate. The coupling factor is proportional to the slot dimensions on the ground plane, as the electric field distribution over the DRA generates an electric and magnetic dipole moment. There are slight variations in the simulated and theoretical values of slot dimensions made on the ground plane. The calculated values are used to optimize the antenna design parameters to achieve the maximum impedance bandwidth. The slot dimension calculations are shown in Table 3. The slot length and width dimensions for the cross slot is uniform and is placed at the center of the coordinate axis. The slot impedance is here uniform to that of characteristics impedance of the feed which helps in enhancing the radiation efficiency of the DRA. The electric field distribution over both XZ and XY plane is shown in Figure 5. The excited characteristics mode here is TE_{1Y1}, the cross slot excitation has increased the gain of the DRA without changing its mode of excitation. The fundamental characteristics mode of excitation here matches with the 50 Ohm input impedance of the DRA. The DRA here is linearly polarized and can be used for unidirectional indoor 5G applications because of its high gain and efficiency. The impedance bandwidth improvement from rectangular slot to cross slot is about 0.8 percentage which is a bandwidth of 0.22 GHz. Figure 6 shows the electric volume magnitude in dB scale.

The gain enhancement occurs by changing the rectangular slot to a cross slot, where a cross slot fed DRA acts as a magnetic dipole. There a slight improvement in the gain values has been achieved from rectangular to cross slot over the ground plane. The cross slot excites the DRA with higher electrical energy coupling resulting in enhancing the gain of the DRA. The impedance variation occurs when a rectangular slot is replaced with a cross slot over the ground plane. The slot dimensions are responsible for the coupling between the DRA and the circular microstrip patch. The gain improvement of 0.35 dB has been recorded here. The cross slot enhances gain with minimum cross polarization. The rectangular slot is placed at the upper edge of the DRA, but the cross slot is placed at the center of the DRA. These slot positions are optimized locations and are studied and fixed to resonate at the desired resonating frequency of the antenna. The study of mode excitation and fields is shown in Figures 5 and 6.

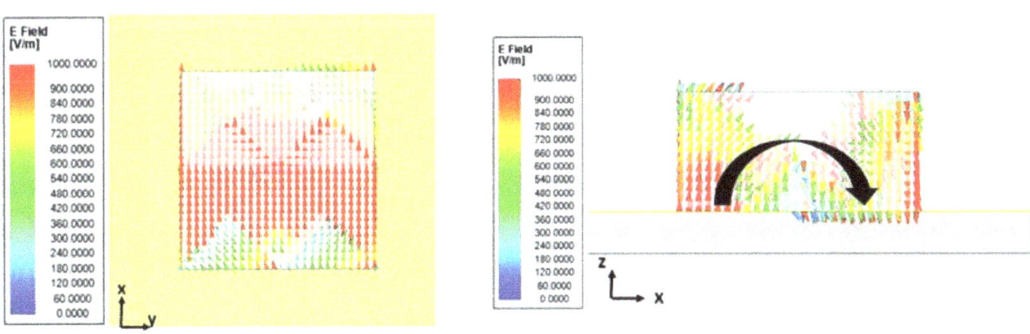

(**a**) (**b**)

Figure 5. Electric field distribution over DRA (**a**) XY Plane (**b**) XZ Plane.

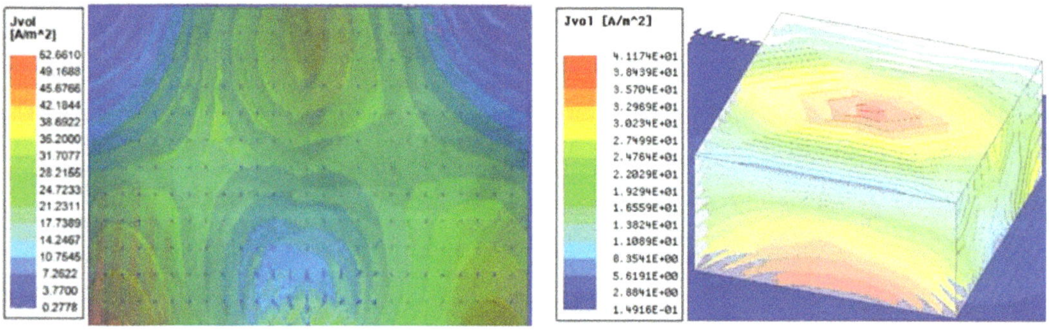

(**a**) (**b**)

Figure 6. Electric field magnitude distribution over DRA (**a**) Field density front view (**b**) 3D View.

3. Evolutionary Study and Analysis

The design of the final antenna has been done from a conventional way of feeding a dielectric resonator antenna which is shown in Figure 2. The DRA is placed over a rectangular slot made on the ground plane. The ground plane is placed above to the substrate and a micro strip line is placed at the other side of the substrate. The gap of 0.254 mm which is the thickness of the substrate is maintained between the micro strip feed line and the rectangular slot or DRA on the ground plane. The length and width of the micro strip line is optimized to match with the characteristic's impedance of the antenna. A general input impedance of 50 ohm is observed to excite the DRA under fundamental modes. The magnitude of electric field distribution over the two different slots are also shown in Figure 7c,d. The cross slot has been prepared with two narrow rectangular slots of equal length and width. The optimization study of different slot dimensions with impedance bandwidth has also presented here. The optimization study is carried out for the rectangular slot and the cross slot over the ground plane. The characteristics mode analysis also has been done at different slot length and width dimensions. For characteristics mode of TE_{1y1} the electric field across the slots follows the similar radiation pattern as per both E-plane and H-plane, which minimizes the back-lobe radiation in the DRA. Figure 7 shows the evolution process of final DRA design from using a rectangular slot to a cross slot and Figure 8 shows the electric field density in dB scale. The uniform slot dimensions also reduce the cross polarized power in the antenna. The feed of either a cross slot or a rectangular slot make the DRA act like a magnetic dipole. As the DRA is placed directly above the ground plane with a cross slot helps also to reduce the back propagation and cross pol power of the antenna. The cross slot offers a minimum cross pol in the design.

In millimeter waves the dimensions of substrate and ground plane are calculated in terms of wavelength of desired resonating frequency and controlled to uniform the field distribution and fringing fields over the DRA and the ground plane. The Electric fields over the metallic patch are radiating outwards and is maximum at the feed point as a conventional radiator. The DRA works like a magnetic dipole under is characteristics mode of excitation as shown in Figure 7c,d. The characteristics of both the rectangular and cross slot are studied and its performance analysis is recorded in Table 4. The cross slot has achieved high gain and wide bandwidth and high isolation as compared to the rectangular slot on the ground plane. Figure 9 shows the reflection coefficient of both the rectangular and cross slot and Figure 10 shows the corresponding input impedances at the resonating frequencies. The dimensions of the slot apertures for both rectangular and cross slot are remained uniform. The center of the cross slot and rectangular slot are coincided with the DRA and the coordinate axes. The co and cross pol radiation pattern for both the rectangular and cross slot are shown in Figures 11 and 12.

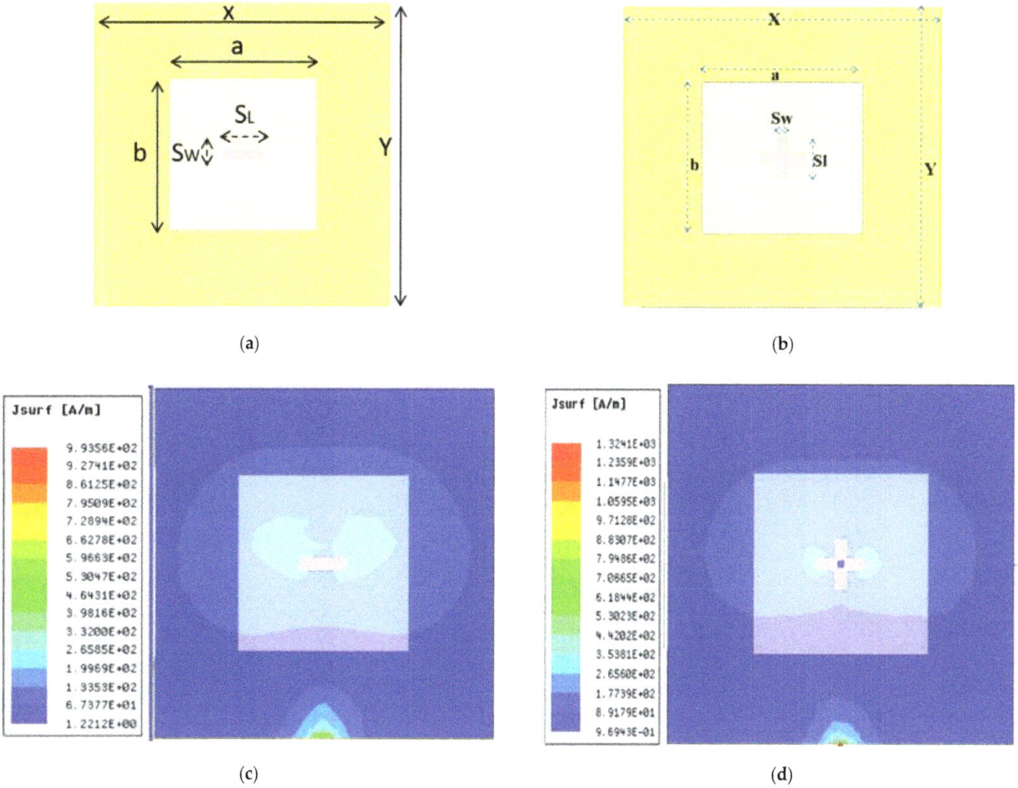

Figure 7. Evolution of the antenna design: (**a**) Micro strip line fed with rectangular slot DRA; (**b**) Circular patch fed cross slot DRA (**c**) Surface current density over a rectangular slot (**d**) Surface Current density over a cross slot.

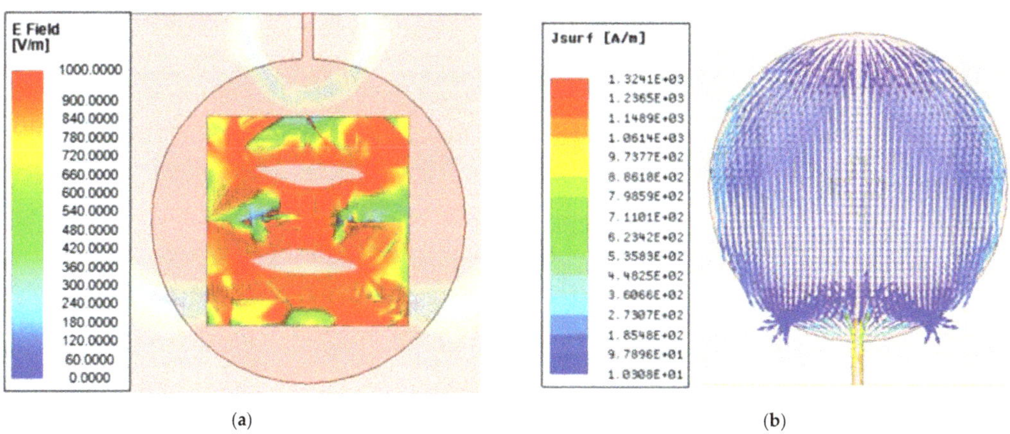

Figure 8. (**a**) Electric field distribution over the circular patch in XY Plane (**b**) Electric field distribution over the circular patch in XY plane.

Table 4. Performance Comparison of different slot apertures.

Slot Type	Bandwidth (GHz)	Bandwidth (%)	S11 Isolation (dB)	Gain Simulated (dB)
Rectangular Slot	0.84 GHz (25.96–26.8 GHz)	3.23%	−24.18 dB	6.8 dB
Cross Slot	1.06 GHz (25.32–26.38 GHz)	4.07%	−32.35 dB	8.04 dB

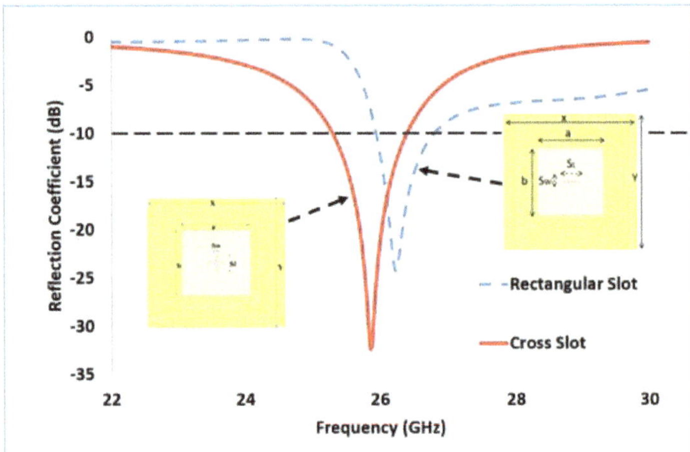

Figure 9. Simulated Reflection coefficient (dB) for different slot structures.

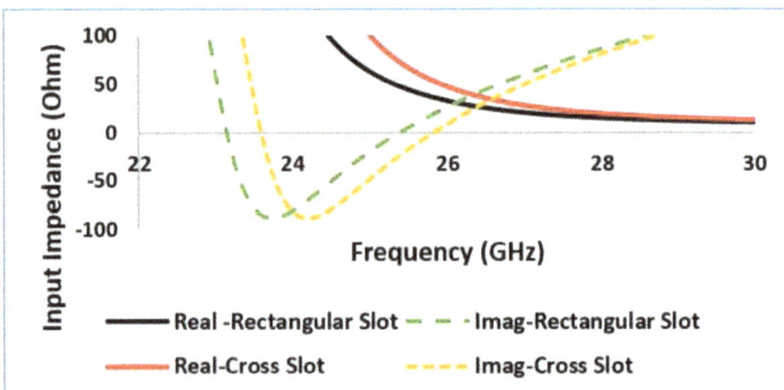

Figure 10. Input Impedance (ohm) at different slot structures.

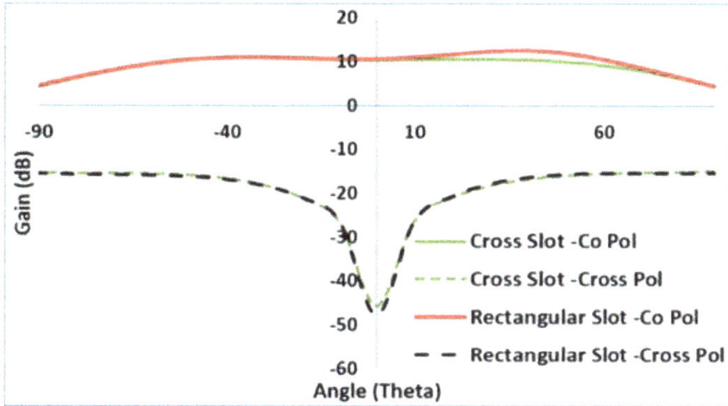

Figure 11. Simulated Radiation pattern in YZ Plane for different slot structures.

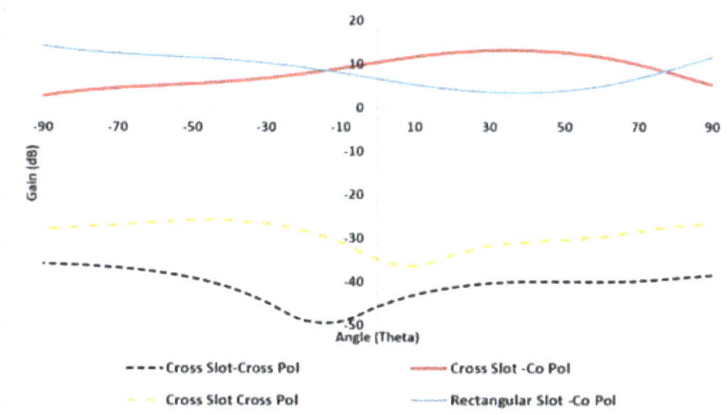

Figure 12. Simulated Radiation pattern in XZ Plane for different slot structures.

The simulated gain values for both rectangular and cross slot in both co and cross polarization is shown in Figures 11 and 12 in both YZ plane and XZ plane, respectively. There is non-uniform radiation pattern with low cross pol power in both the planes. The maximum radiation is along the broadside direction of the antenna. The simulation study has been carried including all the antenna parameters as reflection coefficient, gain, and radiation pattern for both the rectangular and cross slot on the ground plane.

The performance of DRA fed by both rectangular slot and cross slot is shown in Table 4. The gain and bandwidth both have been improved by using a cross slot than a rectangular slot. This gain enhancement is observed without changing the feed dimensions. There is an improvement of 0.22 GHz in bandwidth which is 0.8% from rectangular slot to cross slot. There is also an improvement in gain values. This optimization study has helped in finding the maximum gain, efficiency, and bandwidth of the antenna.

4. Results

The simulated and measured reflection coefficient for cross slot is shown in Figure 13 and the corresponding input impedance at different patch radius is shown in Figure 14. The characteristics impedance of 50 Ohm has been maintained near the patch radius, which has helped in achieving.

1. Delivering maximum power to the antenna which enhances the radiation efficiency further.
2. Minimum cross pol power in both the E and H plane of the antenna.

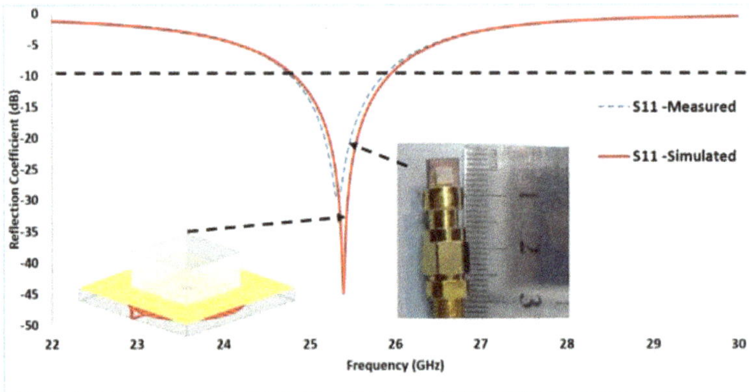

Figure 13. Reflection coefficient (dB) vs. frequency (GHz).

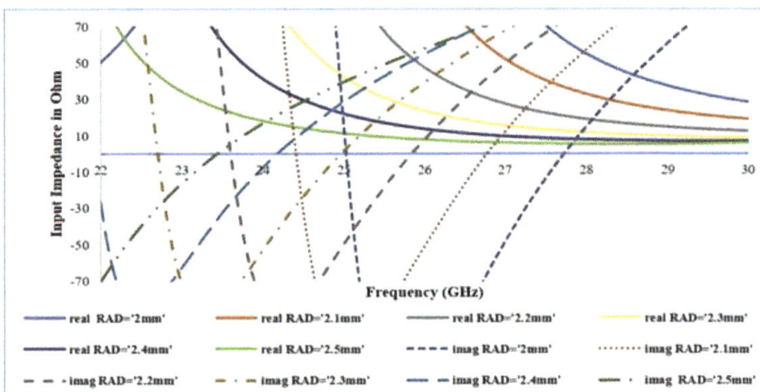

Figure 14. Variation of Characteristic's input impedance in ohm at different radius of circular patch in mm.

The patch radius is varied with different radius to achieve the desired impedance bandwidth. The simulation study of the patch radius is also carried out and its reflection coefficient is shown in Figure 15. At patch radius 2.2 mm it has achieved maximum isolation of −32.4 dB and other values isolation data is shown in Table 4 at different patch radius. The input impedance at 26 GHz with patch radius 2.2 mm is 49 Ohm which matches with the characteristic's impedance of the antenna. The simulated and measured reflection coefficient isolation is −44.82 dB and −29.23 dB, respectively. The DRA dimensions are calculated and are optimized in terms of the aspect ratio with respect to the Q factor. A bandwidth of 1.12 GHz is suitable to use this antenna for indoor 5G small cells. Here the DRA is excited under the characteristic's mode TE $_{1Y1}$. The aspect can be optimized by changing the height of the DRA. Figure 16 shows the reflection coefficient of DRA at different DRA heights.

The physical dimension of the DRA is very small, which makes several fabrication errors and destabilizes the radiation characteristics of the antenna. Here, the input impedance of the strip line to the patch is varied based on the impedance offed by the connector. Moreover, the characteristics impedance is well matched which is around 49 Ohm and has offered maximum radiation efficiency. Figures 17 and 18 shows the optimization of feed length and width of the microstrip line and its corresponding input impedance variation. Table 5 represents the reflection coefficient at different dimensions of the cross slot. The cross slot has been made from a rectangular slot. The variation in the slot structure shows that the cross pol of cross slot is higher than that of rectangular slot maintain a non-variable input impedance. The uniform variation in the slot dimension varies the impedance values.

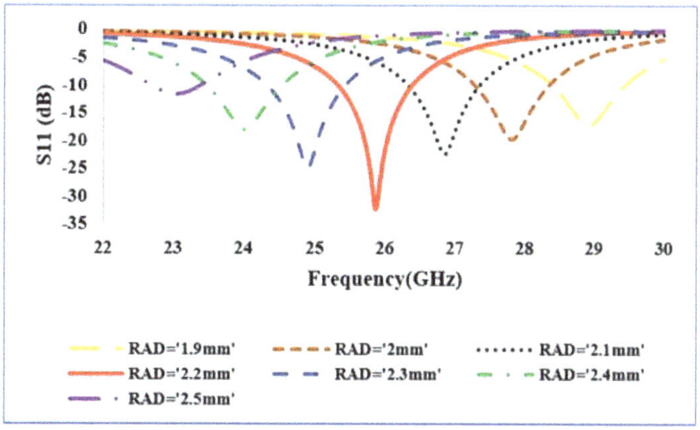

Figure 15. Reflection coefficient in dB at different radius values of the circular patch in mm.

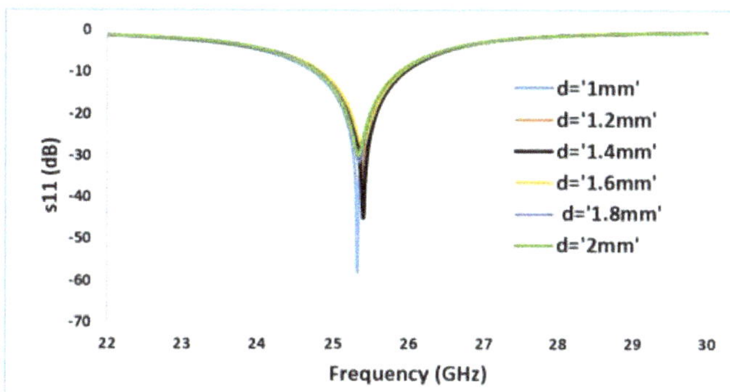

Figure 16. Reflection coefficient in dB at different radius values of the circular patch in mm.

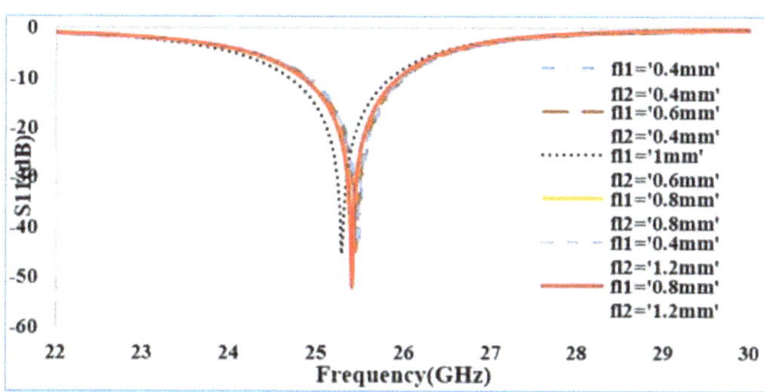

Figure 17. Reflection coefficient (dB) vs. Frequency (GHz) at different feed length (fl).

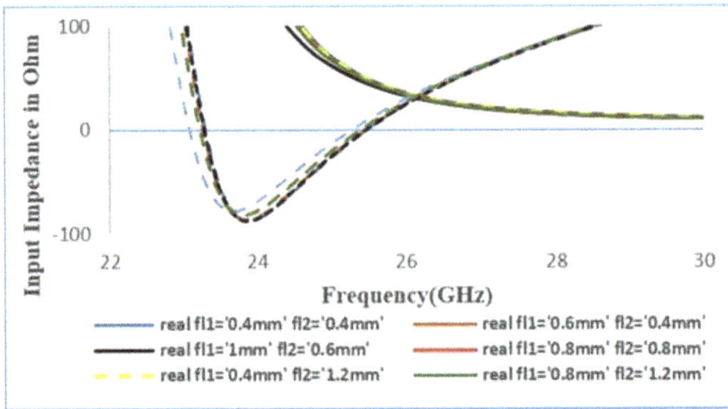

Figure 18. Input Impedance of DRA at different feed dimensions.

Table 5. Return Loss (dB) at different slot dimensions.

Slot Length (fl1,fl2)	Resonating Frequency (GHz)	Return Loss (S11) (dB)
0.4,0.4	28.98	−30.86
0.6,0.4	27.82	−37.50
1.0,0.6	26.84	−45.36
0.8,0.8	25.86	−38.04
0.4,1.2	24.92	−39.78
0.8,1.2	24.06	−52.07
0.4,0.4	23.02	−30.86
0.6,0.4	28.98	−37.50

Different slot dimensions changes the voltage across the terminal load and is mode dependent on the impedance appears at the feed terminal. The Table 6 shows the feed length and the impedance bandwidth isolation of the DRA. The Maximum isolation of −52.07 dB is observed at the slot dimensions of 1.2 mm length and 0.8 mm width. The dimensions of cross slot are similar and is placed at z = 0. At first the rectangular slot is studied and to improvise the characteristics parameters a cross slot has been made on the ground plane. The change in slot structure has not much change the cross-pol power but has enhanced the gain and bandwidth of the antenna. Figure 19 shows the fabrication process and the measurement of antenna parameters in the anechoic chamber.

Table 6. Return Loss (dB) for different patch radius.

Patch Radius in mm (RAD)	Resonating Frequency (GHz)	Return Loss (S11) (dB)
1.9	28.98	−17.09
2.0	27.82	−19.88
2.1	26.84	−22.51
2.2	25.86	−32.35
2.3	24.92	−24.97
2.4	24.06	−18.14
2.5	23.02	−11.07
1.9	28.98	−17.09

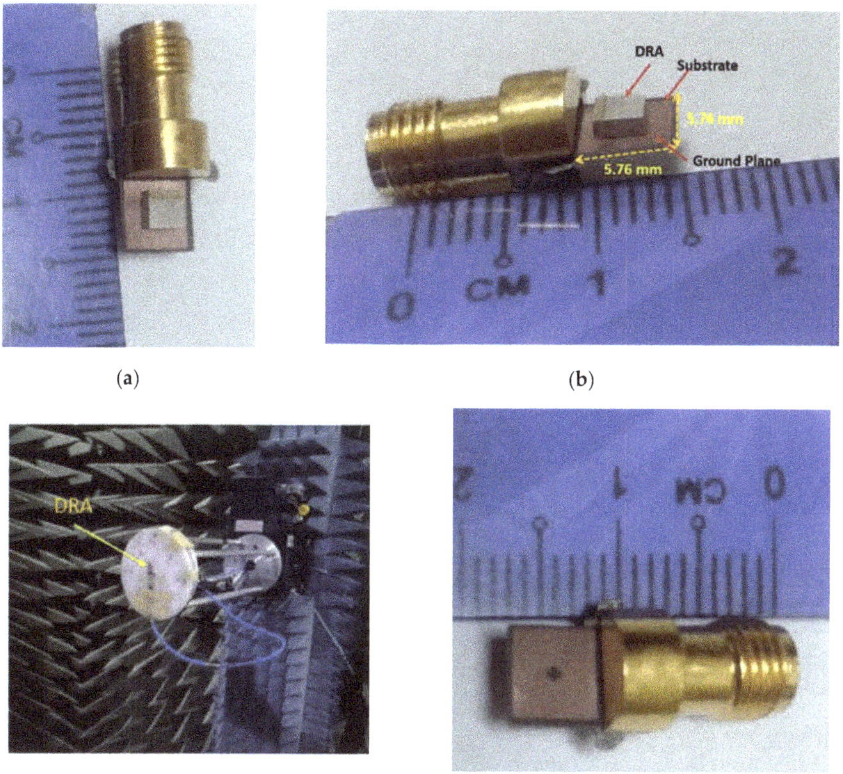

(a) (b) (c) (d)

Figure 19. Fabricated Rectangular DRA at resonating frequency 26 GHz (**a**) Rectangular DRA Top View; (**b**) side view (**c**) Measurement in anechoic chamber (**d**) Cross slot aperture.

Figure 20 represents the gain and efficiency measured at the desired resonating frequency. The DRA exhibits a higher efficiency of 96 percentage and gain improvement to 10 dB. Figures 21 and 22 represents the gain of DRA both in E plane and H plane. The simulated isolation in cross pol power in E plane is −45.62 dB and in H plane is −49.73 dB. The measured values of isolation in cross pol −43.64 dB in E plane and −49.12 dB in H plane. The simulated and measured gain of the DRA is 10.57 dB and 8.04 dB, respectively. The cross slot in the ground plane has reduced the cross-pol power and has enhanced the gain of the DRA. The cross pol reduces the back-lobe radiation of the antenna. There is slight shift in the cross-pol minima by 10 degree in the measured results. Similarly, the co pol and cross pol power in H plane is shifted by 10 degree. The recorded efficiency of the DRA is 96 percent, as the slot impedance is 49 ohm which is very close to the characteristic's impedance of 50 ohm. So maximum power has been delivered to the antenna enhancing its gain and efficiency. In Table 7, the previous work on DRA has been recorded and compared with this work at millimeter wave frequency bands. This singly fed DRA design has more advantages compared to the previous work in terms and gain and efficiency. The DRA designed here can be used as an efficient radiator for 5G indoor small cells. The antenna is linearly polarized and has maximum radiation along the broadside direction which can help in minimizing the path loss component of the propagation. The performance parameters as bandwidth, gain, efficiency, and radiation pattern are shown here.

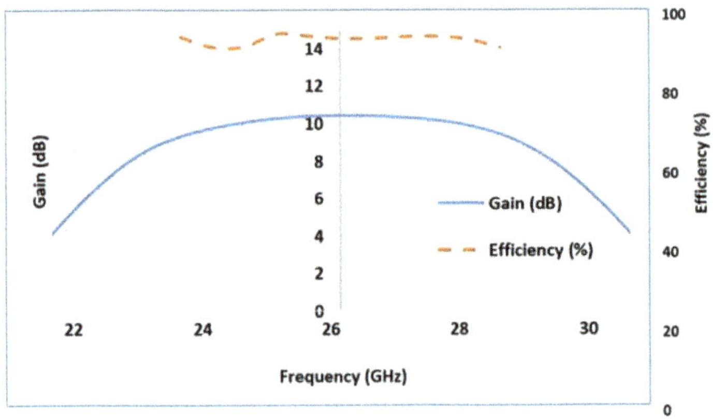

Figure 20. The measured gain and efficiency of the DRA.

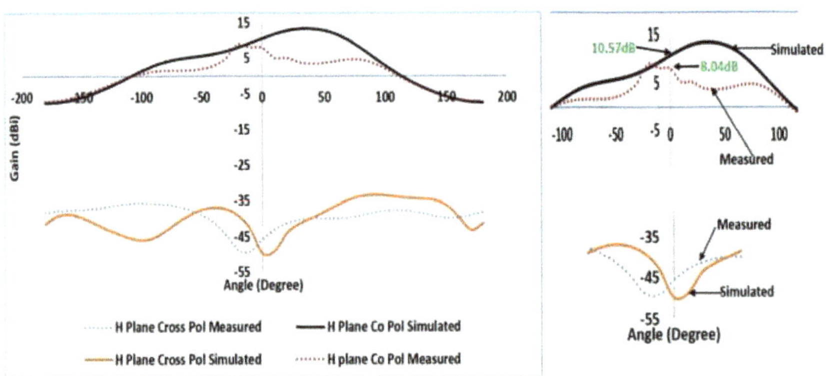

Figure 21. Radiation pattern XZ plane.

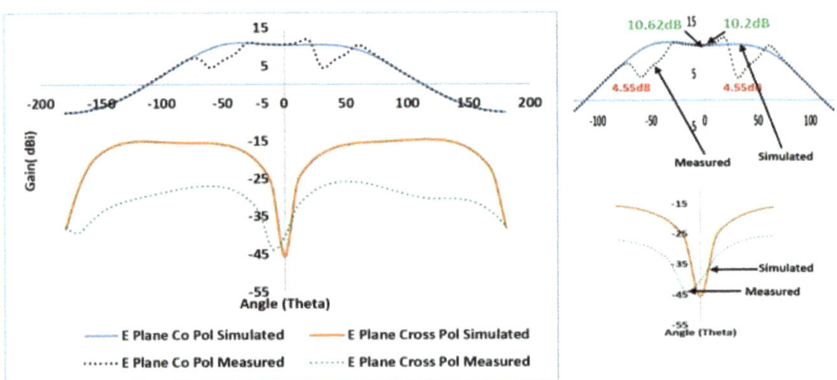

Figure 22. Radiation Pattern YZ plane.

Table 7. Performance Analysis of the proposed work.

Ref No	Frequency GHz	DRA Shape	Permittivity ε_r	Bandwidth %	Electrical Dimensions	Gain dB	Efficiency %
[20]	35 GHz	Cylindrical	10	15.6	$0.14\lambda_0 \times 0.12\ \lambda_0$	6.9	95
[21]	36.22 GHz	Rectangular	10.2	8.6	$0.24\lambda_0 \times 0.3\lambda_0 \times 0.24\lambda_0$	5.51	95
[22]	24 GHz	Rectangular	10	3.74	$0.38\lambda_0 \times 0.51\lambda_0 \times 0.24\lambda_0$	5.9	Not Mentioned
[23]	60 GHz	Rectangular	12.6	6.1	$0.2\lambda_0 \times 0.2\lambda_0 \times 0.2\lambda_0$	6.0	98
[24]	35 GHz	Cylindrical	10.2	11.0	$0.4\lambda_0 \times 0.4\lambda_0 \times 0.07\lambda_0$	5.5	88
[25]	26 GHz	1*2 MIMO Rectangular DRA	10.2	7.3	$0.39\lambda_0 \times 0.39\lambda_0 \times 0.11\lambda_0$	7.1	Not Mentioned
[26]	32 GHz	Grid DRA	17	5.31	$0.3\lambda_0 \times 0.3\lambda_0 \times 0.1\lambda_0$	6.4	89
Proposed	26 GHz	Rectangular	10	4.07	$0.25\lambda_0 \times 0.25\lambda_0 \times 0.12\lambda_0$	8.04	96

In Table 7 the gain, bandwidth, electrical dimensions, and efficiency of the designed antenna is compared with other antennas referred. With permittivity of 10 the DRA has achieved a higher gain value compared to other antennas. The quarter wavelength dimensions of the antenna are electrically larger at desired resonating frequency. Generally, the circular metallic patch used here offers an upward electric field coupling to the DRA through the cross slot with low surface impedances. The reactance value of the antenna is very low and is matched with the input impedance resulting in a broadside radiation.

5. Discussion

Using a simple novel method of aperture coupling a Dielectric resonator antenna with permittivity of 10 has achieved a gain of 8.04 dB. A non-resonating circular patch antenna has been used as a feed to the ceramic based dielectric resonator antenna. Cross polarization power is also minimized with the slot dimensions made on the ground plane. The difficulty was high in the fabrication of the antenna and its placement on the ground plane. The design will be further developed to large array patterns to achieve higher gain and efficiency. There is a shift in the resonating frequency from simulation study to measurement study because of

1. Difficulties in the placement of the DRA on the slot apertures
2. Use of conductive glue to fix the DRA on the ground plane.

Whereas the impact on radiation efficiency was less as the measured cross polarization power was less.

3. Fabrication error.

6. Conclusions

Dielectric Resonator Antennas can be preferred because of wide bandwidth and high efficiency at millimeter wave frequency bands. 5G band 30 GHz from 24.3 to 27.8 GHz band is widely used and can be used for indoor cellular applications. All the design parameters were calculated using MATLAB and all boundary conditions for simulation environment were achieved here. The measurement conditions were satisfied between the DRA and the receiver antenna. The DRA is linearly polarized and is excited under the characteristic's mode TE_{1Y1}. An enhancement in gain and bandwidth has been also done by using a cross slot aperture in ground plane. A singly fed DRA proposed here has achieved high gain (10.57 dB simulated and 8.04 dB Measured), high efficiency (96% Measured and 98% Simulated), and wide bandwidth (1.12 GHz) makes it more suitable antenna for indoor millimeter wave 5G small cell applications.

Author Contributions: Investigation, Performed Experiment, formal analysis, and methodology, A.G.; supervision and project administration, M.H.J. and A.A.A.; data curation and assisted in paper writing I.A. All authors have read and agreed to the published version of the manuscript.

Funding: This work was supported in part by Universiti Teknologi Malaysia Under TDR Grant vote 05G20 and HiCOE Grant vot 4J415.

Institutional Review Board Statement: Not Applicable.

Informed Consent Statement: Not Applicable.

Data Availability Statement: Not Applicable.

Conflicts of Interest: The authors declare no conflict of interest.

References

1. Mongia, R.K.; Bhartia, P. Dielectric resonator antennas—A review and general design relations for resonant frequency and bandwidth. *Int. J. Microw. Millim. Wave Comput. Eng.* **1994**, *4*, 230–247. [CrossRef]
2. Mongia, R.K. Reduced size metallized dielectric resonator antennas. In Proceedings of the IEEE Antenna and Propagation Society International Symposium 1997, Digest, Montreal, QC, Canada, 13–18 July 1997; Volume 4, pp. 2202–2205. [CrossRef]
3. Keyrouz, S.; Caratelli, D. Dielectric Resonator Antennas: Basic Concepts, Design Guidelines, and Recent Developments at Millimeter-Wave Frequencies. *Int. J. Antennas Propag.* **2016**, *2016*. [CrossRef]
4. Mongia, R.K.; Ittipiboon, A. Theoretical and Experimental Investigations on Rectangular Dielectric Resonator Antennas. *IEEE Trans. Antennas Propag.* **1997**, *45*, 1348–1356. [CrossRef]
5. Pozar, D.M. A review of aperture coupled microstrip antennas: History, operation, development, and applications. *Univ. Mass. Amherst* **1996**, 1–9.
6. Huang, C.-Y.; Wu, J.-Y.; Wong, K.-L. Cross-slot-coupled microstrip antenna and dielectric resonator antenna for circular polarization. *IEEE Trans. Antennas Propag.* **2002**, *47*, 605–609. [CrossRef]
7. Altaf, A.; Seo, M. Dual-band circularly polarized dielectric resonator antenna for wlan and wimax applications. *Sensors* **2020**, *20*, 1137. [CrossRef]
8. Guha, D.; Kumar, C. Microstrip Patch versus Dielectric Resonator Antenna Bearing All Commonly Used Feeds: An experiment study to choose the right element. *IEEE Antennas Propag. Mag.* **2016**, *58*, 45–55. [CrossRef]
9. Mongia, R.K.; Ittipiboon, A.; Cuhaci, M. Measurement of Radiation Efficiency of Dielectric Resonator Antennas. *IEEE Microw. Guid. Wave Lett.* **1994**, *4*, 80–82. [CrossRef]
10. Petosa, A.; Mongia, R.K.; Ittipiboon, A.; Wight, J.S. Investigation of various feed structures for linear arrays of dielectric resonator antennas. In Proceedings of the IEEE Antennas and Propagation Society International Symposium 1995, Digest, Newport Beach, CA, USA, 18–23 June 1995; Volume 4, pp. 1982–1985. [CrossRef]
11. Kishk, A.A. Dielectric resonator antenna elements for array applications. In Proceedings of the IEEE International Symposium on Phased Array Systems and Technology, Boston, MA, USA, 14–17 October 2003; pp. 300–305.
12. Gaya, A.; Jamaluddin, M.H.; Ali, I.; Mohamad, H. Dual Band Circularly Polarized Rectangular Dielectric Resonator Antenna for millimeter wave 5G Applications. In Proceedings of the 2019 IEEE Asia-Pacific Conference on Applied Electromagnetics (APACE), Melacca, Malaysia, 25–27 November 2019; pp. 1–5.
13. Kishk, A.A. Wide-Band Truncated Tetrahedron Dielectric Resonator Antenna Excited by a Coaxial Probe. *IEEE Trans. Antennas Propag.* **2003**, *51*, 2913–2917. [CrossRef]
14. Trinh-Van, S.; Yang, Y.; Lee, K.Y.; Hwang, K.C. A wideband circularly polarized pixelated dielectric resonator antenna. *Sensors* **2016**, *16*, 1349. [CrossRef] [PubMed]
15. Lopato, P.; Herbko, M. A circular microstrip antenna sensor for direction sensitive strain evaluation. *Sensors* **2020**, *18*, 310.
16. Trinh-Van, S.; Yang, Y.; Lee, K.Y.; Hwang, K.C. Broadband circularly polarized slot antenna loaded by a multiple-circular-sector patch. *Sensors* **2020**, *18*, 1576.
17. Gaya, A.; Jamaluddin, M.H.; Kamarudin, M.R.; Ali, I. A wideband dielectric resonator antenna with a cross slot aperture for 5G communications. *TELKOMNIKA* **2019**, *17*, 2218–2225. [CrossRef]
18. Ali, I.; Jamaluddin, M.H.; Gaya, A.; Rahim, H.A. A Dielectric Resonator Antenna with Enhanced Gain and Bandwidth for 5G Applications. *Sensors* **2020**, *20*, 675. [CrossRef] [PubMed]
19. Petosa, A.; Ittipiboon, A. Dielectric Resonator Antennas: A Historical Review and the Current State of the Art. *IEEE Antennas Propag. Mag.* **2010**, *52*, 91–116. [CrossRef]
20. Lai, Q.; Almpanis, G.; Fumeaux, C.; Benedickter, H.; Vahldieck, R. Comparison of the radiation efficiency for the dielectric resonator antenna and the microstrip antenna at Ka band. *IEEE Trans. Antennas Propag.* **2008**, *56*, 3589–3592.
21. Abdel Wahab, W.M.; Busuioc, D.; Safavi-Naeini, S. Low cost planar waveguide technology-based dielectric resonator antenna (DRA) for millimeter-wave applications: Analysis, design, and fabrication. *IEEE Trans. Antennas Propag.* **2010**, *58*, 2499–2507. [CrossRef]

22. Pan, Y.M.; Leung, K.W.; Luk, K.M. Design of the millimeter-wave rectangular dielectric resonator antenna using a higher-order mode. *IEEE Trans. Antennas Propag.* **2011**, *59*, 2780–2788. [CrossRef]
23. Ohlsson, L.; Bryllert, T.; Gustafson, C.; Sjoberg, D.; Egard, M.; Arlelid, M.; Wernersson, L.-E. Slot-coupled millimeter-wave dielectric resonator antenna for high-efficiency monolithic integration. *IEEE Trans. Antennas Propag.* **2013**, *61*, 1599–1607. [CrossRef]
24. Gong, K.; Hu, X.H. Low-profile substrate integrated dielectric resonator antenna implemented with PCB process. *IEEE Antennas Wirel. Propag. Lett.* **2014**, *13*, 1023–1026. [CrossRef]
25. Pan, Y.M.; Qin, X.; Sun, Y.X.; Zheng, S.Y. A Simple Decoupling Method for 5G Millimeter-Wave MIMO Dielectric Resonator Antennas. *IEEE Trans. Antennas Propag.* **2019**, *67*, 2224–2234. [CrossRef]
26. Mazhar, W.; Klymyshyn, D.M.; Wells, G.; Qureshi, A.A.; Jacobs, M.; Achenbach, S. Low-Profile Artificial Grid Dielectric Resonator Antenna Arrays for mm-Wave Applications. *IEEE Trans. Antennas Propag.* **2019**, *67*, 4406–4417. [CrossRef]

sensors

MDPI

Article

Design of Minimum Nonlinear Distortion Reconfigurable Antennas for Next-Generation Communication Systems

Germán Augusto Ramírez Arroyave [1,*], Antoni Barlabé [1], Lluís Pradell [1], Javier Leonardo Araque Quijano [2], Bedri A. Cetiner [3] and Luis Jofre-Roca [1,*]

1 Department of Signal Theory and Communications (TSC), School of Telecommunications Engineering, Universitat Politècnica de Catalunya (UPC), Campus Nord, 08034 Barcelona, Spain; antoni.barlabe@upc.edu (A.B.); lluis.pradell@upc.edu (L.P.)

2 Department of Electrical and Electronic Engineering (DIEE), Faculty of Engineering, Universidad Nacional de Colombia, Ciudad Universitaria, Bogotá 111321, Colombia; jlaraqueq@unal.edu.co

3 Electrical and Computer Engineering Department, Utah State University, 4120 Old Main Hill, Logan, UT 84322-4120, USA; bedri.cetiner@usu.edu

* Correspondence: german.augusto.ramirez.a@upc.edu (G.A.R.A.); luis.jofre@upc.edu (L.J.-R.)

Abstract: Nonlinear effects in the radio front-end can degrade communication quality and system performance. In this paper we present a new design technique for reconfigurable antennas that minimizes the nonlinear distortion and maximizes power efficiency through the minimization of the coupling between the internal switching ports and the external feeding ports. As a nonlinear design and validation instance, we present the nonlinear characterization up to 50 GHz of a PIN diode commonly used as a switch for reconfigurable devices in the microwave band. Nonlinear models are extracted through X-parameter measurements supported by accurate calibration and de-embedding procedures. Nonlinear switch models are validated by S-parameter measurements in the low power signal regime and by harmonic measurements in the large-signal regime and are further used to predict the measured nonlinearities of a reconfigurable antenna. These models have the desired particularity of being integrated straightforwardly in the internal multi-port method formulation, which is used and extended to account for the power induced on the switching elements. A new figure of merit for the design of reconfigurable antennas is introduced—the power margin, that is, the power difference between the fed port and the switching elements, which combined with the nonlinear load models directly translates into nonlinearities and power-efficiency-related metrics. Therefore, beyond traditional antenna aspects such as port match, gain, and beam orientation, switch power criteria are included in the design methodology. Guidelines for the design of reconfigurable antennas and parasitic layers of minimum nonlinearity are provided as well as the inherent trade-offs. A particular antenna design suitable for 5G communications in the 3.5 GHz band is presented according to these guidelines, in which the specific switching states for a set of target performance metrics are obtained via a balancing of the available figures of merit with multi-objective separation criteria, which enables good control of the various design trade-offs. Average Error Vector Magnitude (EVM) and power efficiency improvement of 12 and 6 dB, respectively, are obtained with the application of this design approach. In summary, this paper introduces a new framework for the nonlinear modeling and design of reconfigurable antennas and provides a set of general-purpose tools applicable in cases beyond those used as examples and validation in this work. Additionally, the use of these models and guidelines is presented, demonstrating one of the most appealing advantages of the reconfigurable parasitic layer approach, their low nonlinearity.

Keywords: reconfigurable antennas; reconfigurable parasitic layers; antenna optimization; antenna design; nonlinear characterization; behavioral modelling; x-parameters; PIN diode

Citation: Ramírez Arroyave, G.A.; Barlabé, A.; Pradell, L.; Araque Quijano, J.L.; Cetiner, B.A.; Jofre-Roca, L. Design of Minimum Nonlinear Distortion Reconfigurable Antennas for Next-Generation Communication Systems. *Sensors* **2021**, *21*, 2557. https://doi.org/10.3390/s21072557

Academic Editor: Naser Ojaroudi Parchin

Received: 17 February 2021
Accepted: 26 March 2021
Published: 6 April 2021

1. Introduction

Wireless communications systems using medium and high transmitter power and advanced modulation techniques must ensure that nonlinearities are below certain safety levels as harmonic components and intermodulation products can degrade system performance and even damage components.

In particular, for New Radio (NR), the Fifth Generation of Mobile Communications (5G), very stringent performance requirements in terms of signal quality, unwanted emissions, and intermodulation are specified [1–4]. These requirements are accompanied by conducted and radiated testing for verifying conformance to the standard for transmitting and receiving User Equipment (UE)s and Base Station (BS)s [5,6].

As power handling is higher in the BS, attention is focused on this component for which the values specified in [4] are taken as reference. The BS output power limit for conducted tests (defined at antenna connector, or the Transceiver Array Boundary (TAB) connector depending on the BS type) for the cell coverage scenarios considered are (i) Wide Area (No upper limit), (ii) Medium-Range $\leq$38 dBm, (iii) Local Area $\leq$24 dBm.

On the side of signal quality, the metric of interest is the Error-Vector Magnitude (EVM), defined according to the modulation scheme used—QPSK 17.5%, 16QAM 12.5%, 64QAM 8%, 256QAM 3.5%. On the other hand, the limit on unwanted, out-of-band and spurious, emissions is given on terms of two additional metrics, the Adjacent Channel Leakage power Ratio (ACLR) and the Operating Band Unwanted Emissions (OBUE). Of these two, the former is the most stringent with a limit of 45 dB. Finally, on the side of intermodulation, the requirements are that the Inter-Modulation Products (IMP) be attenuated by at least 30 dB.

These metrics and performance requirements imply stringent design goals, but even with a careful engineering some undesired nonlinearities can still persist; therefore, nowadays it is common the use of digital compensation techniques to mitigate the effects of the analog components to achieve conformance to the technical specifications. For these compensation techniques to be effective, a nonlinear model of the analog device to be compensated for is required [7,8].

Even though antennas are traditionally reckoned as inherently linear devices, the recent evolution of the field has relied on introducing nonlinear loads and components such as integrated switches, varactors, and PIN diodes into or in the surroundings of the (re-)radiating structure. This evolution has raised concerns on how these new kinds of nonlinear devices can impact system performance. Consequently, accurate nonlinear antenna modeling is a need that must be addressed, bearing a twofold advantage, as a component design tool and as input for RF system designers to integrate the antenna into the transceiver chain.

The use of PIN diode switches in microwave radio front ends is widespread, and new applications are rapidly emerging, for example in Reconfigurable Antenna (RA) [9,10], reconfigurable Parasitic Layer (PL) [11], reconfigurable surfaces [12,13], reconfigurable phased arrays and a variety of sub-systems for RADAR and Millimeter Waves (mmW) communications. Notwithstanding that PIN diodes are nonlinear components, their system impact on figures of merit such as EVM, and Inter-Modulation Distortion (IMD) is generally overlooked or shadowed by elements conventionally believed to be more critical such as power amplifiers (PA). But when the number of PIN diodes grows or when placed in high power paths, their contribution to system nonlinearities should be accounted for.

Although the initial design and proof of concept of reconfigurable devices can rely on models such as the ideal switch abstraction, linear circuit equivalents, or measured S parameters, the validity and accuracy of these models are rapidly lost as biasing, frequency, and power are changed. Therefore, in stringent applications such as 5G, nonlinear models of the PIN diode should be used to predict system behavior in different operation conditions.

Nonlinear models of microwave devices and circuits [14] can be categorized either as Compact/Physical (based on the physics governing the device) or Black-Box/Behavioral (based on the characteristics of the device from its terminals) [15]. When the internal structure of the Device Under Test (DUT) is not disclosed or is of no interest to the system designer, black-box modeling is the most suitable approach, with various alternatives available [16–21].

The X parameters [22] stand out among the available behavioral modeling choices. They are an appealing option given some of their features such as extraction stability, time-invariance, convergence to the S parameters in the small-signal regime, and their integration into simulation software.

Albeit the nonlinear behavior of the PIN diode is widely reported and mostly attributed to its nonlinear resistance characteristic [23–25], and SPICE models suitable for the microwave range have been proposed [26,27], most manufacturers do not provide the parameters and values needed to be used in compact models at high frequencies.

Consequently, X parameters are proposed in this work to experimentally obtain a behavioral model of a PIN diode commonly used for mmW applications. The extracted diode model is validated through independent S parameters and harmonic distortion measurements.

Reconfigurable antennas in general and parasitic layer based reconfigurable antennas [11] presenting compact size, low complexity, low power consumption, and small nonlinearities, in particular, are an appealing alternative for the deployment of the 5G-NR usage scenarios (Ultra-Reliable and Low Latency Communications (URLLC), Enhanced Mobile Broadband (EMB), and Massive Machine Type Communications (MMTC)).

The extracted nonlinear model of the PIN diode switch is a useful input to gain insight into the nonlinear behavior of reconfigurable antennas using PIN diode switches like the one illustrated in Figure 1, based on [28], which is taken as a case of study in this work to demonstrate a new methodology for the design of minimum nonlinearity reconfigurable antennas and parasitic layers suitable for BS and Customer-Premises Equipment (CPE) in 5G EMB scenarios.

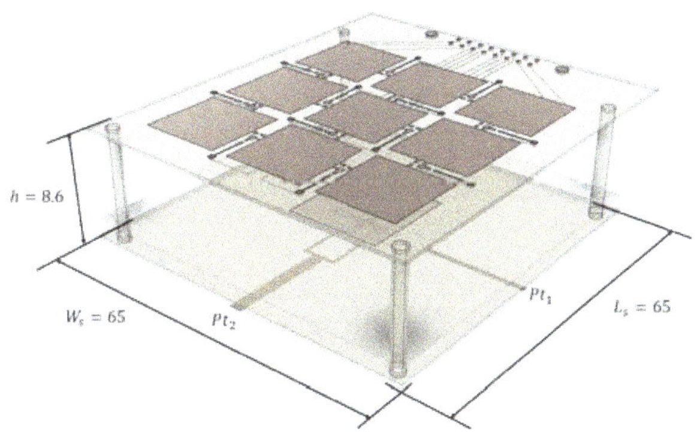

Figure 1. Exploded view of the Parasitic-Layer-based Reconfigurable-antenna used as case study. (Units in mm).

The nonlinear behavior of RAs is scarcely reported in the literature, and design strategies considering nonlinearities are non-existent. Two recent experimentally-oriented works are [29], where Third Order Intercept (TOI) and 1 dB Compression Point (1dB CP) are presented, and [30] which focuses on EVM and IMP measurements. Nevertheless, both cases lack an analytical formulation to explain or estimate the nonlinear behavior of the antenna, which would be required to calculate the measured effects or to consider the nonlinearities in the design stage. This is precisely the gap that this work intends to fill, opening new possibilities for the well-balanced design of reconfigurable antennas for the rigorous 5G performance requirements.

Notwithstanding the lack of theoretical approaches and the scarcity of experimental works studying nonlinear effects in RA, the case of nonlinearly loaded basic antennas and arrays has been tackled to some extent. With few exceptions that deal with the nonlinear problem entirely within the electromagnetic solution, which proves useful in cases where nonlinearities are inherent to the antenna or propagation medium materials, the most reasonable approach to the analysis of nonlinearly loaded antennas is to split the problem into an electromagnetic (linear) and a circuital (nonlinear) problem. The former is commonly dealt with by Full-Wave Electromagnetic (FWEM) numerical techniques, whereas the latter can be tackled by formulations such as the Harmonic Balance (HB) analysis [31].

Nonlinear antenna analysis can be traced back to the works of [32], based on a Time-Domain (TD) formulation of Method of Moments (MoM) that allows the direct inclusion of the nonlinear load model into the calculation. Another approach is presented in [33,34], where the antenna problem is solved by MoM, while a Frequency-Domain (FD) formulation based on a Modified Volterra Series (MVS) is used to solve the resulting nonlinear circuit for single and multiple antenna cases. Likewise, Ref. [35] presents a direct nonlinear space-time solution of the whole antenna current leading to limited applicability and an FD solution of the antenna problem followed by a circuital nonlinear time-marching solution. Harmonic balance to solve the nonlinear problem is introduced in [36] alongside a transformation matrix technique in the frequency domain to deal with the nonlinear harmonic load case. In [37], the reflection algorithm [31] and a variation of HB where the direct solution is replaced by Generalized Minimal Residual Method (GMRES) are used. Likewise, in [38], the Arithmetic Operator Method (AOM) is used to deal with the nonlinearity of the load.

A common drawback in these works is that they assume a given simplified model for the nonlinear switching element $i - v$ characteristic which is not often accurate for real-life devices neither directly measurable for RF and microwave components.

Therefore, given the nonlinear characteristics of the loads, a suitable modeling technique must be used. Measured X-parameters of the PIN diode are developed in this work as a convenient alternative for stating the characteristics of the nonlinear loads.

Likewise, the Internal Multi-Ports Method (IMPM) [39–41] is further extended in this work to efficiently account for the power delivered to the switching elements and the fed ports. The IMPM requires only one FWEM simulation to determine the behavior of all the possible switches configurations, and for the present application, its use involves placing an antenna port during FWEM simulation on the locations where the switching elements may be placed. This approach employs efficient analytical formulations to perform all the subsequent circuital calculations for the switching states of interest, thus allowing to study reconfigurability attending performance both of the port match and the radiated fields. By combining this technique with the nonlinear models of the diode, estimates of system parameters like the EVM and IMD can be made, allowing the design of minimum nonlinearity reconfigurable antennas and parasitic layers.

The remainder of this paper is organized as follows—Section 2 presents the nonlinear diode characterization and validation as well as the application of the extracted models to a particular case, Section 3 introduces the criteria for the minimum nonlinearity design of reconfigurable parasitic layers alongside novel design guidelines for the implicit trade-offs, Section 4 shows the application of the design guidelines for a particular design of a minimum nonlinearity PL-enhanced antenna; finally, the conclusions section summarizes the main contributions of this work and the possible future developments.

2. Nonlinear Switch Characterization

The PIN diode MACOM MA4AGBLP912, operating over the 50 MHz–40 GHz frequency range, was selected for testing. The measurements are performed using a Keysight N5245A PNA-X network analyzer capable of extracting X parameters, connected by means of a Cascade-Microtech 150 µm-pitch Ground-Signal-Ground (GSG) Co-Planar Waveguide (CPW) probe to the test fixture.

2.1. Interpretation of X-Parameters

The notation used in X parameters is based on port, harmonic and propagation direction convention, in which A_{qn} represents the nth harmonic of an incoming signal at port q, while B_{pm} is the mth harmonic of an outgoing signal at port p.

In the general X-parameters formulation, outgoing port waves are expressed as a function of the Large Signal Operating Point Stimuli (LSOPS), which is composed of the device DC Stimulus (DCS), and the RF Stimuli, which in turn is composed of the fundamental magnitude $|A_{11}|$, and the harmonic set of incoming signals A_{qn} which are conveniently referred to the phase of the fundamental $P = e^{j\phi 11}$ ($\phi 11 = \angle A_{11}$).

A common approximation, valid for most devices, that significantly reduces the computational resources required for measurement, is to consider that the fundamental is the only large RF signal entering the DUT, hence, the only one that can drive it into a nonlinear behavior, while the injected/reflected signals at the remaining ports and harmonics combinations are considered small-signal components. Accordingly, the reference LSOPS is solely defined by $refLSOPS = \{\{DCS_q\}, |A_{11}|\}$.

Then, with the aid of linearization and harmonic superposition assumptions, dropping the dependence on $refLSPOS$ for the sake of clarity, the port waves can be related [42] through:

$$B_{pm} = X_{pm}^{(FB)} P^m + \sum_{(q,n)\neq(1,1)} X_{pm,qn}^{(S)} A_{qn} P^{m-n} + \sum_{(q,n)\neq(1,1)} X_{pm,qn}^{(T)} A_{qn}^{*} P^{m+n}, \tag{1}$$

where $X_{pm}^{(FB)}$, $X_{pm,qn}^{(S)}$, and $X_{pm,qn}^{(T)}$ are the X-parameters of type FB, S, and T respectively, whose interpretation is as follows: the $X_{pm}^{(FB)}$ represent the large-signal part of B_{pm}, and are a set of mappings from A_{11} to the output waves at port p and harmonic m for a system perfectly matched at the output port and perfectly matched at each harmonic at all ports. The $X_{pm,qn}^{(S)}$ and $X_{pm,qn}^{(T)}$ terms determine the sensitivity to mismatch of the system, relating the contributions of the small harmonic-signals from the port q at harmonic n to the outgoing wave at port p and harmonic m.

The X-parameters model states that the set of outgoing port waves (for all ports and frequencies combinations) result from a linear mapping of the set of incoming port waves (for all ports and frequencies combinations), but in contrast to S-parameters, not only signal ratios are combined, but additionally this modeling considers the contributions depending on the amplitude of the RF Stimulus (RFS) and the relative phases of the remaining incoming signals (harmonics and intermodulation products) with respect to this reference stimulus.

2.2. CPW Test Fixture

The test-fixture used is supported on an Alumina substrate ($\epsilon_r = 9.6$, $\tan\delta = 0.0004$, stable up to 67 GHz), and consists of a CPW transmission line with strip width $w = 0.1$ mm, gap $g = 0.05$ mm, and length $L_{Line} = 1.175$ mm from the GSG probe contact plane to the device reference plane, wherein the diode under test is mounted with its cathode connected to the CPW strip and anode connected to ground as is shown in Figure 2. This setup is representative of a PIN diode switch on a reconfigurable antenna when modelled as an internal port where reflection coefficient is of interest.

In order to proceed with the extraction of the diode model, a broadband Short-Open-Load-Thru (SOLT) calibration is first performed, followed by input power calibration and harmonics phase reference calibration as is required for the measurement of X parameters [42]. The diode is biased using an external DC-power supply connected to a built-in biasing-tee of the network analyzer. Subsequently, the diode S parameters are measured, retaining only the reflection coefficient.

(**a**) Test fixture detail.

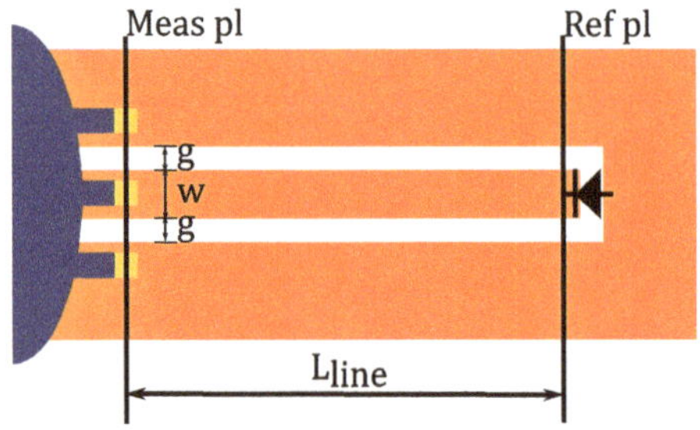

(**b**) Diode mounting schematic.

Figure 2. X-parameters measurement setup.

As the SOLT calibration reference plane is defined on the GSG probe-CPW line contact, a de-embedding procedure is applied to remove the contribution of the CPW line from the measured diode S parameters and retrieve the actual diode reflection coefficient. A model for the CPW-*even* mode accounting for dispersion, obtained with the aid of accurate 2.5D electromagnetic simulation using Keysight Momentum, is used for that purpose.

Considering the diode mounting and that the X-parameters measurement system used in this work requires the calibration and registering of both ports, some comments on the validity of the measurements are deemed convenient.

Although in the case of S parameters, it is clear that two- and one-port S-parameter measurements would yield the same reflection coefficient, this may not be evident in the case of X-parameters, as in general, the nonlinear functions involved depend on the incoming signals from all the ports/harmonics combinations. However, as in the test fixture used for diode characterization both ports are physically uncoupled, the waves injected at port 2 do not affect the measurement at port 1, hence, inter-port wave dependence vanishes. This is expressed by the reduced one port X-parameters

$$B_{1k} = X_{1m}^{(FB)} P^m + \sum_{n \neq 1} X_{1m,1n}^{(S)} A_{1n} P^{m-n} + \sum_{n \neq 1} X_{1m,1n}^{(T)} A_{1n}^* P^{m+n}, \tag{2}$$

which can be posed in matrix form, allowing the integration of the results into further circuital formulations such as harmonic balance calculations.

$$B_1 = PX_1^{(FB)} + PX_1^{(S)} (P')^{-1} A_1 + PX_1^{(T)} P' A_1^*, \tag{3}$$

where $B_1 = [B_{11} \cdots B_{1N}]^T$, $P = diag(P^m)$, $P' = diag(P^{m+1})$, $X_1^{(FB)} = [X_{11}^{(FB)} \cdots X_{1N}^{(FB)}]^T$, $A_1 = [A_{12} \cdots A_{1N}]^T$, and $X_1^{(S)}, X_1^{(T)}$ are matrices whose elements are of the form $X_{1m,1n}^{(\cdot)}$.

2.3. *Measurements*

Extensive measurements of S- and X-parameters of diode MA4AGBLP912 were performed in the 1–50 GHz range, using different biasing levels. Note that in the case of the PIN diode, the biasing current IDC_1 is equivalent to the DCS_q term (part of the LSOPS) in the X-parameters notation.

Measurements of S_{11} from 45 MHz to 50 GHz are shown in Figure 3. From this figure, it is clear that in the forward bias condition varying DC level has an important impact on the diode behavior. Likewise, it can be observed that for biasing currents higher than 15 mA the diode achieves a stable behavior. Therefore, 0, 1 and 20 mA bias are taken as representative of the diode "OFF", "NL" (nonlinear) and "ON" states. Measurement results are validated against S-parameter files provided by the manufacturer (up to 30 GHz) ascertaining a very good agreement.

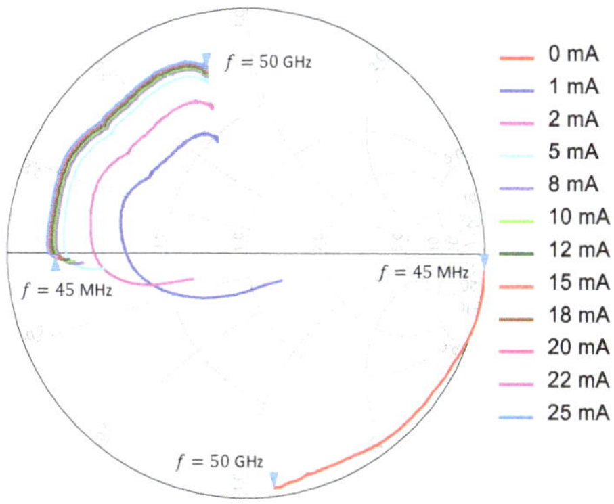

Figure 3. Measured S-parameter (de-embedded) vs. Bias.

One-port X-parameter measurements were carried out at different biasing levels $I_b = \{0, 1, 2, 5, 10, 20\}$ mA. For each of these DCS a power-frequency grid for $|A_{11}|$ is defined, varying input power in the range $P_{in} = \{-10, -6, -2, 2, 6, 10\}$ dBm and sweeping fundamental frequency from 1 to 16 GHz with $n_f = 16$ points, considering up to the third harmonic (total frequency expanding from 1 to 48 GHz). That is, in total there are 6 sets of measurements each with a total of 3 X parameters over a $6 \times 16 \times 3$ sampling grid. In contrast, a conventional S-parameter measurement would only contain 6 sets of n_f points each.

To ease the comparison of X parameters for the different diode biasing levels, slices of $X_{1m}^{(FB)}$, $X_{11,1n}^{(S)}$, and $X_{11,1n}^{(T)}$ can be taken either at fixed frequencies or input powers of the fundamental A_{11}. For example, slices representing the variation with input power at a fundamental frequency $f_1 = 5$ GHz are illustrated in Figure 4 for the diode "ON" and "OFF" states.

Valuable information about the nonlinearity of the DUT can be extracted from the variation with RFS power of the large signal response $X_{1m}^{(FB)}$, represented in Figure 4a, where it can be observed that for the "OFF" and "ON" states the fundamental and the harmonic tones linearly increase with input power. This figure also demonstrates that the diode is not driven into a high nonlinearity region for the powers used in the model extraction, as in both cases the second and third harmonics are at least 40 dB below the fundamental level. Besides, the $X_{1m}^{(FB)}$ terms allow an estimation of the Third Harmonic Intercept Point (IP3) that is ∼35 dBm in the case of the "OFF" state, and ∼43 dBm for the "ON" state.

Another common approach to estimate the nonlinearity of the DUT is by defining the nonlinear transition as the input power resulting in the 1dB CP of the reflected fundamental wave; this value can be extracted directly from Figure 4a or by representing the large signal reflection coefficient $X_{11,11}^{(S)} \equiv X_{11}^{(FB)}/|A_{11}|$. In the diode case, it is apparent that for the input powers used, compression does not occur for the "OFF" nor the "ON" states. In contrast, for the case of the "NL" state, 1 dB CP occurs at around 6 dBm.

On the other hand, the variation with input power of the set of functions $X_{11,1n}^{(S)}$ illustrated in Figure 4b and $X_{11,1n}^{(T)}$ illustrated in Figure 4c, indicate a small contribution of the reflected second and third harmonics to the total response at fundamental for both the "OFF" and "ON" diode states.

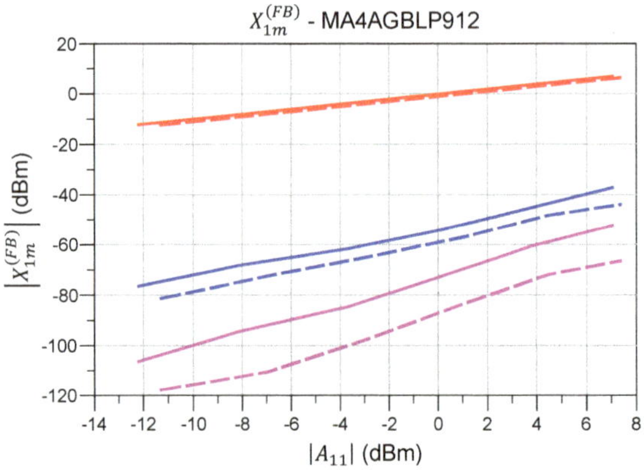

(a) $X^{(FB)}$ vs. input power ($f_1 = 5$ GHz).

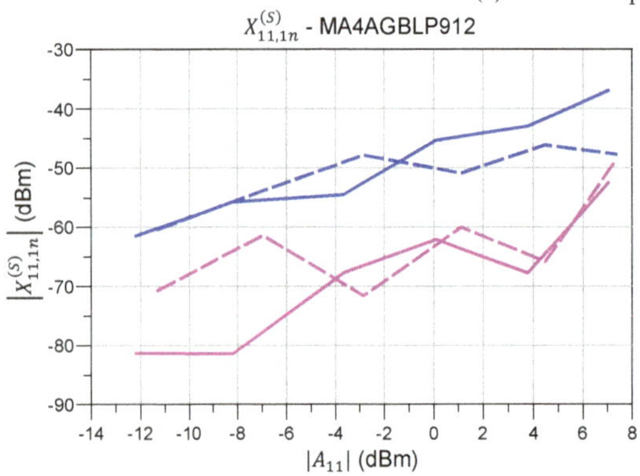

(b) $X^{(S)}$ vs. input power ($f_1 = 5$ GHz).

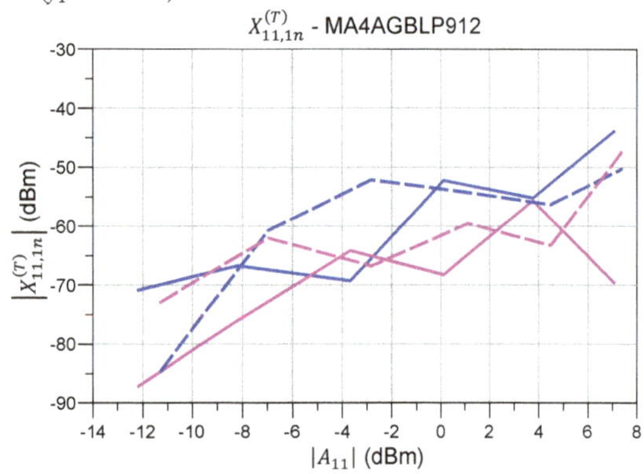

(c) $X^{(T)}$ vs. input power ($f_1 = 5$ GHz).

Figure 4. X-Parameters variation with input power at fundamental $f_1 = 5$ GHz. (— f_1, — $2f_1$, — $3f_1$, — "Off", --- "On").

The X parameters convergence to the S parameters in the small-signal regime is verified for both the "ON" and "OFF" states. Direct S-parameter measurements are compared to X-parameter reduced to S parameters using circuit simulation in Keysight ADS. Results are reported in Figure 5. There is a very good agreement between the two kinds of parameters with maximum magnitude and phase deviation of 0.5 dB and 2.5° respectively.

A further validation was performed contrasting $X_{1m}^{(FB)}$ with an independent measurement using a spectrum analyzer at ($DCS_q = \{0, 1, 20\}$ mA, $P_{in} = \{-2.5, 5\}$ dBm, f_1 =1–16 GHz). A fundamental to second harmonic rejection above $\{45, 13, 55\}$ dB for the three biasing levels was observed at $P_{in} = 5$ dBm, these values compare well with those observed from $X_{1m}^{(FB)}$ of $\{49, 13.5, 52\}$ dB at $P_{in} = 6$ dBm for the same biasing levels.

2.4. Nonlinear Assessment of RA and PL

The extracted diode model based on the X-parameters measurements, once validated, becomes a fundamental input to assess the possible nonlinear behavior of reconfigurable antennas and parasitic layers from an analytical viewpoint, and ultimately, a valuable piece of information to be further used as a design tool in the case of minimum nonlinearities reconfigurable antennas and parasitic layers.

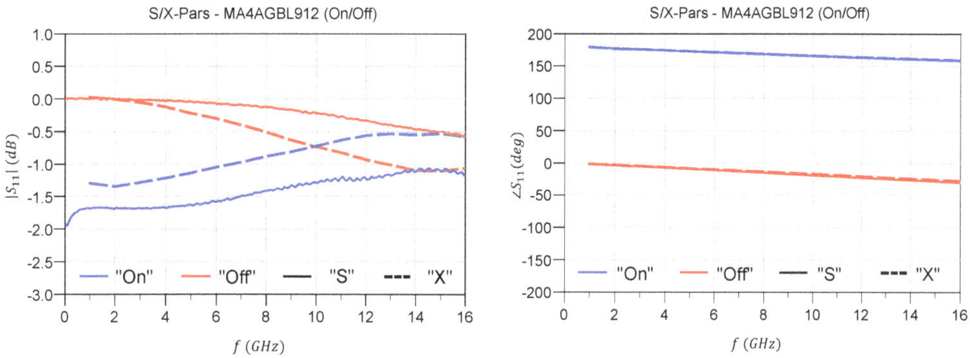

(**a**) Magnitude. (**b**) Phase.

Figure 5. *S*- and *X*- parameters small signal convergence.

To illustrate the use of the X-parameters in a particular case, the antenna reported in [30] and reproduced in simplified manner in Figure 6a is taken as case of study. This antenna uses a total of 17 switches based on the MA4AGFCP910 PIN diode, of which the most critical is the one located in the main feed path (S_6 in the diagram); likewise, the set of 12 diodes in the intermediate PL (S_5) are jointly (de)activated, hence, although individually not as significant as S_6, when combined, their impact on nonlinearity is similar or can be even larger. The switches S_{1-4} providing beam-steering are loosely coupled to the fed power, therefore, their impact on antenna nonlinearity is marginal.

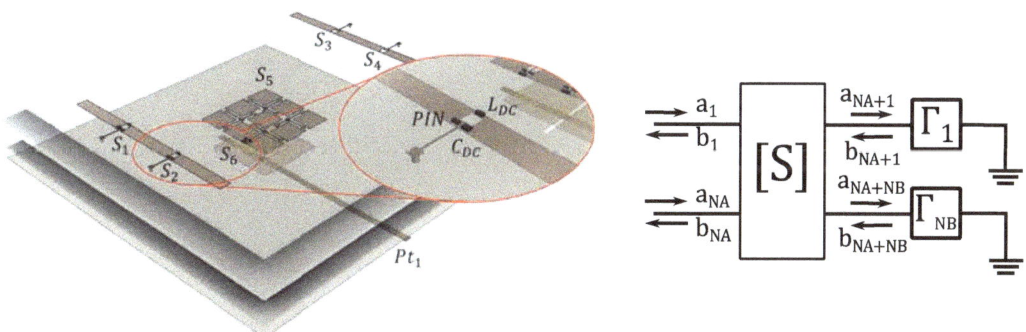

(**a**) Parasitic-Layer-based Reconfigurable-antenna. (**b**) Internal multi-ports method.

Figure 6. Parasitic-Layer-based Reconfigurable-antenna and antenna equivalent circuit.

Notwithstanding that this antenna has 6 diverse modes of operation allowing for beam-steering and bandwidth reconfigurability, given the antenna symmetries and the expected small contribution to nonlinearity of the switches on the upper PL, only modes M_1, M_4 providing bandwidth reconfigurability are considered for analysis. Switch states for these modes are summarized in Table 1.

Table 1. Switch states for the antenna modes of interest.

Mode	S_1	S_2	S_3	S_4	S_5	S_6
M_1	0	0	0	0	0	1
M_4	0	0	0	0	1	0

As mentioned in the introduction, a common approach for the analysis of nonlinearly loaded antennas is to split the problem into an electromagnetic (linear) and a circuital network (nonlinear) problem. The first one can be dealt by numerical techniques such as MoM, Finite Elements Method (FEM), Finite-Differences Time-Domain (FDTD), whereas the second one can be tackled by nonlinear formulations like the Harmonic Balance (HB).

The internal multi-port method, illustrated in Figure 6b, which proceeds by replacing each switch location by a simulation port that can be later loaded with the proper switch parameters, and that has been extensively used for the analysis and design of RA and PL, is naturally fitted to translate the electromagnetic problem into a circuital one, in which the use of X-Parameters models of the loads, in combination with nonlinear circuital simulation, allow to calculate the nonlinear metrics of interest. After performing the FWEM simulation using the time domain solver in CST Studio, Keysight ADS is used to carry on the corresponding circuital S-parameters and harmonic balance simulations.

Figure 7 shows that in the small-signal regime, the antenna port matching predicted by the X-parameters is quite similar to that predicted by the S-Parameters, ratifying the validity of the approach.

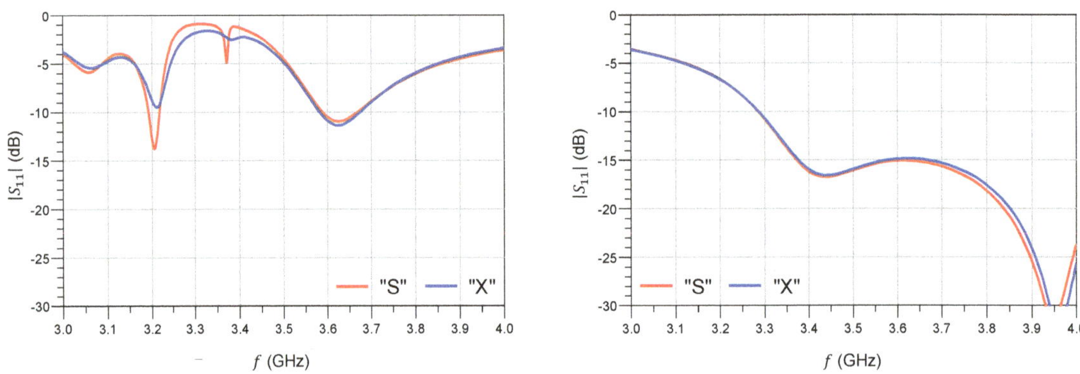

(**a**) Antenna mode M_1. (**b**) Antenna mode M_4.

Figure 7. Port match in a test reconfigurable antenna.

Afterwards, two relevant metrics to assess the nonlinearity of the antenna are determined by means of an HB simulation in which the input power is swept from −10 to 30 dBm. Figure 8 shows the harmonic components generated by the switches as seen from the antenna input port. From this variation of fundamental and harmonics with input power, the TOI of the antenna can be estimated to be $IP_3 = 55$ dBm and 41 dBm for the M_1 and M_4 operation modes, respectively.

One step further, the Error Vector Magnitude can also be derived with the aid of the X-parameters nonlinear model. The EVM is a communications performance metric [43] that gained relevance in favor of traditional ones such as Bit-Error Ratio (BER) because of its independence on the receiver (no need to decode the received bit-stream), and its independence on the modulation scheme used [44] in either high- or low-distortion environments.

The EVM is related to transmitter imperfections [45] such as modulator carrier leakage, IQ mismatch, nonlinearity, local oscillator phase noise and frequency error. An approach to calculate the EVM on nonlinear devices is presented in [46], assuming one system impairment is dominant whereas the remaining are modelled as part of the Gaussian noise.

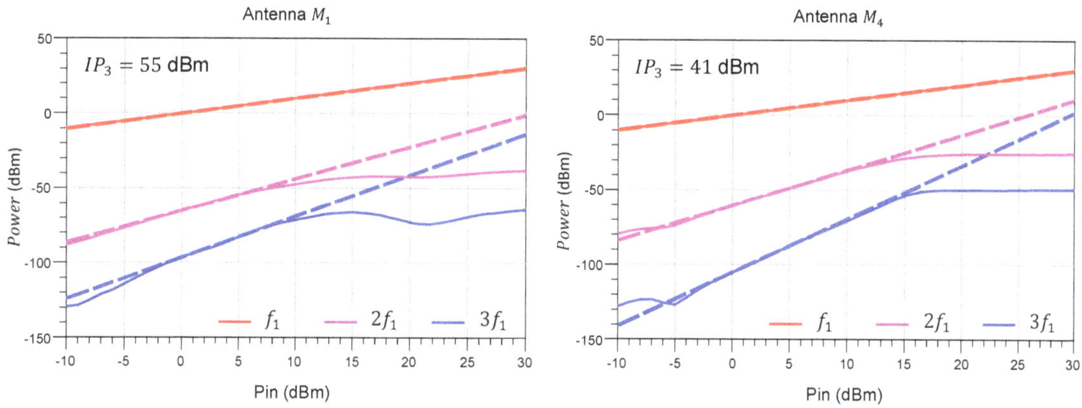

Figure 8. Harmonic components generated by the test reconfigurable antenna, with indication of IP_3. (**left**) M_1, (**right**) M_4.

In general, the EVM is defined as

$$EVM = \frac{\frac{1}{N_S}\sum_{n=1}^{N_S}|S_r(n) - S_t(n)|^2}{P_0}$$
$$P_0 = \frac{1}{N_S}\sum_{n=1}^{N_S}|S_t(n)|^2, \quad (4)$$

where $S_t(n)$ and $S_r(n)$ are the nth transmitted and received symbols, respectively.

Although it is valid only in "Data-aided receivers" [47], the Signal-to-Noise Ratio (SNR)-based approximation for EVM is proven to hold well for high SNR setups. It can also be used as an upper bound for the real EVM when the received symbols are estimated.

Using a base-band signal model and reasoning similar to that in [46], the EVM can be expressed in terms of the SNR and the in-band nonlinear distortion (HD). Therefore, acknowledging that the major contribution to the in-band nonlinear distortion comes from the third order intermodulation products [45], the approximated EVM can be expressed as:

$$EVM \approx \sqrt{\frac{1}{SNR} + HD}$$
$$EVM \approx \sqrt{\frac{N_0}{P_0} + \left(\frac{P_0}{IP_3}\right)^2}, \quad (5)$$

which states that in low input power setups, the main driver of EVM is the system noise, while in high power setups the harmonic distortion is the main responsible for error.

The approximated EVM is calculated using the results of the HB simulation of the PL-based RA under test, and assuming a moderately high noise floor of $N = -60$ dBm. Results are shown in Figure 9. The observed results are in agreement with the trends and values observed in the measurements reported in [30], where the antenna operation mode M_4 presents a higher EVM than the M_1, and the EVM is below 8% for input powers up to $P_{in} = 30$ dBm, for all the antenna states of interest. This results also demonstrate that the PL-approach is a low-nonlinearity alternative to achieve complex antenna functions in accordance with the performance limits established for 5G.

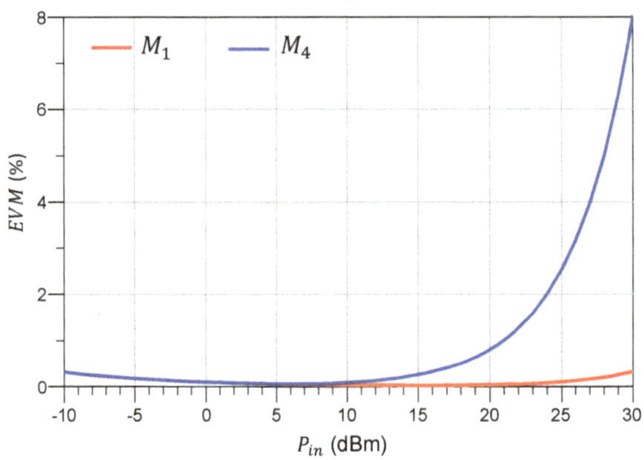

Figure 9. Error Vector Magnitude (EVM) of test reconfigurable antenna.

3. Minimum Nonlinearity PL Design Guidelines

The advanced design of parasitic layers, and reconfigurable antennas in general, heavily relies on optimization techniques to find an adequate set of switching states to achieve a target set of figures of merit. These figures of merit are usually specified in terms of fundamental antenna parameters for example, port parameters, such as match and coupling, and pattern features such as direction of maximum, gain and polarization.

Although nothing prevents the inclusion of system-related performance metrics impacted by the antenna, such as coverage, throughput, BER or EVM, or any other parameter of interest in particular applications, the use of basic antenna properties is often the most efficient option, as the computational cost of calculating these derived quantities make them impractical when evaluating beyond a few thousands of switch states. This complexity is exacerbated if structural variations of the basic radiator are to be considered in a co-optimization process as for example in [48].

Therefore, the use of basic antenna parameters and system-level performance estimators directly derived through simple calculations may be used whenever possible to predict the better performing configuration when comparing among several different alternatives.

As the main interest of this work is the design of reconfigurable antennas and parasitic layers with minimum nonlinearities and impact on related system-level metrics such as EVM, a partitioning of the design process into three stages is proposed aiming to reduce the computation time and achieving controllable results. The stages of the design methodology are:

- Choose an optimal trade-off starting point for the antenna geometry.
- Determine the optimum switch configurations using the performance metrics vector defined in terms of fundamental antenna parameters derived from small-signal formulations.
- Based on nonlinear load models, accurately calculate the nonlinearities and system-related parameters for the optimum configurations.

A high level flowchart diagram illustrating the main elements of the proposed design methodology is presented in Figure 10.

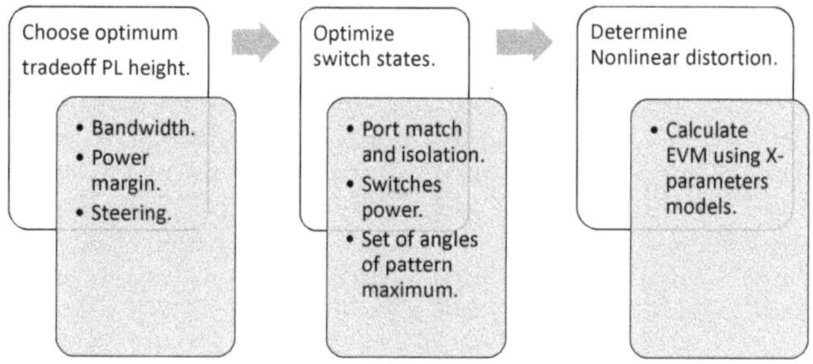

Figure 10. Flow chart with the main elements of the design methodology.

3.1. Adding New Design Criteria

A general optimization problem is defined as a search in an n-dimensional decision space, mapped through a cost function f to a lower k-dimensionality objective space, in which a comparison of elements x in the original space is performed through metrics y defined in the destination space. In the case of RA design, diverse antenna parameters compose the comparison metrics, hence, a multiobjective optimization problem is implicit.

$$\begin{aligned} f : \mathbb{R}^n &\rightarrow \mathbb{R}^k \\ y &= f(x). \end{aligned} \tag{6}$$

Common RA design approaches mix multiple parameters to obtain a single performance indicator to facilitate decision, however, if the cost function is defined in a way such that $k = 1$, there is a risk of over-simplifying the problem as there exists the possibility that f be a surjective projection of $\mathbb{R}^n \rightarrow \mathbb{R}$ having several individuals with very similar cost values and quite different configuration and performance in the individual criteria; this prevents the discovery of the trade-offs involved in the design process, and may result in the optimization being biased towards the easier of the individual performance metrics, neglecting the harder ones.

Adding dimensions to the objective space allows the creation of Pareto fronts, representing the best objective combination, enabling the visualization of trade-offs and exploring the interrelation between objectives. Furthermore, adding new decision axes to the objective space creates new separation criteria for configurations with very similar performance on the other axes.

In this work, a new power-related metric is defined in addition to the port parameters and far-field related metrics allowing to account for antenna port match, isolation, beam pointing, gain, power efficiency and nonlinearity.

3.2. Extension of the IMPM to Account for Nonlinearities

The Internal Multi-Ports Method (IMPM) is an efficient way to characterize the behavior of a reconfigurable device for a set of switch states, requiring only one FWEM simulation at setup. After setup, formulations yielding results equivalent to circuital simulations allow to determine the effects of the terminating loads on the fundamental antenna parameters. A simple extension to the IMPM is added to account for the power delivered to the fed ports and to the switches, becoming a useful indicator of power efficiency and possible excitation of nonlinearities. Expressions to calculate loaded port parameters, radiated field and port power are briefly summarized.

The setup FWEM simulation is performed to obtain the reference network scattering parameters $S \in \mathbb{C}^{N_p \times N_p}$, and the field radiated when the nth port is fed with unit amplitude (and the others are terminated in matched loads) $\vec{E}_n(\theta, \phi)$, $n \in \{1 \dots N_P\}$, where $N_p = N_A + N_B$, N_A, and N_B are the number of total, fed (external), and loaded (internal) ports, respectively. The matrices and vectors involved in the formulation can be partitioned to index the external or internal ports, A, B subscripts are used to denote this partitioning.

From the defining relation of the total S-matrix $\mathbf{b} = \mathbf{S}\,\mathbf{a}$ in terms of the power wave amplitudes $a_m = v_m^+/\sqrt{2Z_{cm}}$, $b_m = v_m^-/\sqrt{2Z_{cm}}$ ($v_m^{\pm}$ is the peak amplitude of the incoming/outgoing voltage wave at the mth port) for ports with possibly different characteristic impedances, the block partition described above results in:

$$\begin{bmatrix} \mathbf{b_A} \\ \mathbf{b_B} \end{bmatrix} = \begin{bmatrix} \mathbf{S_{AA}} & \mathbf{S_{AB}} \\ \mathbf{S_{BA}} & \mathbf{S_{BB}} \end{bmatrix} \begin{bmatrix} \mathbf{a_A} \\ \mathbf{a_B}. \end{bmatrix} \tag{7}$$

Noting that only input ports have a source signal $\mathbf{a_A}$, while for the loaded ports the incoming signals are actually reflections on the load terminations, obtainable from $\mathbf{\Gamma_B}$, the diagonal matrix telling the reflection coefficient of the terminating loads, we can write:

$$\begin{bmatrix} \mathbf{b_A} \\ \mathbf{b_B} \end{bmatrix} = \begin{bmatrix} \mathbf{S_{AA}} & \mathbf{S_{AB}} \\ \mathbf{S_{BA}} & \mathbf{S_{BB}} \end{bmatrix} \begin{bmatrix} \mathbf{a_A} \\ \mathbf{\Gamma_B b_B} \end{bmatrix} \tag{8}$$

From the expressions above, we can compute all the unknown wave amplitudes from the known input amplitudes $\mathbf{a_A}$:

$$\mathbf{a_B} = \mathbf{\Gamma_B b_B} \tag{9}$$

$$\mathbf{b_A} = \left\{ \mathbf{S_{AA}} + \mathbf{S_{AB}} \left(\mathbf{\Gamma_B}^{-1} - \mathbf{S_{BB}} \right)^{-1} \mathbf{S_{BA}} \right\} \mathbf{a_A} \tag{10}$$

$$\mathbf{b_B} = (\mathbf{I} - \mathbf{S_{BB}}\,\mathbf{\Gamma_B})^{-1} \mathbf{S_{BA}}\,\mathbf{a_A}. \tag{11}$$

Now, the aggregated radiated field for a particular excitation and load condition can be calculated as follows from both $\mathbf{a_A}$ and $\mathbf{a_B}$:

$$\bar{E}_{tot}(\theta,\phi) = \sum_{n \in pts} a_n \bar{E}_n(\theta,\phi). \tag{12}$$

A number of quantities of interest can now be obtained from these wave amplitudes:

$$P_{in} = ||\mathbf{a_A}||^2 - ||\mathbf{b_A}||^2 \tag{13}$$

$$P_{sw} = ||\mathbf{b_B}||^2 - ||\mathbf{a_B}||^2 \tag{14}$$

$$P_{ant} = ||\mathbf{a_A}||^2 + ||\mathbf{a_B}||^2 - ||\mathbf{b_A}||^2 - ||\mathbf{b_B}||^2 \tag{15}$$

$$\eta_{tot} = \frac{\eta_{ant} P_{ant}}{P_{in}}, \tag{16}$$

where P_{in} is the net input power to the reconfigurable antenna system, P_{sw} is the power dissipated by the switches, P_{ant} is the net input power to the elementary radiator (the passive/linear portion of the system), and η_{tot}, η_{ant} are the efficiencies for the full reconfigurable antenna system and the passive antenna element, respectively.

Under a pure tone excitation, and considering that the antenna building materials and surrounding environment have a linear behavior for all operating circumstances, the only source of nonlinear distortion are the internal port loads and their frequency conversion characteristics.

Although the presented formulation relates only one input frequency to the same output frequency, it can be extended to account for harmonics induced by the nonlinear loads by properly adapting the vectors and matrices involved in the calculations. In particular, if N_H is the number of harmonics to consider, the vector quantities in the presented formulation can be expressed as composed by sub-vectors of N_H terms each, forming a longer extended vector of size $N_p N_H$. This implies also the extension of the involved matrices to include the relationships at the harmonic frequencies, therefore each matrix entry will be a sub-matrix of size $N_H \times N_H$, totaling a size of $(N_p N_H) \times (N_p N_H)$ for a full N_p-ports matrix.

In this case, $\mathbf{S}$, which involves the linear antenna portion, will be a matrix composed by diagonal matrices, that is, for any pair of ports (p, q), and harmonics (m, n), $S_{pq}(m, n) = 0$, $\forall m \neq n$.

On the other hand, given the non-linearity of the terminating loads (switches), $\mathbf{\Gamma_B}$ will be a block diagonal matrix with each block representing a given switch, and all blocks being fully populated; therefore the reflection coefficient of a single load Γ^l must provide a representation for harmonic frequencies conversion, taking the form

$$\Gamma^l = \begin{bmatrix} \Gamma^l(1,1) & \dots & \Gamma^l(1,N_H) \\ \vdots & \ddots & \vdots \\ \Gamma^l(N_H,1) & \dots & \Gamma^l(N_H,N_H). \end{bmatrix} \tag{17}$$

From this expression, the convenience of the X-parameter formulation for the loads modelling results evident as it embeds the required frequency conversion as a function of the input power.

As an additional remark, it must be noted that when input power is low enough, the cross-frequency components are negligible and consequently the Γ^l reduce to diagonal matrices, thus the harmonics calculated are only affected by signals in their same frequency, and the harmonic-extended system is reduced to a linear relationship.

As mentioned earlier, for the optimization stage the interest is on efficiently comparing several thousands of configurations; therefore, a valid indicator of the nonlinear performance of a given switch configuration is adopted instead of solving the nonlinear extended system. For this purpose, the power margin (ΔP) is defined as the difference between the power delivered to the fed port and the power delivered to the switches, over the frequency range where the excited port is well matched:

$$\begin{aligned} f_r &: |S_{ii}(f_r)| < -10\text{dB} \\ \Delta P &= \left\langle \frac{P_{in}(f_r)}{P_{sw}(f_r)}. \right\rangle \end{aligned} \tag{18}$$

The definition and use of this metric is supported on the fact that for the "ON" and "OFF" switch states, up to moderate input powers, the power of the fundamental tone is much higher than that of the harmonics, as was confirmed by the X-parameters measurements performed. Therefore, the nonlinearities can be predicted by the switch power at the fundamental frequency. Likewise, as was evidenced for the antenna presented as case of study, although the maximum power through a switch is a significant driver of nonlinearities, the total power delivered to the switches set may be a more determinant factor.

3.3. *PL Design Trade-Offs*

From the expressions obtained above, it can be observed that the nonlinear perturbation comes from the waves reflected by the loads, given by the term $\mathbf{S_{AB}}\left(\mathbf{\Gamma_B}^{-1} - \mathbf{S_{BB}}\right)^{-1}\mathbf{S_{BA}}\,\mathbf{a_A}$. Accordingly, if the nonlinear effects are to be minimized there are two approaches:

1. Reducing the $\mathbf{S_{AB}}$, $\mathbf{S_{BA}}$ terms, that is, the coupling between external and internal ports. This is the underlying philosophy of the PL as there is a physical separation between the feed ports and the switches. Note that this coupling can be controlled by the PL height above the basic antenna, adding one degree of freedom to the design. However, changing the PL height over the basic antenna also impacts the radiated field and port match, requiring a delicate balance.
2. Reducing the $\left(\mathbf{\Gamma_B}^{-1} - \mathbf{S_{BB}}\right)^{-1}$ harmonic terms. This could be attained by a careful antenna design that provides an adequate matching between the terms in the subtraction, however, given that normally the minuend will be dominant, this is mostly technology dependent, reinforcing the need for accurately characterizing the switch nonlinearities and the benefits of using these models as a detailed design tool.

To examine the effects of the PL height 'h' over one basic antenna and its effects on the reconfigurability of parameters like port matching, radiated field, and nonlinearities, the parasitic-layer-based reconfigurable-antenna illustrated in Figure 1 is taken as a reference.

A parametric variation of 'h' is performed on this reference antenna, aiming to determine a feasible trade-off region which can be used as design guideline, even for other PL pixellations and number of switches, as well as other antenna configurations. Therefore, the interest is in average or expected behavior rather than in particular switch states behavior. Aggregated port match bandwidth, maximum steering angle, and average power margin are used as metrics of comparison.

The PL based RA is composed of a basic 2-ports aperture-coupled microstrip patch radiator matched at $f_0 = 3.5$ GHz, on a $th = 3$ mm thick Rogers RO4003 substrate $\epsilon_r = 3.55$. The PL is a 3×3 pixellation supported on a $th = 0.51$ mm thick Rogers RO4003 substrate, placed at a distance h from the basic antenna, with size and switches numbering as illustrated in Figure 11. This design is based on the single-port presented in [28], which was fabricated and measured with good agreement between simulations and measurements, working from 2.4 to 2.5 GHz, providing steering in nine directions $\theta_0 \in \{-30°, 0°, 30°\}$, $\phi_0 \in \{0°, 45°, 90°, 135°\}$ with an average gain of 6.5 dB.

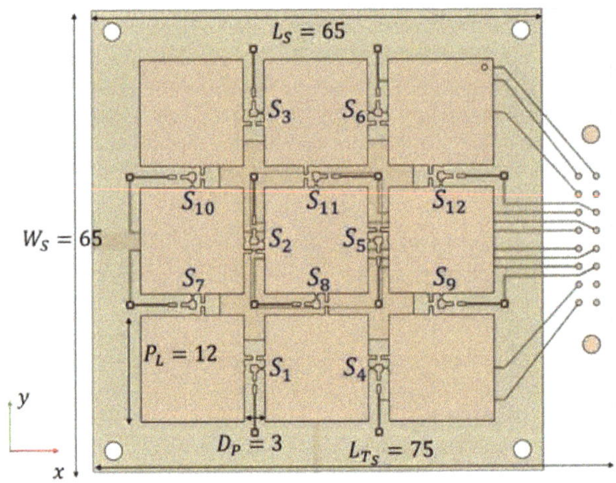

Figure 11. Parasitic-Layer-based Reconfigurable-antenna.

The parametric sweep is performed using the frequency-domain solver of CST studio, for $h = \lambda_0/\{40, 20, 15, 12, 10, 9, 8, 6, 5, 4\}$, retaining for each case the 'unloaded' S-parameters file and the radiated fields generated by each 'port'. In a later step, the IMPM is applied to determine the metrics of interest. As the number of switch configurations is relatively low (4096), exhaustive search is feasible in a short time for each height value. The sweep results for antenna port 1 are summarized in the set of plots of Figure 12 which represent the probability distribution of configurations for a given figure of merit in terms of the variations of the parameter h. Results for port 2 are almost identical and are omitted.

The first that can be observed in Figure 12a is that, as expected, the PL loading alters the port impedance in a way that when the separation is smaller, the obtainable bandwidth is higher, and when separation grows, convergence to the bandwidth of the simple antenna is achieved. This is in a good agreement with the experimental results in [40], for a higher complexity PL where a statistical sweep is performed for practical reasons. There can also be appreciated that most configurations are matched at a frequency above that of the basic antenna, suggesting that when starting a new design, the base antenna should resonate at a lower frequency than that of the intended central frequency. An important remark from this figure, is that a reconfigurable bandwidth greater than 10% can be achieved for separations lower than $\lambda/8$.

On the other hand, Figure 12b, illustrates the power margin variation with h. Although high power margins ~9 dB can be achieved for some configurations with very narrow separations, average margins in excess of 6 dB are achieved for separations beyond $\lambda/12$. However, there must be noted that the average Δ_P increment is not linear, and beyond $\lambda/8$ growth is slower.

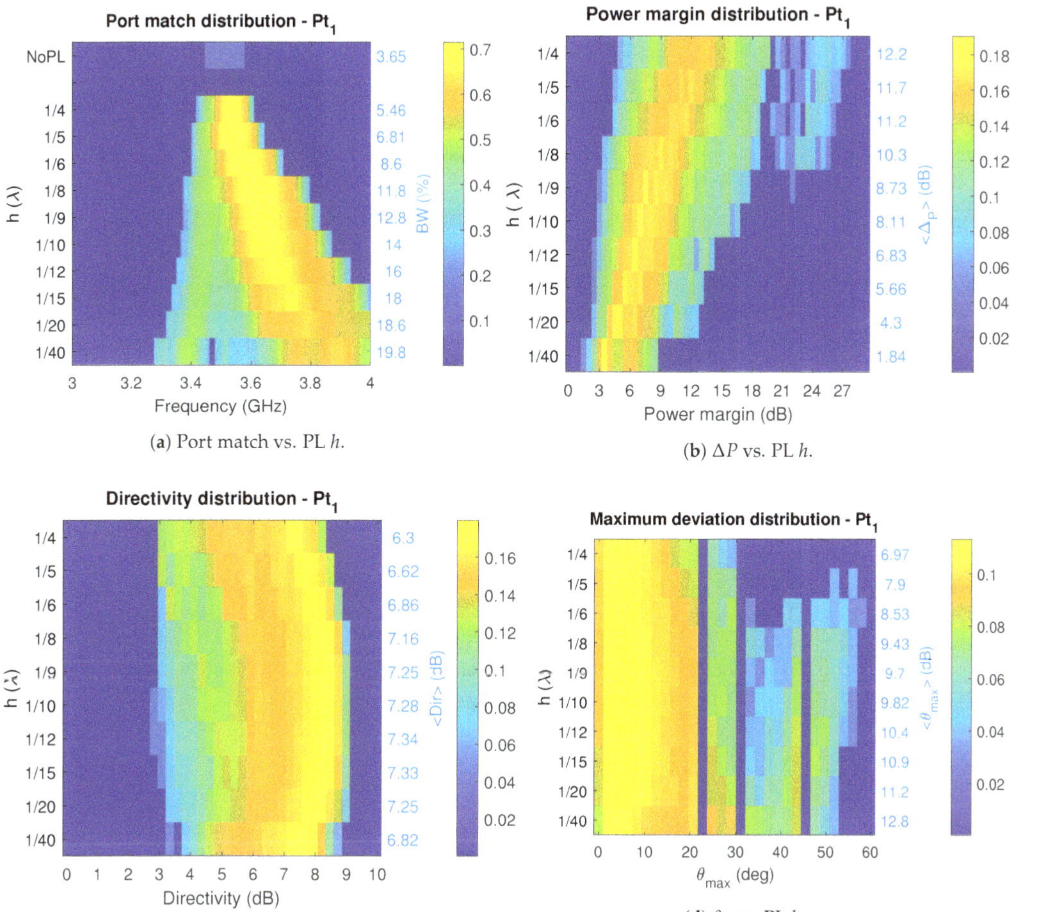

(**a**) Port match vs. PL h.

(**b**) ΔP vs. PL h.

(**c**) Directivity vs. PL h.

(**d**) θ_0 vs. PL h.

Figure 12. Parasitic-Layer-based Reconfigurable-antenna tradeoffs.

Variation of directivity with h is shown in Figure 12c. Directivity is preferred over gain to isolate the effect of the PL on the shape of the radiated field, without including the effects of port matching. This figure shows that directivity can be improved up to a pair of dBs with respect to that of the basic antenna. Most importantly, for a wide range of height variations ($\lambda/20 - \lambda/6$), on average, the effect of the PL does not lead to a degradation of directivity when compared to that of the base antenna.

Finally, the angular variation of the pattern maximum with the PL height over the basic antenna, allows to determine the feasibility of finding patterns with maximum radiation pointing towards specific directions (θ_0, ϕ_0). A 2D-histogram of the (θ_0, ϕ_0) distribution was created for each value of h to get an exact description of the angular distribution of pattern maximum. From this depiction a trend was observed, and is that the maximum deviation tends to cluster at $(\theta_0, \phi_0) = (45^\circ, \{60^\circ, 120^\circ, 240^\circ, 300^\circ\})$, and $(30^\circ, \{0^\circ, 180^\circ, 360^\circ\})$. Taking this into consideration, a simpler diagram of the maximum θ deviation, for all the possible ϕ values, is created as illustrated in Figure 12d, where configurations with $D < 6$ dB have been filtered out. From this figure, it is apparent that the maximum deviation changes little with height, and that up to $\lambda/5$ deviations around 50° are achievable. However, the angular distribution shows that there are more configurations with larger steering for PL separations in the range $\lambda/20 - \lambda/10$.

Finally, aiming to capture the overall reconfigurability potential of the PL-based-RA, and as a general purpose design guideline, the average metrics encountered are expressed by means of a tradeoffs surface as illustrated in Figure 13, showing how much bandwidth reconfigurability, power margin and steering can be expected for a particular setup.

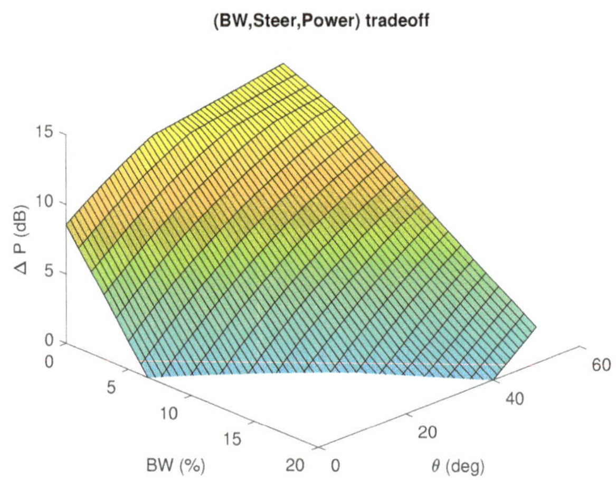

Figure 13. Design Trade-Offs of a Parasitic Layer (PL) based Reconfigurable Antenna (RA).

The set of plots of Figures 12 and 13 have presented a general description of the effects of the PL on the fundamental parameters of a basic antenna, and important guidelines for the design as the optimum height region for placing the PL to achieve a desired compromise of objectives.

4. Parasitic Layer Design Example

It is foreseen that a large portion of the initial deployments of the 5G-NR will be concentrated on the Freuency Range 1 (FR1) (sub-6 GHz portion of the spectrum) [49], and special interest is placed in the 3.5 GHz band where moderately large bandwidths up to 100 MHz can be allocated. A particular reconfigurable antenna example for the 3.5 GHz band is presented in this section to illustrate the minimum nonlinearities design method developed in the previous section.

The most computationally expensive part of the design process is the initial FWEM simulation to get the 'unloaded' antenna parameters, this simulation must be performed for any candidate value of h as there is not a straightforward way to reuse simulation results for other h values. Accordingly, the presented guidelines for determining the optimum h are intended to greatly reduce the design time.

After the fundamental design trade-offs of the reconfigurable parasitic layer are established, an optimum region for the separation between the basic antenna and the PL can be determined based on the compromise that must be made between bandwidth, linearity, and beam-steering. For instance, for 5G applications covering the n78 band at 3.5 GHz, a good starting point should be $h = \lambda/9 - \lambda/12$ that can achieve around

12–15% bandwidth, an average 7.3 dB directivity, and a rich beam-steering capability with maximum deviations up to 55°, joined with average power margins in the range 7–9 dB.

However, beyond appropriate region for PL placement, particular indications on how to determine the switch states to achieve the optimum compromise of the objectives are not yet given. The optimum switch states determination is demonstrated by means of a particular design example.

In this case, the same topology used to determine the basic trade-offs is employed with some variations, first the basic antenna is modified to resonate at a lower frequency aiming to have a lower cutoff frequency and a richer number of well-matched configurations at the central frequency of 3.5 GHz, also, the pixel size is increased to 14 mm aiming to have finer control of steering.

The switch state determination is posed as a multidimensional optimization problem, in which a 3D cost function considering S-parameters, beam pointing and switches power, is defined and explored using Pareto fronts to obtain antenna realizations with an optimum objective compromise. The cost function vector is defined as:

$$\begin{aligned} J(k) &= \{J_S(k), J_{\Delta P}(k), J_{\Delta(\theta,\phi)}(k)\} \\ J_S(k) &= \max_f \left(\max_{i,j} (|S_{ij}(k,f)|) \right) \\ J_{\Delta P}(k) &= -\Delta P(k) \\ J_{\Delta(\theta,\phi)}(k) &= \Delta(\theta,\phi)(k, f_c, \theta_{0i}, \phi_{0i}), \end{aligned} \tag{19}$$

where each component is normalized and mapped to the range $[0,1]$ by the limits *target* $S_{ij} \in \{-10, 0\}$ dB, *target* $\Delta P \in \{-10, -3\}$ dB, *target* $\Delta(\theta,\phi) \in \{0°, 10°\}$. Imposing requirements on port match and coupling, power margin, and angular deviation from target. The set of desired angles is $(\theta_0, \phi_0) = \{(0°, 0°)$, and $\{(40°, \{0°, 45°, 90°, 135°, 180°, 225°, 270°, 315°\})\}$, therefore $J_{\Delta(\theta,\phi)}(k) \in \mathbb{R}^9$. Further restrictions on results are imposed discarding those configurations with $Dir < 6$ dB.

After each individual is assigned a cost vector, weak Pareto optimality criterion is used to retain only the better performing configurations for each of the desired modes of operation. The process is illustrated for the $(\theta,\phi) = (0°, 0°)$ target in Figure 14. The cost of all the possible configurations is shown in Figure 14a, while the reduced set of non-dominated individuals is illustrated in Figure 14b, this set represents the configurations with better performance in any of the objectives defined.

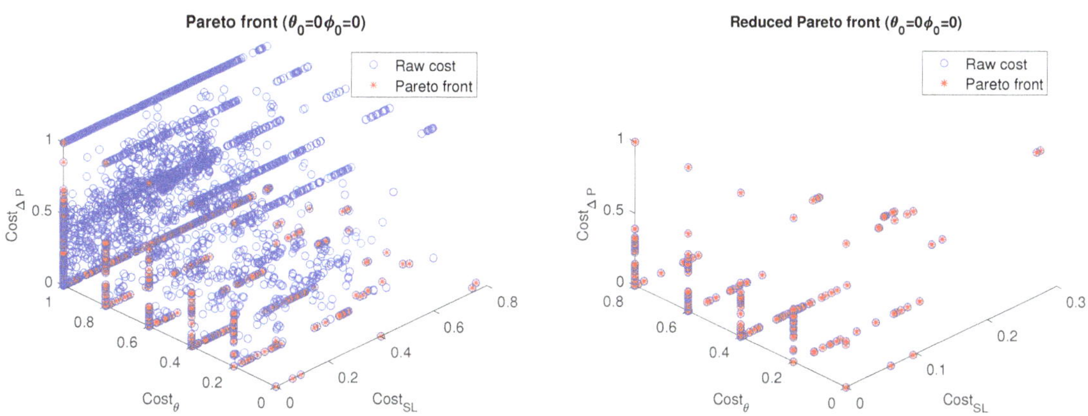

(**a**) 3D Cost function—all configurations. (**b**) Reduced 3D Cost—Pareto optimal configurations.

Figure 14. Optimization process.

Depending on the reachable cost vector for a particular target mode of operation, and whether a small mismatch, miss-pointing, or higher power going to the switches is tolerable, the designer can evaluate a subset of the configurations on the non-dominated set and determine, based on features like for instance pattern shape, impedance bandwidth, or nonlinear metrics, the configurations that better suits the application.

The procedure illustrated in Figure 14, is carried out for each of the 9 beam directions intended, and for both ports. Although this process can be seen as performing 18 independent optimizations, it is very fast as cost vector is calculated only once for all the switches configurations and the remaining problem is dealt as a classification task repeated for the intended 18 configurations. Results are shown and discussed only for port 1 noting that the values obtained for port 2 are very similar.

Once the better performing configurations are found, these are validated by simulation with CST Design Studio, finding an almost exact match. Similarly, the port parameters calculated using the measured S- and X-parameters models of the loads are verified to be in a very good agreement.

The final configuration of the antenna have a satisfactory performance characterized by an aggregated bandwidth of 10.5% centered at the intended $f_c = 3.5$ GHz. Additionally, the optimum configurations providing beam-steering, denoted as M_{1-9}, are well matched and isolated at the central frequency, albeit slight shifting of minimum is allowed as the balancing criteria privileges accurate pointing over port impedance. The port match and isolation at f_c are summarized in Table 2.

Table 2. Antenna port parameters at f_c.

Param	M_1	M_2	M_3	M_4	M_5	M_6	M_7	M_8	M_9
S_{11}	−11.95	−7.80	−16.81	−13.19	−12.18	−8.59	−11.95	−12.24	−11.67
S_{21}	−14.63	−19.49	−13.32	−10.41	−10.47	−19.39	−11.62	−10.73	−10.25
S_{22}	−11.30	−11.53	−13.32	−9.82	−10.24	−12.29	−9.95	−10.39	−11.37

On the other hand, the power margin for the optimum configurations are (13.37, 5.50, 7.66, 7.75, 6.98, 5.52, 7.04, 7.66, 6.94) dB, demonstrating a mean value of 8.34 dB, what is a 6 dB improvement when compared with a design strategy that only considers port match and radiation pattern characteristics, and is in accordance to the expected value from the design guidelines, this means also that a small fraction of the power entering the antenna is dissipated on switches and thus a low nonlinearity is expected.

Likewise, Figure 15 shows the radiation diagrams for the final antenna configurations for port 1. From this figure, a very good pointing towards the target angles defined is observed. Also, a mean gain of 6.13 dB is found which is in accordance with previous PL designs.

Finally, the IP3 and further EVM nonlinear metrics of the antenna optimum configurations are calculated using the diode nonlinear diode models extracted by X-parameters measurements. Results for EVM are illustrated in Figure 16 showing that up to 30 dBm the EVM is below 6% for all the antenna states, and that up to 35 dBm the mean EVM is below 4%. In comparison with a design strategy that only considers port match and radiation pattern characteristics, where EVM values up to 20% are observed, this is a 12 dB improvement. Note that the modes with the lower power margin present the higher EVM as expected from the design assumptions.

This result confirms the suitability of this parasitic-layer-based reconfigurable-antenna design for 5G NR mobile broadband systems, and at the same time shows the benefits of the new design approach considering the switches power to minimize the nonlinear distortion.

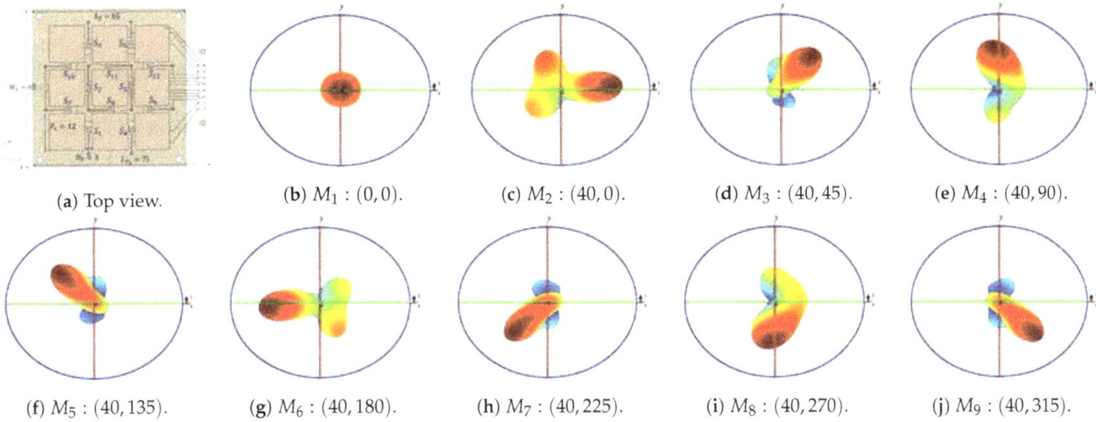

Figure 15. Radiation diagrams at Port 1, for operating mode M_i, $(i \in 1 \ldots 9)$, pointing towards (θ_0, ϕ_0).

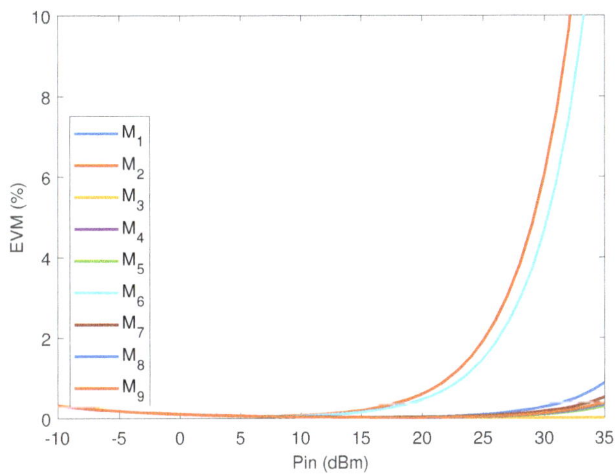

Figure 16. EVM of the Minimum Nonlinearity Parasitic-Layer-based Reconfigurable Antenna.

Even though, to the best of the authors' knowledge, this is the first effort to minimize the nonlinear distortion of a reconfigurable antenna, and the design technique has implications beyond the particular results presented as example, a comparison with similar works reported in the literature, in terms of basic antenna parameters, size, and nonlinearities when available, is presented in Table 3.

Table 3. Comparison between the designed antenna and other antennas reported on literature. (#*Sw*: Number of switches, #*Pt*: Number of ports, BW: Bandwidth (GHz), Av. G: Average Gain (dB), # Beams: Number of beams, Steer: Beam steering angle (deg), Size: Occupied volume (wavelengths at central frequency), EVM: Mean EVM (%) at a given input power.

Ref.	#*Sw*	#*Pt*	BW	Av G	#Beams	Steer	Size	EVM
This work	12	2	3.3–3.8	6.5	9	40	$0.76 \times 0.87 \times 0.10$	4 at 35 dBm
[30]	17	1	3.1–3.9	9	3	30	$1.05 \times 1.05 \times 0.25$	6 at 30 dBm
[29]	49	1	2.4–2.6	7.5			$1.38 \times 0.71 \times 0.007$	16 at 5 dBm
[28]	12	1	2.4–2.5	6.5	9	30	$0.80 \times 0.74 \times 0.10$	NA
[40]	60	1	2.4–3.0	4	9	30	$2.16 \times 1.08 \times 0.05$	NA

5. Conclusions

A method for the nonlinear characterization of a mmW PIN diode has been proposed. It is based on X-parameters measurements obtained using a test set composed of a nonlinear network analyzer, a one-port on-wafer probe station and a CPW test fixture.

The proposed method has been applied to characterize a commercial PIN diode commonly used in reconfigurable antenna applications spanning a frequency range from 1 to 48 GHz.

The measured X parameters have been validated by good agreement obtained between the harmonic rejection derived from X parameters to independent harmonic measurements, as well as from measurements in the small signal regime, where nonlinear X-parameters converge to measured S parameters.

The X parameters model is used to assess the possible nonlinearities on a reconfigurable antenna, showing consistency with experimental results.

Guidelines for the design of Reconfigurable Antennas in general and particular application to Parasitic Layers have been presented accounting for a new performance metric related to the power efficiency and possible emergence of nonlinearities.

Optimum tradeoffs for the separation of the basic antenna and a reconfigurable parasitic layer to attaining particular objectives related to port match, nonlinearity, and beam-steering have been discussed.

A complete methodology for the design of minimum nonlinearity reconfigurable antennas has been presented, accounting also for port match and beam-steering.

The methodology is based on (i) Determining the optimal PL distance to the basic antenna. (ii) Determine the optimum switch configurations. (iii) Based on nonlinear load models, accurately calculate the nonlinearities and system-related parameters only for the optimum configurations.

A design example of a parasitic-layer-based reconfigurable-antenna suitable for 5G NR communications in the 3.5 GHz band, from the port match, beam-steering and nonlinear distortion is presented.

The particular antenna design is a suitable alternative for 5G-NR EMB usage scenarios in the 3.5 GHz band, bearing 2 ports with a 11% bandwidth, 9 beams pointing in $(\theta_0, \phi_0) = \{(0^\circ, 0^\circ)$, and $\{(40^\circ, \{0^\circ, 45^\circ, 90^\circ, 135^\circ, 180^\circ, 225^\circ, 270^\circ, 315^\circ\})\}$, and an average realized gain of 6.1 dB. The whole structure fits within a compact size of $0.76 \times 0.87 \times 0.1$ wavelengths. The presented antenna has an average Error Vector Magnitude (EVM) of 4% at an input power of 35 dBm.

With the application of this new design methodology an improvement of 12 and 6 dB of the EVM and power efficiency, respectively, were obtained when compared with conventional reconfigurable antenna design strategies.

Author Contributions: Conceptualization, All authors; methodology, All authors; software, G.A.R.A., J.L.A.Q. and L.P.; validation, L.P., A.B., J.L.A.Q., B.A.C. and L.J.-R.; formal analysis, G.A.R.A., A.B. and L.P.; investigation, All authors; resources, L.P., A.B., B.A.C.; data curation, G.A.R.A.; writing—original draft preparation, G.A.R.A.; writing—review and editing, All authors; visualization, G.A.R.A.; supervision, L.J.-R.; project administration, L.J.-R.; funding acquisition, L.J.-R. All authors have read and agreed to the published version of the manuscript.

Funding: This research was supported by the Spanish "Comision Interministerial de Ciencia y Tecnologia" (CICYT) under projects TEC2013-47360-C3-1-P/AEI/10.13039/501100011033, TEC2016-78028-C3-1-P/AEI/10.13039/501100011033, and MDM2016-O6OO, and Catalan Research Group 2017 SGR 219, and by the Colombian "Departamento Administrativo de Ciencia, Tecnología e Innovaci6n" (Colciencias) through convocatoria 727 de 2015.

Data Availability Statement: The datasets generated during the current study are available from the corresponding authors on reasonable request.

Acknowledgments: To the technical staff of the AntennaLAB for their manufacturing and assembling of prototypes.

Conflicts of Interest: The authors declare no conflict of interest. The funders had no role in the design of the study; in the collection, analyses, or interpretation of data; in the writing of the manuscript, or in the decision to publish the results.

Abbreviations

The following abbreviations are used in this manuscript:

mmW	Millimeter Waves
RA	Reconfigurable Antenna
PL	Parasitic Layer
DUT	Device Under Test
TOI	Third Order Intercept
IP3	Third Harmonic Intercept Point
1dB CP	1 dB Compression Point
IMD	Inter-Modulation Distortion
EVM	Error-Vector Magnitude
IMPM	Internal Multi-Ports Method
LSOPS	Large Signal Operating Point Stimuli
RFS	RF Stimulus
DCS	DC Stimulus

References

1. 3GPP TS 38.101-1 V16.6.0. *User Equipment (UE) Radio Transmission and Reception; Part 1: Range 1 Standalone (Release 16)*; Technical Specification; 3rd Generation Partnership Project: Valbonne, France, 2021.
2. 3GPP TS 38.101-2 V16.6.0. *User Equipment (UE) Radio Transmission and Reception; Part 2: Range 2 Standalone (Release 16)*; Technical Specification; 3rd Generation Partnership Project: Valbonne, France, 2021.
3. 3GPP TS 38.101-4 V16.3.0. *User Equipment (UE) Radio Transmission and Reception; Part 4: Performance Requirements (Release 16)*; Technical Specification; 3rd Generation Partnership Project: Valbonne, France, 2021.
4. 3GPP TS 38.104 V16.6.0. *Base Station (BS) Radio Transmission and Reception (Release 16)*; Technical Specification; 3rd Generation Partnership Project: Valbonne, France, 2021.
5. 3GPP TS 38.141-1 V16.6.0. *Base Station (BS) Conformance Testing Part 1: Conducted Conformance Testing (Release 16)*; Technical Specification; 3rd Generation Partnership Project: Valbonne, France, 2021.
6. 3GPP TS 38.141-2 V16.6.0. *Base Station (BS) Conformance Testing Part 2: Radiated Conformance Testing (Release 16)*; Technical Specification; 3rd Generation Partnership Project: Valbonne, France, 2021.
7. Hausmair, K.; Gustafsson, S.; Sánchez-Pérez, C.; Landin, P.N.; Gustavsson, U.; Eriksson, T.; Fager, C. Prediction of Nonlinear Distortion in Wideband Active Antenna Arrays. *IEEE Trans. Microw. Theory Tech.* **2017**, *65*, 4550–4563. [CrossRef]
8. Teodoro, S.; Silva, A.; Dinis, R.; Barradas, F.M.; Cabral, P.M.; Gameiro, A. Theoretical Analysis of Nonlinear Amplification Effects in Massive MIMO Systems. *IEEE Access* **2019**, *7*, 172277–172289. [CrossRef]
9. Christodoulou, C.G.; Tawk, Y.; Lane, S.A.; Erwin, S.R. Reconfigurable Antennas for Wireless and Space Applications. *Proc. IEEE* **2012**, *100*, 2250–2261. [CrossRef]
10. Haupt, R.L.; Lanagan, M. Reconfigurable Antennas. *IEEE Antennas Propag. Mag.* **2013**, *55*, 49–61. [CrossRef]
11. Rodrigo, D.; Damgaci, Y.; Unlu, M.; Cetiner, B.A.; Romeu, J.; Jofre, L. Antenna Reconfigurability based on a Novel parasitic pixel layer. In Proceedings of the 5th European Conference on Antennas and Propagation (EUCAP), Rome, Italy, 11–15 April 2011; pp. 3497–3500.
12. Zhang, L.; Chen, X.Q.; Liu, S.; Zhang, Q.; Zhao, J.; Dai, J.Y.; Bai, G.D.; Wan, X.; Cheng, Q.; Castaldi, G.; et al. Space-Time-Coding Digital Metasurfaces. *Nat. Commun.* **2018**, *9*, 4334–4344. [CrossRef]
13. Basar, E.; Di Renzo, M.; De Rosny, J.; Debbah, M.; Alouini, M.; Zhang, R. Wireless Communications Through Reconfigurable Intelligent Surfaces. *IEEE Access* **2019**, *7*, 116753–116773. [CrossRef]
14. Pedro, J.C.; Maas, S.A. A comparative overview of microwave and wireless power-amplifier behavioral modeling approaches. *IEEE Trans. Microw. Theory Tech.* **2005**, *53*, 1150–1163. [CrossRef]
15. Golio, M.; Dunleavy, L.; Gneiting, T. History and state-of-the-art in large signal modeling for RF/microwave power amplifier development. In Proceedings of the 2015 IEEE MTT-S International Microwave Symposium, Phoenix, AZ, USA, 17–22 May 2015; pp. 1–4. [CrossRef]
16. Verspecht, J. Large-signal network analysis. *IEEE Microw. Mag.* **2005**, *6*, 82–92. [CrossRef]
17. Verspecht, J.; Root, D.E. Polyharmonic distortion modeling. *IEEE Microw. Mag.* **2006**, *7*, 44–57. [CrossRef]
18. Cai, J.; King, J.B.; Merrick, B.M.; Brazil, T.J. Padé-Approximation-Based Behavioral Modeling. *IEEE Trans. Microw. Theory Tech.* **2013**, *61*, 4418–4427. [CrossRef]
19. Cai, J.; Brazil, T.J. Reduced-Complexity Polynomial Based Nonlinear Behavioral Modeling. *IEEE Microw. Wirel. Compon. Lett.* **2014**, *24*, 496–498. [CrossRef]

20. Cai, J.; King, J.; Yu, C.; Liu, J.; Sun, L. Support Vector Regression-Based Behavioral Modeling Technique for RF Power Transistors. *IEEE Microw. Wirel. Compon. Lett.* **2018**, *28*, 428–430. [CrossRef]
21. Xu, J.; Jones, R.; Harris, S.A.; Nielsen, T.; Root, D.E. Dynamic FET model—DynaFET—For GaN transistors from NVNA active source injection measurements. In Proceedings of the 2014 IEEE MTT-S International Microwave Symposium (IMS2014), Tampa, FL, USA, 1–6 June 2014; pp. 1–3. [CrossRef]
22. Root, D.E.; Verspecht, J.; Sharrit, D.; Wood, J.; Cognata, A. Broad-band poly-harmonic distortion (PHD) behavioral models from fast automated simulations and large-signal vectorial network measurements. *IEEE Trans. Microw. Theory Tech.* **2005**, *53*, 3656–3664. [CrossRef]
23. Caverly, R.H.; Hiller, G. Distortion In p-i-n Diode Control Circuits. *IEEE Trans. Microw. Theory Tech.* **1987**, *35*, 492–501. [CrossRef]
24. Caverly, R.H.; Hiller, G. The frequency-dependent impedance of p-i-n diodes. *IEEE Trans. Microw. Theory Tech.* **1989**, *37*, 787–790. [CrossRef]
25. Gatard, E.; Sommet, R.; Bouysse, P.; Quere, R. An Improved Physics-Based Formulation of the Microwave p-i-n Diode Impedance. *IEEE Microw. Wirel. Compon. Lett.* **2007**, *17*, 211–213. [CrossRef]
26. Caverly, R.H.; Quinn, N. A SPICE model for simulating the impedance-frequency characteristics of high frequency PIN switching diodes. In Proceedings of the 1999 IEEE International Symposium on Circuits and Systems (ISCAS), Orlando, FL, USA, 30 May–2 June 1999; Volume 6, pp. 282–285.
27. Caverly, R.H.; Drozdovski, N.V.; Drozdovskaia, L.M.; Quinn, M.J. SPICE modeling of microwave and RF control diodes. In Proceedings of the 43rd IEEE Midwest Symposium on Circuits and Systems (Cat. No. CH37144), Lansing, MI, USA, 8–11 August 2000; Volume 1, pp. 28–31.
28. Li, Z.; Ahmed, E.; Eltawil, A.M.; Cetiner, B.A. A Beam-Steering Reconfigurable Antenna for WLAN Applications. *IEEE Trans. Antennas Propag.* **2015**, *63*, 24–32. [CrossRef]
29. Lotfi, P.; Soltani, S.; Murch, R.D. Printed Endfire Beam-Steerable Pixel Antenna. *IEEE Trans. Antennas Propag.* **2017**, *65*, 3913–3923. [CrossRef]
30. Towfiq, M.A.; Khalat, A.; Blanch, S.; Romeu, J.; Jofre, L.; Cetiner, B.A. Error Vector Magnitude, Intermodulation, and Radiation Characteristics of a Bandwidth- and Pattern-Reconfigurable Antenna. *IEEE Antennas Wirel. Propag. Lett.* **2019**, *18*, 1956–1960. [CrossRef]
31. Maas, S. *Nonlinear Microwave and RF Circuits*, 2nd ed.; Artech House: Morristown, NJ, USA, 2003; p. 603.
32. Schuman, H. Time-domain scattering from a nonlinearly loaded wire. *IEEE Trans. Antennas Propag.* **1974**, *22*, 611–613. [CrossRef]
33. Sarkar, T.; Weiner, D. Scattering analysis of nonlinearly loaded antennas. *IEEE Trans. Antennas Propag.* **1976**, *24*, 125–131. [CrossRef]
34. Sarkar, T.K.; Weiner, D.D.; Harrington, R.F. Analysis of Nonlinearly Loaded Multiport Antenna Structures Over an Imperfect Ground Plane Using the Volterra-Series Method. *IEEE Trans. Electromagn. Compat.* **1978**, 278–287. [CrossRef]
35. Liu, T.; Tesche, F. Analysis of antennas and scatterers with nonlinear loads. *IEEE Trans. Antennas Propag.* **1976**, *24*, 131–139. [CrossRef]
36. Huang, C.-C.; Chu, T.-H. Analysis of wire scatterers with nonlinear or time-harmonic loads in the frequency domain. *IEEE Trans. Antennas Propag.* **1993**, *41*, 25–30. [CrossRef]
37. Lee, K.-C. Two efficient algorithms for the analyses of a nonlinearly loaded antenna and antenna array in the frequency domain. *IEEE Trans. Electromagn. Compat.* **2000**, *42*, 339–346. [CrossRef]
38. Sheshyekani, K.; Sadeghi, S.H.H.; Moini, R. A Combined MoM-AOM Approach for Frequency Domain Analysis of Nonlinearly Loaded Antennas in the Presence of a Lossy Ground. *IEEE Trans. Antennas Propag.* **2008**, *56*, 1717–1724. [CrossRef]
39. Quijano, J.L.A.; Vecchi, G. Optimization of an Innovative Type of Compact Frequency-Reconfigurable Antenna. *IEEE Trans. Antennas Propag.* **2009**, *57*, 9–18. [CrossRef]
40. Rodrigo, D.; Cetiner, B.A.; Jofre, L. Frequency, Radiation Pattern and Polarization Reconfigurable Antenna Using a Parasitic Pixel Layer. *IEEE Trans. Antennas Propag.* **2014**, *62*, 3422–3427. [CrossRef]
41. Ramírez, G.A.; Araque, J.L.; Ballesteros, C.; Blanch, S.; Romeu, J.; Cetiner, B.; Jofre, L. Study of interconnecting switch currents in reconfigurable parasitic layer antennas. In Proceedings of the 2019 IEEE International Symposium on Antennas and Propagation and USNC-URSI Radio Science Meeting, Atlanta, GA, USA, 7–12 July 2019; pp. 313–314. [CrossRef]
42. Root, D.E.; Verspecht, J.; Horn, J.; Marcu, M. *X-Parameters. Characterization, Modeling, and Design of Nonlinear RF and Microwave Components*; Cambridge University Press: Cambridge, UK, 2013.
43. Shafik, R.A.; Rahman, M.S.; Islam, A.R.; Ashraf, N.S. On the error vector magnitude as a performance metric and comparative analysis. In Proceedings of the 2006 International Conference on Emerging Technologies, Peshawar, Pakistan, 13–14 November 2006; pp. 27–31. [CrossRef]
44. McKinley, M.; Remley, C.A.; Myslinski, M.; Kenney, J.S. EVM Calculation for Broadband Modulated Signals. In Proceedings of the 64th ARFTG Microwave Measurements Conference, Fall 2004, Orlando, FL, USA, 2–3 December 2004; pp. 45–52. [CrossRef]
45. Liu, R.; Li, Y.; Chen, H.; Wang, Z. EVM estimation by analyzing transmitter imperfections mathematically and graphically. *Analog Integr. Circuits Signal Process.* **2006**, *48*, 257–262. [CrossRef]
46. Gharaibeh, K.M.; Gard, K.G.; Steer, M.B. Accurate estimation of digital communication system metrics—SNR, EVM and /spl rho/ in a nonlinear amplifier environment. In Proceedings of the 64th ARFTG Microwave Measurements Conference, Fall 2004, Orlando, FL, USA, 2–3 December 2004; pp. 41–44. [CrossRef]

47. Mahmoud, H.A.; Arslan, H. Error vector magnitude to SNR conversion for nondata-aided receivers. *IEEE Trans. Wirel. Commun.* **2009**, *8*, 2694–2704. [CrossRef]
48. Araque Quijano, J.L.; Vecchi, G. Optimization of a Compact Frequency- and Environment-Reconfigurable Antenna. *IEEE Trans. Antennas Propag.* **2012**, *60*, 2682–2689. [CrossRef]
49. GSMA. *3.5 GHz in the 5G Era Preparing for New Services in 3.3–4.2 GHz*; Technical Report; GSM Association: London, UK, 2021.

sensors

MDPI

Article

A Highly Compact Antipodal Vivaldi Antenna Array for 5G Millimeter Wave Applications

Amruta Sarvajeet Dixit [1], Sumit Kumar [1,*], Shabana Urooj [2] and Areej Malibari [3,4]

1 Symbiosis Institute of Technology, Symbiosis International Deemed University, Pune 412115, India; amrutaamalode@gmail.com
2 Department of Electrical Engineering, College of Engineering, Princess Nourah bint Abdulrahman University, Riyadh 84428, Saudi Arabia; smurooj@pnu.edu.sa
3 College of Engineering, Princess Nourah bint Abdulrahman University, Riyadh 84428, Saudi Arabia; aamalibari@pnu.edu.sa
4 Department of Computer Science, Faculty of Computing and IT, King Abdulaziz University, Jeddah 80200, Saudi Arabia
* Correspondence: er.sumitkumar21@gmail.com

Abstract: This paper presents a compact 1 × 4 antipodal Vivaldi antenna (AVA) array for 5G millimeter-wave applications. The designed antenna operates over 24.19 GHz–29.15 GHz and 30.28 GHz–40.47 GHz frequency ranges. The proposed antenna provides a high gain of 8 dBi to 13.2 dBi and the highest gain is obtained at 40.3 GHz. The proposed antenna operates on frequency range-2 (FR2) and covers n257, n258, n260, and n261 frequency bands of 5G communication. The corrugations and RT/Duroid 5880 substrate are used to reduce the antenna size to 24 mm × 28.8 mm × 0.254 mm, which makes the antenna highly compact. Furthermore, the corrugations play an important role in the front-to-back ratio improvement, which further enhances the gain of the antenna. The corporate feeding is optimized meticulously to obtain an enhanced bandwidth and narrow beamwidth. The radiation pattern does not vary over the desired operating frequency range. In addition, the experimental results of the fabricated antenna coincide with the simulated results. The presented antenna design shows a substantial improvement in size, gain, and bandwidth when compared to what has been reported for an AVA with nearly the same size, which makes the proposed antenna one of the best candidates for application in devices that operate in the millimeter frequency range.

Keywords: antipodal Vivaldi antenna (AVA); millimeter wave; compact; 5G applications; corrugations

Citation: Dixit, A. S.; Kumar, S.; Urooj, S.;Malibari, A. A Highly Compact Antipodal Vivaldi Antenna Array for 5G Millimeter Wave Applications. *Sensors* **2021**, *21*, 2360. https://doi.org/10.3390/s21072360

Academic Editor: Naser Ojaroudi Parchin

Received: 1 February 2021
Accepted: 8 March 2021
Published: 29 March 2021

1. Introduction

Millimeter-wave applications are used in medical imaging, the military, satellites, and 5G communication. The current 4G communication for mobile phones provides a moderate data rate and capacity and suffers from spectrum shortage [1]. The evolutionary 5G technology is the panacea of current mobile communication issues, and it can be used to provide various revolutionary services. The international telecommunication union (ITU) has specified various specifications of 5G technology; for example, spectrum efficiency should be up to 9 bits/s/Hz, there should be a high data rate from 2 Gbps to 20 Gbps, connection density should be 1 million/km^2, mobility should be up to 500 km/h, and operating frequency should be below 6 GHz (3 GHz to 5.5 GHz) and above 6 GHz (28 GHz, 38 GHz, and 70 GHz) [2,3]. The main requirement for the deployment of 5G technology is that devices should work in the millimeter-wave frequency range [2], but working in this range can cause devices to suffer from higher path losses [4]. Such path losses at higher frequencies can be reduced by using the high-gain antenna array. This fact provoked the design of a compact antenna that can work at such a high-frequency band with enhanced gain and a stable radiation pattern. The antipodal Vivaldi antenna (AVA), which was invented by Gazit [5], is one of the best candidates as a 5G antenna that works in the millimeter-wave range. The Vivaldi antenna, which was invented by Gibson [6], operates at high frequencies and provides a wide bandwidth. the AVA is preferred over the Vivaldi antenna because it provides a wider bandwidth, high gain, a nearly-constant radiation pattern, and it is easy for fabrication. A single AVA cannot meet all the requirements of a 5G antenna, such as high gain (above 8 dBi), efficiency (above 85%), a wide bandwidth (above 6 GHz), and stable radiation patterns; thus, it is necessary to design a highly compact AVA array that fulfils the above-mentioned requirements [7,8].

Various antenna performance enhancement techniques have been investigated by researchers [3]. The gain of the AVA can be increased by using an array [9–11]. In [12], a 1 × 4 AVA array with substrate integrated waveguide (SIW) and corrugations was implemented to obtain a high gain of 23 dBi. Furthermore, the bandwidth of the AVA can be increased by using slots in AVA flare [13,14]. In [15], a wideband antenna was designed to operate from 3.4 GHz to 40 GHz by incorporating slots in the AVA. The finalization of the dimension and position of the slots is a challenging task [16]. Additionally, the AVA can be made compact by incorporating corrugations [17,18]. The corrugations are of a triangular, sinusoidal, rectangular, or square shape. Out of these shapes, the rectangular shape is frequently used in the literature. Some researchers have implemented the AVA with metamaterial [10], dielectric lens [19], and balanced AVA [20] to improve the gain. The balanced AVA consists of three layers: the top and bottom layers function as conductors, and the middle layer functions as a ground. This structure provides equal electric field distribution on both sides to the ground, but the antenna design is complex [21]. Similarly, the design and placement of the metamaterial unit cells are also complex. Furthermore, after incorporating a dielectric lens, the size of the antenna increases [22]. Moreover, some researchers have implemented the AVA with parasitic patches to improve the gain, but this enhancement is not substantial, and the parasitic patch adversely affects the size of the antenna [23]. Hence, it is a challenging task to design a compact antenna with the required performance parameters.

This paper presents a 1 × 4 AVA array with corrugations to achieve a high gain, wide bandwidth, and size reduction. Due to the non-ideal directional characteristics of the practical antennas, the energy radiated from each array antenna element is received by other array elements. The partial amount of this energy may be re-scattered in a different direction, which is known as mutual coupling [24]. If the distance among the array elements is minimal, then the interchange of energy (energy scattering) increases, thereby increasing mutual coupling. Furthermore, this re-scattered energy radiates in any direction other than the intended one, which gives rise to increases in side-lobe levels (SLL). Hence, because of the basic structure of the antenna array, there are more side-lobe levels (SLLs) and mutual coupling of array elements [4]. More SLLs affect the radiation pattern, while mutual coupling alleviates the antenna bandwidth. Both of these issues are mitigated by incorporating corrugations in AVA flares. The corrugations reduce back- and side-lobe levels, and they also improve the bandwidth [25]. Out of the various performance enhancement methods, corrugation is easy to implement, and it is a very effective technique. Furthermore, in [26], comb-shaped (rectangular) corrugations are etched on the AVA flare to identify voids in concrete beams. In [27], rectangular corrugations with a constant, increasing, and decreasing size are demonstrated. In [27], it is proved that AVA performance can be effectively improved by using corrugations with either a constant or decreasing size. In this paper, the constant size corrugation is incorporated in the AVA.

The array elements can be fed by using either series or corporate feeding. In series feeding, different antenna elements are excited differently to alleviate the SLLs. Even though the array with a series feed is compact, it provides a narrow bandwidth, which is not desirable [28]. In corporate feeding, different antenna elements are excited equally, and it enhances the bandwidth [29]. This paper presents the design of a 1 × 4 AVA array with corporate feeding to achieve a wide bandwidth for 5G applications that operate in the millimeter-wave frequency range. The detailed design and parametric study of the

AVA array with corrugations are given in Section 2. Section 3 proves the importance of the proposed antenna with the help of various simulated and measured results. Finally, the overall findings of this paper are summarized in the Conclusion Section 4.

2. Design of Proposed AVA Array

2.1. Single AVA Design

The single-patch conventional AVA and the proposed AVA with corrugation are shown in Figure 1. They are fabricated on RT/duroid 5880 substrate, which has 2.2 dielectric permittivity and a loss tangent of 0.001. The antenna was designed using a high-frequency structure simulator (HFSS) version 2020 R2. Its length is 20 mm, its width is 6 mm, and its substrate thickness is 0.254 mm. The proposed AVA has a top patch that acts as a radiator and a bottom patch that acts as a ground. Both the ground and radiator patches are mirror images of each other. These patches are formed by using an inner circular arc and an outer exponential curve whose equation is given below [30].

$$Y_1 = \pm(C1e^{350x} - C2)\ (8.65 \leq x \leq 17.25) \tag{1}$$

where $C1$ = 0.00678 mm and $C2$ = 3.4 mm

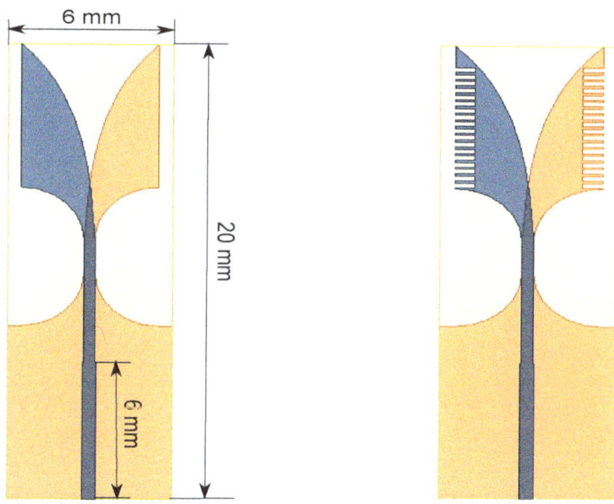

(**a**) Conventional AVA without corrugations. (**b**) AVA with rectangular corrugations.

Figure 1. Design of the proposed single-patch antenna.

The input impedance plot of the single-patch AVA with and without corrugations is shown in Figure 2. The slots of corrugations introduce the RLC circuit, which results in a change in impedance and resonance frequency of the antenna. As shown in Figure 2, the input impedance plot is shifted to the lower frequency range after incorporating corrugations in the AVA. This shifting in input impedance results in a decrease in lower cut-off frequency, as shown in Figure 3. The lower cut-off frequency of the conventional AVA is 26.33 GHz, whereas the lower cut-off frequency of the proposed AVA with corrugations is 25 GHz. Hence, corrugations result in a change in the frequency response of the antenna. By optimizing the dimensions of corrugations, bandwidth and input impedance enhancement can be achieved. The optimized corrugations depth is 0.7 mm, the width of the corrugation teeth is 1mm, and the width of the corrugation slot is 2 mm.

Next, the vital role of corrugations in electric field enhancement is proved in Figure 4. In this figure, region A shows that the concentration of the electric field is less than that of the electric field concentration of region B. Thus, corrugations also contribute to enhancing radiation, resulting in a substantial gain improvement, as shown in Figure 5. The range of gain for the AVA without corrugation is 3.2 dBi to 4.45 dBi, whereas the range of gain for the AVA with corrugations is 4.2 dBi to 4.5 dBi, which is almost constant. Thus, corrugations improve the gain of an antenna.

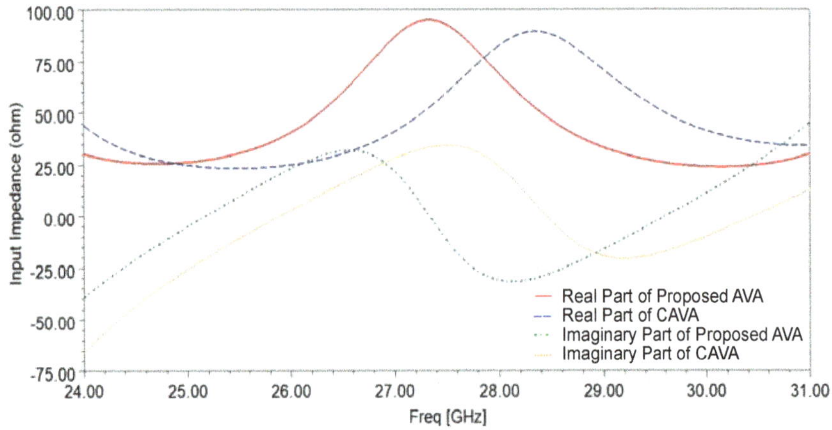

Figure 2. Input impedance of the single-patch AVA with and without corrugations.

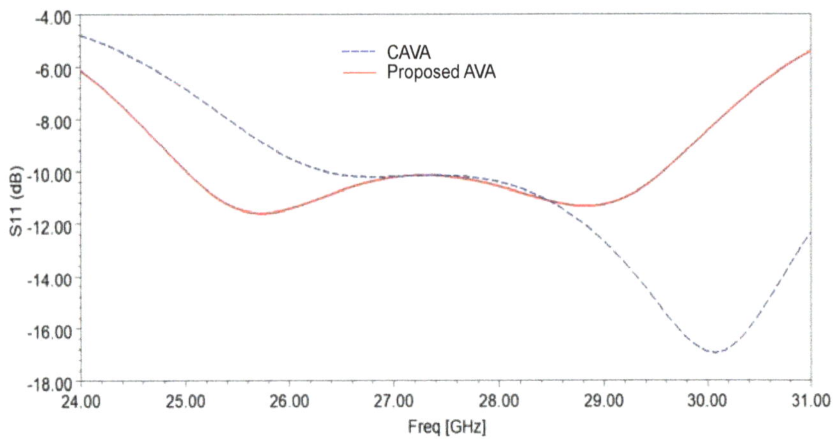

Figure 3. Reflection coefficient of the single-patch AVA with and without corrugations.

2.2. 1 × 4 AVA Array Design

The 1 × 4 AVA array was formed by using the AVA with the corrugations shown in Figure 6a,b. The length of the proposed AVA is 28.8 mm and its width is 24 mm. The rectangular slots are incorporated in the ground plane to improve the frequency response of the AVA array depicted in Figure 7. The dotted graph is S11 of the AVA array without slots and corrugations, whereas the continuous graph is S11 of the AVA array with slots in the ground plane and without corrugations. This figure shows an improvement in the frequency response of the AVA, particularly in the frequency range of 25 GHz to 33 GHz. The slot structure changes the RLC circuit, which results in a change in the frequency response of the antenna.

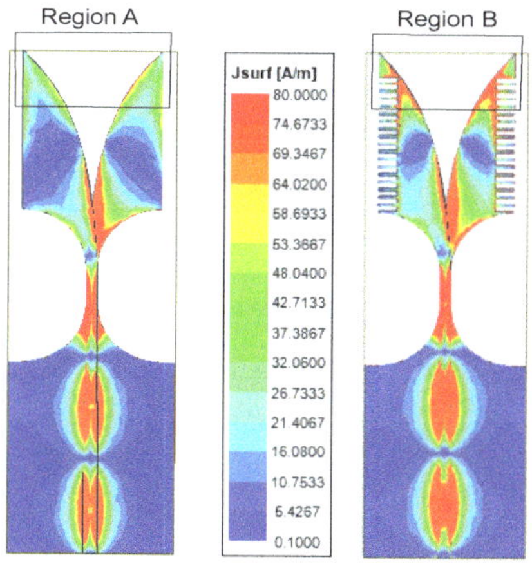

Figure 4. Electric field distribution of the single-patch AVA with and without corrugations at 27.5 GHz.

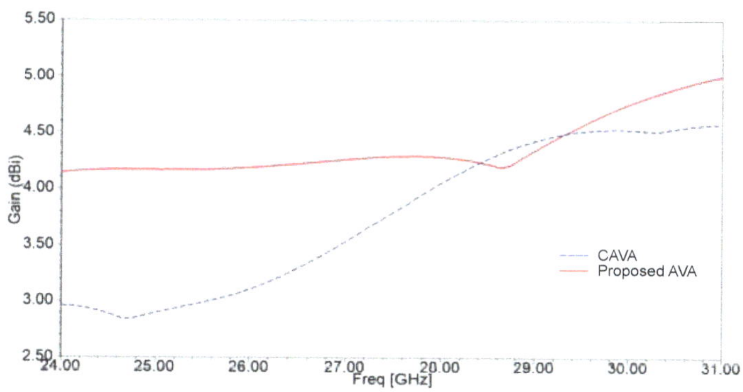

Figure 5. Gain versus frequency plot of the single-patch AVA with and without corrugations.

The mutual coupling between array elements is shown in Figure 8. As depicted in Figure 8, the mutual coupling of the 1 × 4 AVA array is in the acceptable range. The mutual coupling improved after the introduction of slots in the ground plane, and it was further enhanced by incorporating corrugations in the AVA array. The slots in the ground plane provided isolation enhancement of 24 dB at 34 GHz, while corrugations improved the isolation by 36.4 dB at 33.56 GHz. Figure 9 indicates that the input impedance of the AVA array varied from 27 Ω to 100 Ω, whereas the input impedance of the AVA array with corrugations changed from 40 Ω to 63 Ω over the frequency range of 24.2 GHz to 40.5 GHz. Thus, the corrugations also contributed to improving input impedance matching.

Input impedance was also enhanced by using a corporate feeding network. The corporate feeding is present at the top to feed the top patches with equal power. The impedance of each branch of the feeding network is given in Figure 10. Initially, a 50 Ω feeding line was used, and then it was bifurcated into two 100 Ω feeding lines. This structure of the two 100 Ω feeding lines and the single 50 Ω feeding line resembles a 'T' junction. This 'T' junction was repeated to obtain four feeding lines for the four top patches. Additionally, the $\lambda/4$ transformer was used between the 50 Ω and 100 Ω feeding line for good impedance matching. The feeding network provided nearly 50 ± 10 Ω impedance, and it was optimized using HFSS.

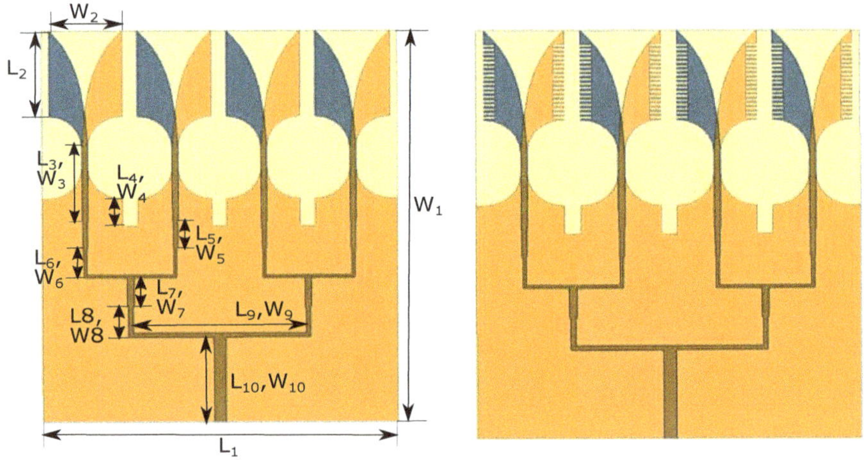

(**a**) AVA array without corrugations (**b**) AVA array with rectangular corrugations

Figure 6. Design of proposed antenna.

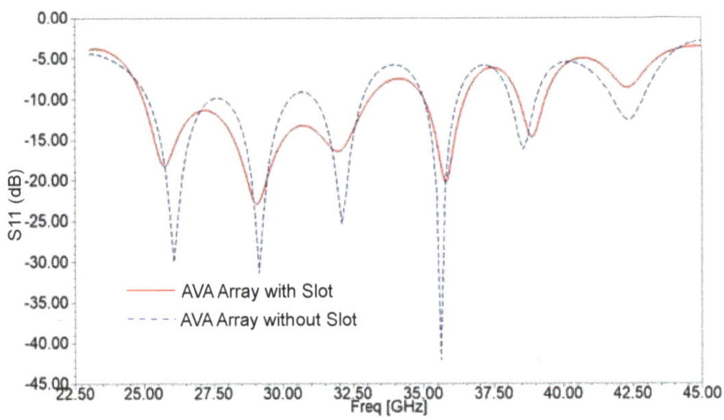

Figure 7. Reflection coefficient of the AVA array with and without slots in the ground.

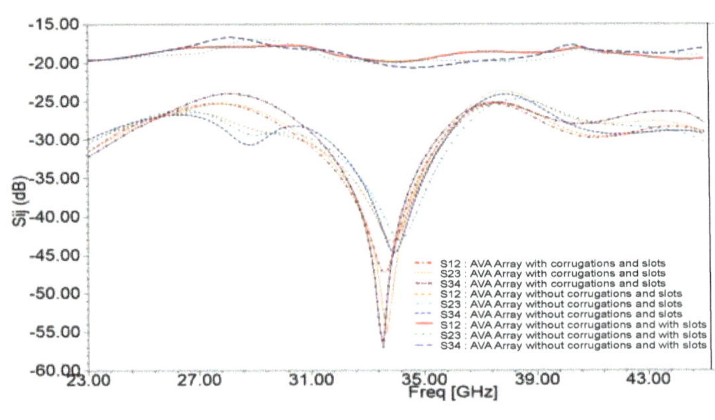

Figure 8. Simulated mutual coupling between array elements after removal of 1×4 power divider.

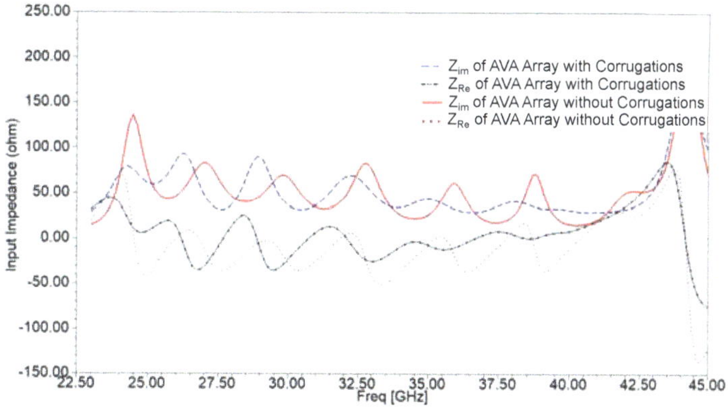

Figure 9. Input impedance of the AVA array with and without corrugations.

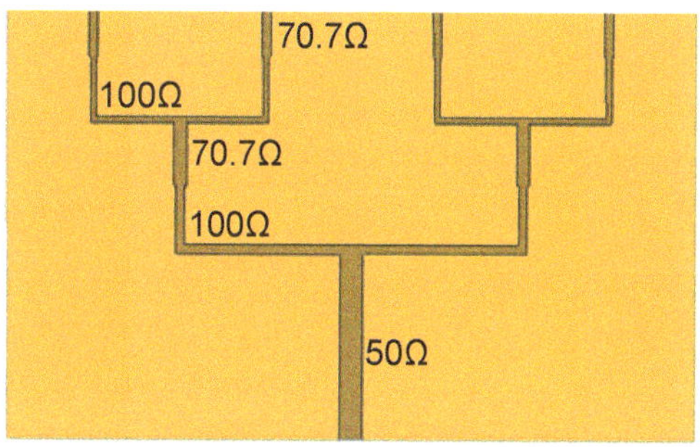

Figure 10. Feeding network of the AVA array.

Corrugations were also introduced in both the top and bottom patches. The structure of corrugation resembles a comb shape, which means that its structure is rectangular, and the size of the rectangle is uniform over the complete top and bottom patches. The corrugations change the capacitance, inductance, and resistance of an antenna, which positively reflects on bandwidth enhancement. The dimension of corrugation's length slot was optimized using HFSS, as shown in Figure 11a. This figure demonstrates that the best reflection coefficient result was obtained for a corrugation length of 0.7 mm. The width of the slot and teeth of the corrugations are 2 mm and 1 mm, respectively. Furthermore, the vital role of corrugations in the bandwidth improvement is shown in Figure 11b. This figure displays the reflection coefficient plots of the AVA array with and without corrugations. It shows that there are tri-bands in the AVA array, which are 25.09 GHz–32.9 GHz, 34.97 GHz–36.24 GHz, and 38.03 GHz–39.07 GHz, whereas a dual-band from 24.19 GHz–29.15 GHz and 30.28 GHz–40.47 GHz was achieved after incorporation of corrugations in the AVA array. This wide frequency band contains three important frequency bands of 5G, communications which are 24.25 GHz–29.5 GHz, 31.8 GHz–33.4 GHz, and 37.5 GHz–40.5 GHz. The fabricated antenna is shown in Figure 12, and the optimized dimensions are given in Table 1.

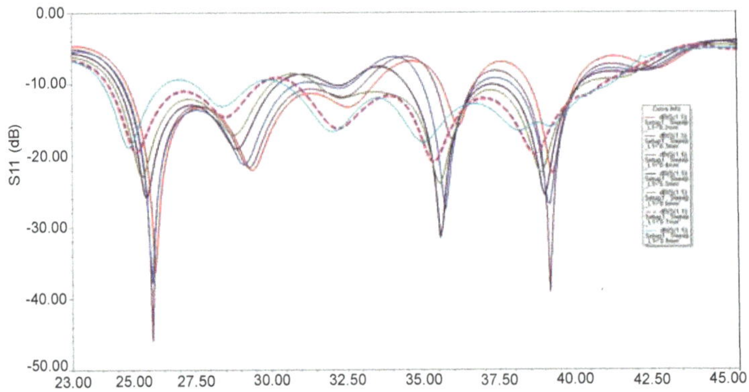

(**a**) S11 for different corrugation length.

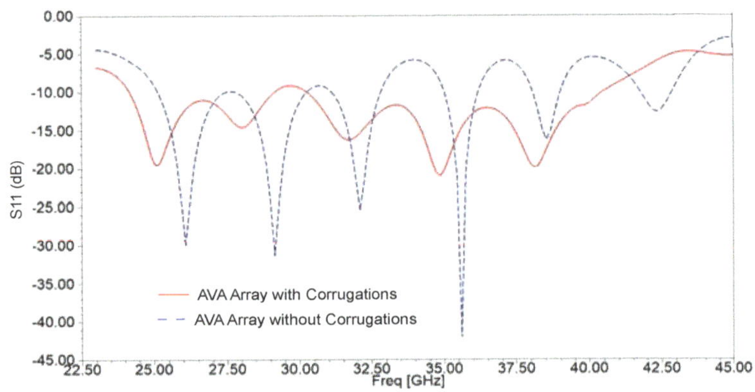

(**b**) S11 of the AVA array with and without corrugations.

Figure 11. Reflection coefficient of the designed antenna.

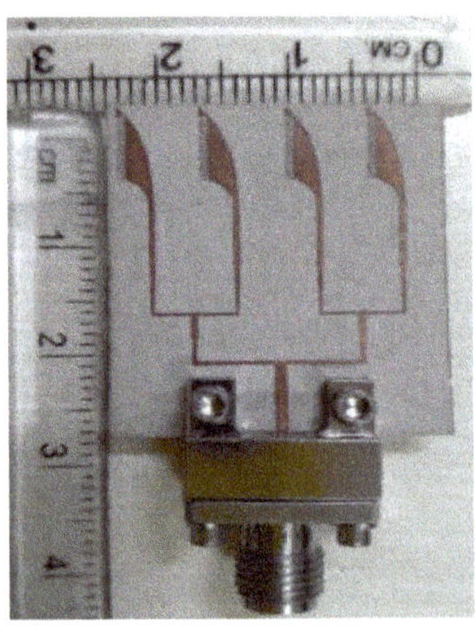

(**a**) Top Patch

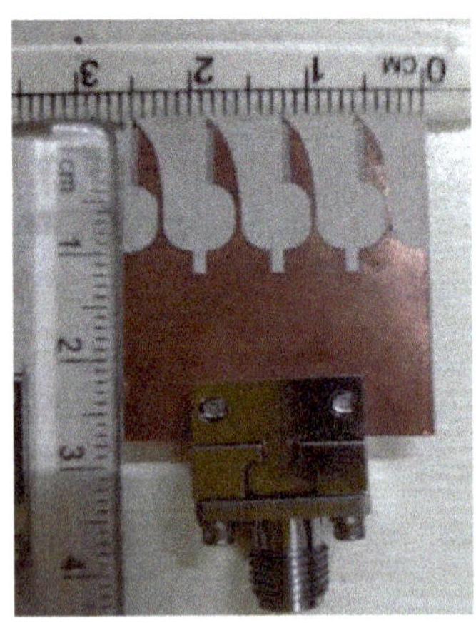

(**b**) Bottom Patch

Figure 12. Fabricated proposed antenna.

Table 1. 1 × 4 AVA array dimensions.

Parameters	Dimensions (mm)	Parameters	Dimensions (mm)
W_1	28.8	L_1	24
W_2	5	L_2	6.3
W_3	0.4	L_3	5.4
W_4	1	L_4	2
W_5	0.3	L_5	2
W_6	0.2	L_6	2
W_7	0.45	L_7	2
W_8	0.3	L_8	2
W_9	0.3	L_9	12
W_{10}	0.8	L_{10}	6.23

3. Results and Discussion

The fabricated antenna was tested using an N5224A performance network analyzer (PNA). The simulated and measured reflection coefficient (S11) results are shown in Figure 13. Because of unavoidable fabrication drawbacks, there were minor dissimilarities between the measured and simulated results. The simulated frequency range is from 24.19 GHz–29.15 GHz to 30.28 GHz–40.5 GHz, whereas the measured operating frequency range is from 24.19 GHz–29.02 GHz to 30.20 GHz–40.33 GHz. The frequency ranges of both simulated and measured results where S11 is above −10 dB are not allotted for 5G applications. Hence, the designed antenna covers 24.25 GHz–29.5 GHz, 31.8 GHz–33.4 GHz, and 37.5 GHz–40.5 GHz frequency bands of 5G communications.

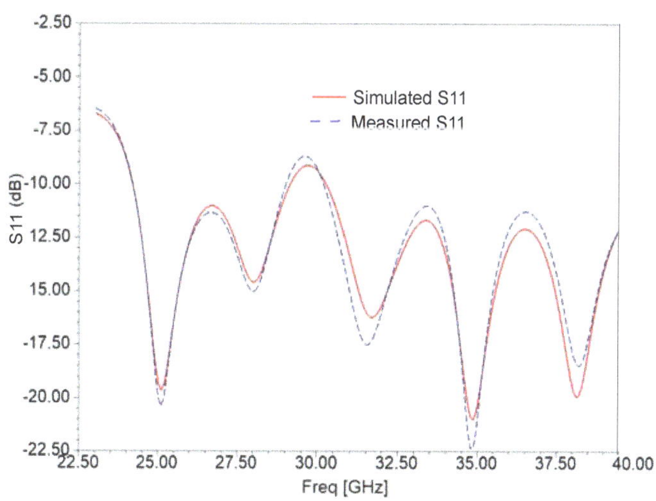

Figure 13. Simulated and measured S11 of the AVA array with corrugations.

The co- and cross-polarization of elevation (H-plane) and azimuth (E-plane) planes at 25.5 GHz, 31.5 GHz, and 38 GHz are shown in Figure 14. This figure proves that the radiation patterns at different frequencies are almost the same, and, thus, the antenna performance is frequency independent. Therefore, the designed antenna provides a stable radiation pattern. In both the E and H planes, the front lobe of the AVA array with corrugations is higher than the front lobe of the AVA array without corrugations at various frequencies. Furthermore, the back- and side-lobe levels are also alleviated in the AVA array with corrugations as compared to the AVA array without corrugations. Hence, this figure depicts the enhancement in the front-to-back ratio and gains by employing corrugations in the AVA flares. Next, as the element spacing is 5 mm, which is less than the center wavelength, grating lobes are absent in all radiation patterns. Importantly, the cross-polarization of the AVA array with corrugations is lower than the cross-polarization of the AVA array without corrugations.

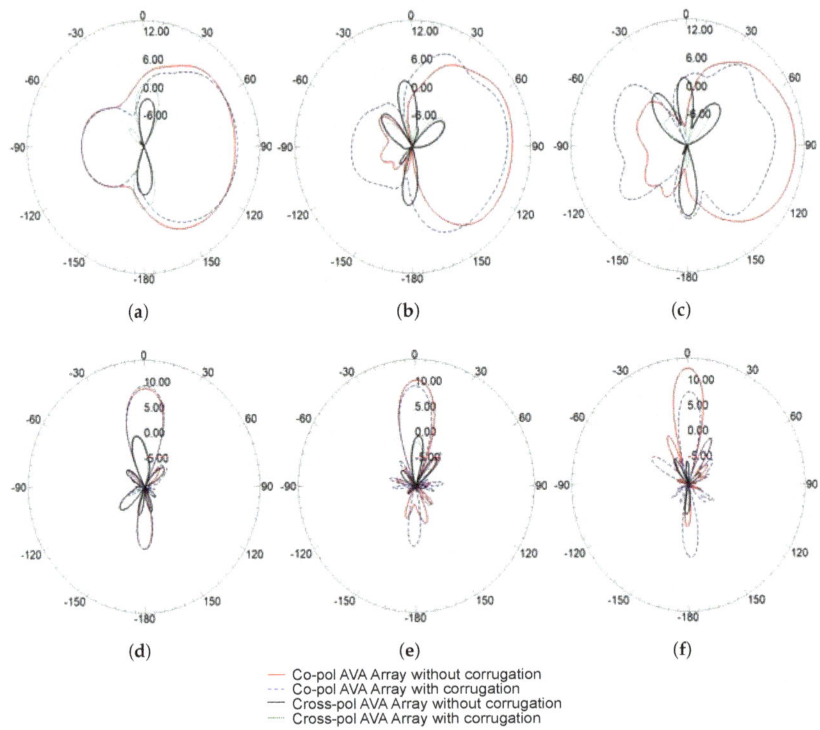

Figure 14. Simulated co-polarization and cross-polarization of E plane (**a**–**c**) and H plane (**d**–**f**).

The importance of corrugations for the improvement of the front-to-back ratio (FBR) can be clearly observed in Figure 15. The FBR of the AVA without corrugations is in the range of 5.7 dB to 10.5 dB, whereas the FBR of the AVA with corrugations is from 8 dB to 24 dB. The highest FBR of the AVA with corrugation is 24 dB at 40.45 GHz, whereas the highest FBR of the AVA without corrugations is 10.5 dB at 27.54 GHz. As a consequence of this, it is proved that the corrugation increases the front lobe and reduces the back lobe, which results in the enhancement of FBR and gain.

Figure 16 shows the simulated electric field distribution of the AVA array with and without corrugations at 25 GHz (upper) and 34 GHz (lower). It shows that the power divider networks are optimized correctly to distribute the input power equally to all four antenna elements. The AVA array structure provided plane-like waves. In a 1 × 4 AVA array with corrugations, the concentration of the electric field is higher at the edges as compared to the AVA array without corrugations. Furthermore, the AVA array with corrugation enhanced directivity as compared to the AVA array without corrugations, which resulted in gain enhancement.

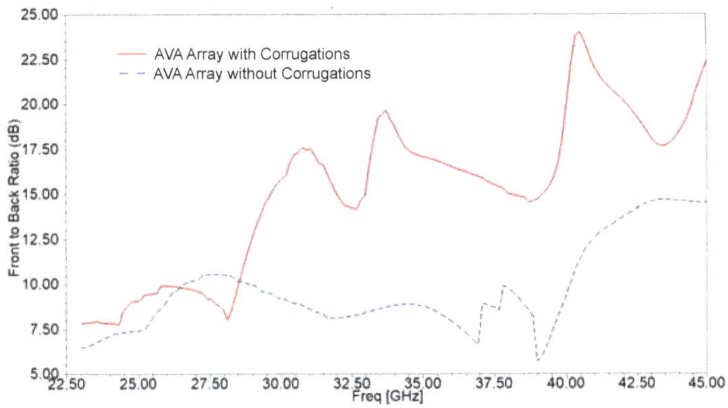

Figure 15. Simulated front-to-back ratio of the AVA array with and without corrugations.

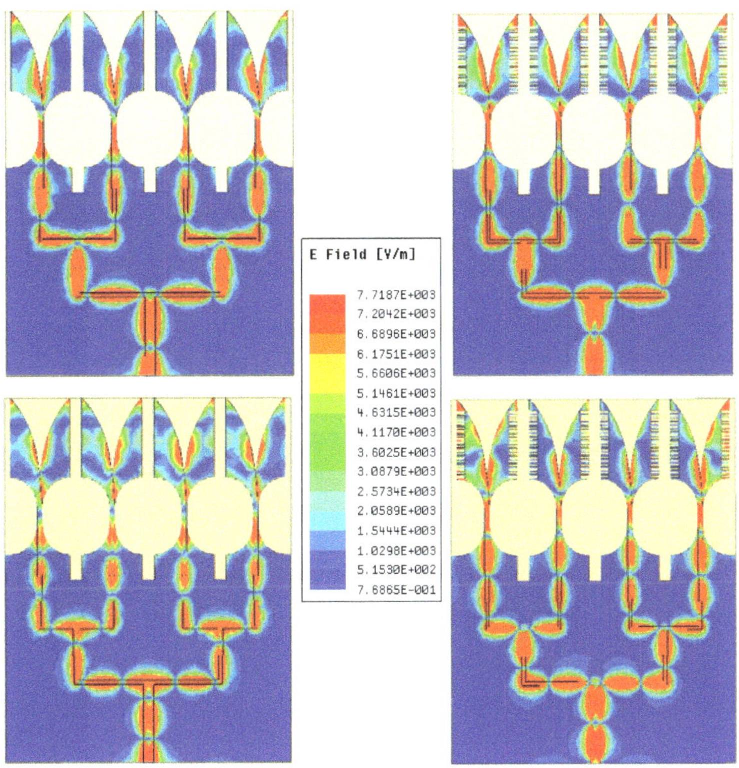

Figure 16. Electric field distribution of the AVA array with and without corrugations at 25 GHz (**upper**) and 34 GHz (**lower**).

The contribution of corrugation to gain enhancement is also shown in Figure 17. The gain of the AVA array without corrugations is from 8.47 dBi to 12.63 dBi, whereas the gain of the AVA array with corrugations is from 8 dBi to 13.2 dBi. Thus, the peak gain was enhanced by 0.57 dBi, and the gain variation was reduced to some extent. The gain of the AVA with corrugations is low at 27.5 GHz, which is due to the losses of slits and mutual coupling between array elements. The efficiency versus frequency graph is shown in Figure 18. The efficiency of the AVA without corrugations varies from 89.64% to 93.37%, and the efficiency of the AVA with corrugations changes from 91.97% to 94.15%. Thus, the proposed antenna provides good efficiency over the desired operating frequency range.

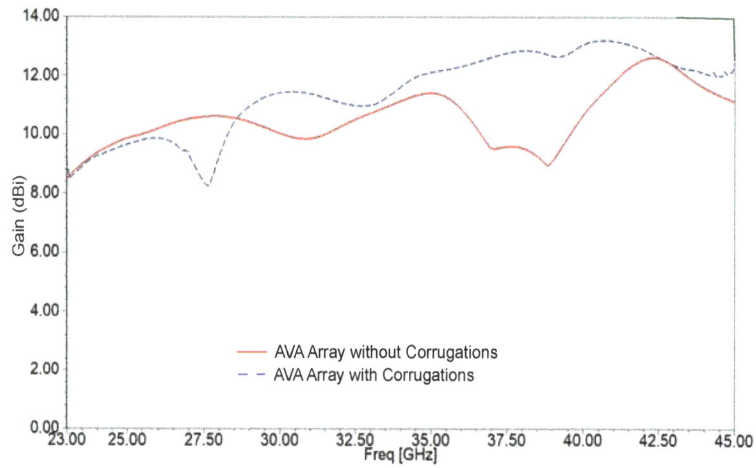

Figure 17. Simulated gain plots of the AVA array with and without corrugations.

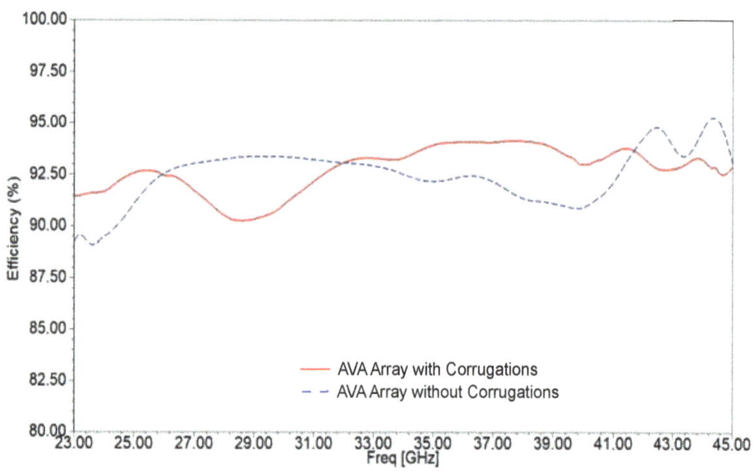

Figure 18. Simulated efficiency of the AVA array with and without corrugations.

The importance of the proposed 1 × 4 AVA array was evaluated by comparing it with other recent AVAs, which are given in Table 2. These antennas were compared by focusing on important parameters, such as relative permittivity, dimensions, gain, and frequency bands of antennas. Dimensions are represented in terms of the center frequency of the antenna. The antenna designed in [31,32] provides a very good bandwidth, but its gain variation is higher, and the antenna size is also very large. In [33,34] an antenna is designed with a wide bandwidth and moderate antenna size, but its gain is small. Further, the antennas designed in [1,35,36] are of moderate size and nearly constant gain, but their bandwidth is very small. Moreover, the antenna designed in [37] is compact, but its gain is low. Additionally, in [38] a moderate-sized antenna with a low gain is designed. Recently, a new antenna was designed in 2021 [39], which is of moderate size and moderate gain for 5G applications. The antenna size is smaller in [40] as compared to the proposed antenna, but its gain and bandwidth are lower. Finally, the antenna proposed in [41–43] provides a wide bandwidth, but it is considerably large. As compared to these antennas, the proposed antenna is very compact and provides a wide bandwidth and a high and nearly constant gain. Hence, the designed antenna is more appropriate for integration into mm-wave-based 5G communication devices than the above-mentioned antennas.

Table 2. The comparison of the proposed AVA with other AVAs.

Ref. No.	Techniques Employed	ϵ_r	Dimensions ($\lambda_c{}^3$)	Gain (dB)	Freq. Band (GHz)
[31]	Fractal	2.94	24.57 × 8.71 × 0.2	0–13.9	4.2–42
[37]	SIW	2.2	2.89 × 1.26 × 0.099	2.15–5.75	11.02–40
[1]	Metamaterial	2.2	7.81 × 3.75 × 0.102	9.2–11.9	24.15–28.5
[36]	DGS	2.2	7.85 × 3.77 × 0.103	9.42–11.44	24.44–28.5
[33]	SIW, DL, Corrugations	2.2	7.4 × 5.27 × 0.076	5–8	58–64
[38]	EBG	4.4	6.796 × 3.398 × 0.063	3–7.25	5–13
[39]	EBG	2.2	9.17 × 5.59 × 0.033	10.4–12.8	26.5–40
[32]	Exponential Strip Lines	2.5	7.37 × 5.84 × 0.047	1–12.5	0.72–17
[41]	Slots	3.38	14.296 × 5.69 × 0.0672	4–8.5	3.2–40
[40]	Kernel Regression	4.4	4.21 × 2.44 × 0.039	−1–6	1–6
[34]	Parasitic Patch	3.5	5.49 × 5.13 × 0.022	2.5–9.8	2.2–12
[35]	Corrugations	2.65	3.24 × 2.52 × 0.036	9–11.5	2.3–11
[42]	Corrugations	3.55	7.91 × 3.165 × 0.061	NG–8	6–18
[43]	Corrugations	3.38	12.16 × 7.22 × 0.03	6.7–15	1.65–18
Proposed AVA Array	Corrugations	2.2	4.6 × 3.83 × 0.04	8–13.2	24.19–29.15, 30.28–40.5

SIW—substrate-integrated waveguide, DGS—defected ground structure, DL—dielectric lens, EBG—electromagnetic bandgap, NA—not applicable, NG—not given

4. Conclusions

A compact 1 × 4 AVA array for 5G mm-wave application is implemented in this paper. The results prove the importance of corrugations in the enhancement of the front-to-back ratio, gain, and bandwidth of antennas. The designed antenna operates in the range of from 24.19 GHz–29.15 GHz to 30.28 GHz–40.5 GHz, which includes three important frequency bands of 5G applications. Moreover, it provides a high and almost constant gain of 8 dBi to 13.2 dBi. Importantly, the improved antenna parameters are achieved without compromising the antenna size, which is 24 mm × 28.8 mm × 0. 254 mm. Furthermore, the radiation patterns of the presented antenna are stable over the operating frequency range. The significance of the proposed antenna is proved by comparing it with the most recent AVAs. This comparison and other antenna results show that the proposed AVA array is suitable for integration in 5G millimeter wave-based communication devices.

Author Contributions: Conceptualization, A.S.D. and S.K.; methodology, A.S.D.; software, A.S.D.; validation, S.U. and A.M.; formal analysis, S.K.; resources, S.U. and A.M.; writing—original draft preparation, A.S.D. and S.K.; writing—review and editing, S.U. and A.M.; supervision, S.K.; project administration, S.K.; funding acquisition, S.U. and A.M. All authors have read and agreed to the published version of the manuscript.

Institutional Review Board Statement: Not applicable.

Informed Consent Statement: Not applicable.

Data Availability Statement: No database required.

Acknowledgments: This research was funded by the Deanship of Scientific Research at Princess Nourah bint Abdulrahman University through the Fast-track Research Funding Program.

Conflicts of Interest: The authors declare no conflict of interest.

References

1. Zhu, S.; Liu, H.; Wen, P. A new method for achieving miniaturization and gain enhancement of vivaldi antenna array based on anisotropic metasurface. *IEEE Trans. Antennas Propag.* **2019**, *67*, 1952–1956. [CrossRef]
2. ITU-R. *Minimum Requirements Related to Technical Performance for IMT-2020 Radio Interface(s)*; Report ITU-R M.2410-0; ITU-R: Geneva, Switzerland, 2017; pp. 1–11.
3. Kumar, S.; Dixit, A.S.; Malekar, R.R.; Raut, H.D.; Shevada, L.K. Fifth Generation Antennas : A Comprehensive Review of Design and Performance Enhancement Techniques. *IEEE Access* **2020**, *8*, 163568–163593. [CrossRef]
4. Zhu, S.; Liu, H.; Chen, Z.; Wen, P. A compact gain-enhanced vivaldi antenna array with suppressed mutual coupling for 5G mmwave application. *IEEE Antennas Wirel. Propag. Lett.* **2018**, *17*, 776–779. [CrossRef]
5. Gazit, E. Improved design of the Vivaldi antenna. *IEE Proc. H Microw. Antennas Propag.* **1988**, *135*, 89–92. [CrossRef]
6. Gibson, P. The Vivaldi Aerial. In Proceedings of the 9th European Microwave Conference, Brighton, UK, 17–20 September 1979; pp. 101–105. [CrossRef]
7. Dixit, A.S.; Kumar, S. A Miniaturized Antipodal Vivaldi Antenna for 5G Communication Applications. In Proceedings of the 7th International Conference on Signal Processing and Integrated Networks (SPIN), Noida, India, 27–28 February 2020; pp. 800–803.
8. Raut, H.D.; Shevada, L.; Malekar, R.; Kumar, S. High gain wideband antennas for 5G applications: A review. In *Inventive Communication and Computational Technologies, Lecture Notes in Networks and Systems*; Ranganathan, G., Chen, J., Rocha, Á., Eds.; Springer: Singapore, 2021; Volume 145, pp. 777–787.
9. Wang, Y.W.; Yu, Z.W. A Novel Symmetric Double-Slot Structure for Antipodal Vivaldi Antenna to Lower Cross-Polarization Level. *IEEE Trans. Antennas Propag.* **2017**, *65*, 5599–5604. [CrossRef]
10. Sun, M.; Chen, Z.N.; Qing, X. Gain enhancement of 60-GHz antipodal tapered slot antenna using zero-index metamaterial. *IEEE Trans. Antennas Propag.* **2013**, *61*, 1741–1746. [CrossRef]
11. Kazemi, R.; Member, S.; Fathy, A.E.; Sadeghzadeh, R.A. Dielectric Rod Antenna Array With Substrate Integrated Waveguide Planar Feed Network for Wideband Applications. *IEEE Trans. Antennas Propag.* **2012**, *60*, 1312–1319. [CrossRef]
12. Tiwari, N.; Rama Rao, T. Substrate Integrated Waveguide Based High Gain Planar Antipodal Linear Tapered Slot Antenna with Dielectric Loading for 60 GHz Communications. *Wirel. Personal Commun.* **2017**, *97*, 1385–1400. [CrossRef]
13. Bai, J.; Shi, S.; Prather, D.W. Modified compact antipodal Vivaldi antenna for 4-50-GHz UWB application. *IEEE Trans. Microw. Theory Tech.* **2011**, *59*, 1051–1057. [CrossRef]
14. Moosazadeh, M.; Kharkovsky, S.; Case, J.T. Microwave and millimetre wave antipodal Vivaldi antenna with trapezoid-shaped dielectric lens for imaging of construction materials. *IET Microw. Antenna Propag.* **2015**, *10*, 1–9. [CrossRef]
15. Moosazadeh, M.; Kharkovsky, S. A Compact High-Gain and Front-to-Back Ratio Elliptically Tapered Antipodal Vivaldi Antenna With Trapezoid-Shaped Dielectric Lens. *IEEE Antennas Wirel. Propag. Lett.* **2015**, *15*, 552–555. [CrossRef]
16. Islam, M.T.; Samsuzzaman, M.; Islam, M.T.; Kibria, S.; Singh, M.J. A homogeneous breast phantom measurement system with an improved modified microwave imaging antenna sensor. *Sensors* **2018**, *18*, 2962. [CrossRef]
17. Briqech, Z.; Sebak, A.; Denidni, T.A. High Gain 60 GHz Antipodal Fermi Tapered Slot Antenna With Sine Corrugation. *Microw. Opt. Technol. Lett.* **2014**, *57*, 6–9. [CrossRef]
18. Hesari, S.S.; Bornemann, J. Antipodal Vivaldi antenna arrays fed by substrate integrated waveguide right-angled power dividers. *Appl. Sci.* **2018**, *8*, 2625. [CrossRef]
19. Moosazadeh, M. High-Gain Antipodal Vivaldi Antenna Surrounded by Dielectric for Wideband Applications. *IEEE Trans. Antennas Propag.* **2018**, *66*, 4349–4352. [CrossRef]
20. Wang, N.N.; Fang, M.; Chou, H.T.; Qi, J.R.; Xiao, L.Y. Balanced Antipodal Vivaldi Antenna with Asymmetric Substrate Cutout and Dual-Scale Slotted Edges for Ultrawideband Operation at Millimeter-Wave Frequencies. *IEEE Trans. Antennas Propag.* **2018**, *66*, 3724–3729. [CrossRef]
21. Sarkar, C.; Saha, C.; Shaik, L.A.; Siddiqui, J.Y.; Antar, Y.M.M. Frequency notched balanced antipodal tapered slot antenna with very low cross- polarised radiation. *IET Microw. Antenna Propag.* **2018**, *12*, 1859–1863.
22. Kaboli, M.; Molaei, A.; Abrishamian, M.S.; Mirtaheri, S.A. Dielectric lens balanced antipodal Vivaldi antenna with low cross-polarisation for ultra-wideband applications. *IET Microw. Antenna Propag.* **2014**, *8*, 1137–1142. [CrossRef]
23. Dixit, A.S.; Kumar, S. A Survey of Performance Enhancement Techniques of Antipodal Vivaldi Antenna. *IEEE Access* **2020**, *8*, 45774–45796. [CrossRef]
24. Balanis, C.A. *Antenna Theory; Analysis and Design*, 3rd ed.; John Wiley and Sons: Hoboken, NJ, USA, 2008; Volume 72, pp. 989–990.
25. Eichenberger, P.J.; Yetisir, E.; Ghalichechian, N. High-Gain Antipodal Vivaldi Antenna with Pseudoelement and Notched Tapered Slot Operating at 2.5-57 GHz. *IEEE Trans. Antennas Propag.* **2019**, 1–11. [CrossRef]
26. Moosazadeh, M.; Kharkovsky, S.; Case, J.T.; Samali, B. Antipodal Vivaldi antenna with improved radiation characteristics for civil engineering applications. *IET Microw. Antenna Propag.* **2017**, *11*, 796–803. [CrossRef]
27. Phalak, K.D.; Briqech, Z.; Sebak, A. Ka-Band Antipodal Fermi-Linear Tapered Slot Antenna With a Knife. *Microw. Opt. Technol. Lett.* **2014**, *57*, 485–489. [CrossRef]

28. Zhu, Q.; Ng, K.B.; Chan, C.H.; Luk, K.M. Substrate-Integrated-Waveguide-Fed Array Antenna Covering 57-71 GHz Band for 5G Applications. *IEEE Trans. Antennas Propag.* **2017**, *65*, 6298–6306. [CrossRef]
29. Kumar, G.; Ray, K.P. *Broadband Microstrip Antennas*; Artech House Antennas and Propagation Library: London, UK, 2003; pp. 1–432.
30. Dixit, A.S.; Kumar, S. The enhanced gain and cost-effective antipodal Vivaldi antenna for 5G communication applications. *Microw. Opt. Technol. Lett.* **2020**, *62*, 1–10.
31. Karmakar, A.; Bhattacharjee, A.; Saha, A.; Bhawal, A. Design of a fractal inspired antipodal vivaldi antenna with enhanced radiation characteristics for wideband applications. *IET Microw. Antennas Propag.* **2019**, *13*, 892–897. [CrossRef]
32. Honari, M.M.; Ghaffarian, M.S.; Mirzavand, R. Miniaturized antipodal vivaldi antenna with improved bandwidth using exponential strip arms. *Electronics* **2021**, *10*, 83. [CrossRef]
33. Horst, M.J.; Member, S.; Ghasr, M.T.; Member, S.; Zoughi, R. Design of a Compact V—Band Transceiver and Antenna for Millimeter-Wave Imaging Systems. *IEEE Trans. Instrum. Meas.* **2019**, *68*, 4400–4411. [CrossRef]
34. Li, Z.; Kang, X.; Su, J.; Guo, Q.; Yang, Y.L.; Wang, J. A Wideband End-Fire Conformal Vivaldi Antenna Array Mounted on a Dielectric Cone. *Int. J. Antennas Propag.* **2016**, *2016*, 1–11. [CrossRef]
35. Liu, C.; Yan, A.; Yu, C.; Xu, T. Improvement on a 2 × 2 Elements High-Gain Circularly Polarized Antenna Array. *Int. J. Antennas Propag.* **2015**, *2015*,1–8, 252717. [CrossRef]
36. Zhu, S.; Liu, H.; Wen, P.; Chen, Z.; Xu, H. Vivaldi Antenna Array Using Defected Ground Structure for Edge Effect Restraint and Back Radiation Suppression. *IEEE Antennas Wirel. Propag. Lett.* **2020**, *19*, 84–88. [CrossRef]
37. Deng, J.Y.; Cao, R.; Sun, D.; Zhang, Y.; Yong, T.; Guo, L.X. Bandwidth Enhancement of an Antipodal Vivaldi Antenna Facilitated by Double Ridge Substrate Integrated Waveguide. *IEEE Trans. Antennas Propag.* **2020**, 1–4. [CrossRef]
38. Hirano, T.; Hirose, A. Wideband and Low Direct-Coupling Tapered Slot Antenna Using Electromagnetic Bandgap Structures. *IEEE Trans. Antennas Propag.* **2019**, *67*, 2272–2279. [CrossRef]
39. Elabd, R.H.; Abdullah, H.H.; Abdelazim, M. Compact Highly Directive MIMO Vivaldi Antenna for 5G Millimeter-Wave Base Station. *J. Infrared Millim. Terahertz Waves* **2021**, *42*, 173–194. [CrossRef]
40. Lee, S.; Hur, J.; Heo, M.B.; Kim, S.; Choo, H.; Byun, G. A Suboptimal Approach to Antenna Design Problems with Kernel Regression. *IEEE Access* **2019**, *7*, 17461–17468. [CrossRef]
41. Moosazadeh, M. Sidelobe level reduction using Teflon for a microwave and millimetre-wave antipodal Vivaldi antenna. *IET Microw. Antennas Propag.* **2020**, *14*, 474–478. [CrossRef]
42. Manoochehri, O.; Farzami, F.; Erricolo, D.; Chen, P.y. Design of a corrugated antipodal Vivaldi antenna with stable pattern. In Proceedings of the 2019 United States National Committee of URSI National Radio Science Meeting (USNC-URSI NRSM), Boulder, CO, USA, 9–12 January 2019; pp. 1–2.
43. Moosazadeh, M. Comb-Shaped Slit Antipodal Vivaldi Antenna and Its Application for Detection of Void Inside Concrete Specimens. In *Antipodal Vivaldi Antennas for Microwave Imaging of Construction Materials and Structures*; Springer: Berlin/Heidelberg, Germany, 2019; pp. 113–130. [CrossRef]

sensors

MDPI

Communication

Chiral Dielectric Metasurfaces for Highly Integrated, Broadband Circularly Polarized Antenna

Bruno Ferreira-Gomes [1], Osvaldo N. Oliveira, Jr. [2] and Jorge Ricardo Mejía-Salazar [1,*]

[1] National Institute of Telecommunications (Inatel), Santa Rita do Sapucaí 37540-000, MG, Brazil; bruno.gomes@mtel.inatel.br
[2] São Carlos Institute of Physics, University of São Paulo, P.O. Box 369, São Carlos 13560-970, Brazil; chu@ifsc.usp.br
* Correspondence: jrmejia@inatel.br

Abstract: We report on the design of a low-profile integrated millimeter-wave antenna for efficient and broadband circularly polarized electromagnetic radiation. The designed antenna comprises a chiral dielectric metasurface built with a 2 × 2 arrangement of dielectric cylinders with slanted-slots at the center. A broadbeam high-gain with wide axial ratio (AR) < 3 dB bandwidth was reached by pairing the electric and magnetic resonances of the dielectric cylinders and the slanted slots when excited by an elliptically polarized driven-patch antenna. This electric-magnetic pairing can be tuned by varying the cylinders diameter and the tilting and rotation angles of the slanted slots. The simulation results indicate impedance-matching bandwidths up to 22.6% (25.3–31.6 GHz) with 3-dB AR bandwidths of 11.6% (26.9–30.2 GHz), which in terms of compactness ($0.95\lambda_0 \times 0.95\lambda_0$) and performance are superior to previous antenna designs. Since the simulations were performed by assuming materials and geometries easily implementable experimentally, it is hoped that circularly polarized antennas based on chiral metasurfaces can be integrated into 5G and satellite communications.

Keywords: chirality; dielectric resonator antennas; metasurfaces

Citation: Ferreira-Gomes, B.; Oliveira, O.N., Jr.; Mejía-Salazar, J.R. Chiral Dielectric Metasurfaces for Highly Integrated, Broadband Circularly Polarized Antenna. *Sensors* **2021**, *21*, 2071. https://doi.org/10.3390/s21062071

Academic Editor: Naser Ojaroudi Parchin

Received: 22 February 2021
Accepted: 10 March 2021
Published: 16 March 2021

1. Introduction

The quest for antennas that are cost-effective, compact, and efficient for a broad bandwidth with a specific radiation pattern relies on the ability to manage millimeter-wave (mm-wave) electromagnetic field interactions. To transfer mm-wave antennas from research laboratories to the market of mobile infrastructures (for 5G communication) and satellite communication, it is crucial to address polarization mismatches and suppress multipath interferences. This may in principle be reached with circularly polarized (CP) antennas working at different ranges in the GHz regime [1–7], but two major limitations must be addressed. The first is to reach high-quality CP radiation with high gain within a broad bandwidth (BW). The quality of a CP electromagnetic wave is defined in terms of its axial-ratio (AR), i.e., the ratio between the major and minor axes of the polarization ellipse. The closer AR is to unity (0 dB) the higher the CP quality, and in practice electromagnetic fields have been considered to be CP for AR below 3 dB [1–7]. As for the second drawback, one has to reduce the size and Ohmic losses (inherent in metallic inclusions) of mm-wave antennas to produce them in high-throughput, integrable platforms. Dielectric resonator antennas (DRAs) made with high-permittivity ($\varepsilon \sim 10 \cdots 10^2$) materials (e.g., ceramics) can concentrate and radiate CP electromagnetic fields with small losses [8]. However, the integration of these CP-DRAs into modern wireless systems is hampered by the need for multiple resonators and a multi-feed mechanism to achieve wide bandwidths [9–13]. Another possibility to address these limitations is to employ chiral systems in the design of CP antennas [14,15]. Chirality refers to the handedness of an object which cannot be superimposed with its mirror image as in left- (LCP) and right-handed circularly polarized

(RCP) light. The interaction of chiral electromagnetic fields with a chiral scatterer depends on this handedness. This may be used to simultaneously improve the handedness of an electromagnetic source while enhancing its electromagnetic field intensity, which can be reached through the proper engineering of suitable chiral structures [16]. In fact, chirality concepts have been used to control electromagnetic wave propagation in the microwave and optical domains [17–19], which serves as inspiration for CP antenna design.

In this communication, we demonstrate theoretically a new concept to produce integrated, highly efficient mm-wave CP antennas in which we employ metasurfaces [20,21]. These types of surfaces were proven useful for several applications, including gain enhancement [22] and phase rectifying [23]. In contrast to previous antenna designs using bulky metallic-based chiral metasurfaces [14,15], we exploit the pairing of electric and magnetic resonances of high-permittivity ($\varepsilon = 16$) cylindrical resonators with slanted slots at the center. An analogous concept in nanophotonics yielded an enhanced broadband near- and far-field chiroptical activity [18]. To reach optimized radiation patterns with high gain and wide AR-BW, we used a 2×2 arrangement of slotted cylinders. Through numerical calculations we show that this concept can be readily applied in antenna arrays by using a metal cavity surrounding each unit cell. The results for the system with metal cavity show promising improvements in gain and AR amplitudes and bandwidths. For comparison purposes, simulations were made using two commercial software packages for antenna design, namely CST Studio and ANSYS HFSS, and the results were in excellent agreement. CST Studio uses a time domain (transient) solver based on the finite integration technique (FIT) with a hexahedral mesh, while HFSS employs the finite element method (FEM) with tetrahedral mesh elements. As we shall demonstrate, antennas made with a chiral dielectric metasurface (CDM) are promising for broadband and efficient CP mm-wave radiation.

2. Antenna Design

2.1. Configuration

Figure 1 shows a zoomed view of the antenna designed to have the center frequency at 28.5 GHz to enable high-throughput applications in 5G and satellite communications.

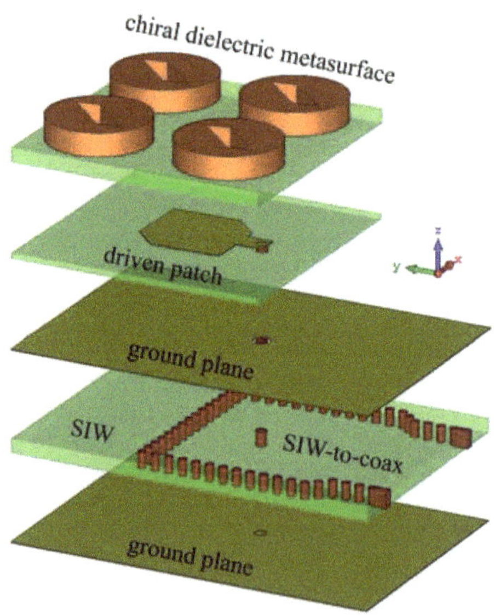

Figure 1. 3D geometric view of the proposed antenna element fed by a SIW ideal connector.

The antenna consists of five parts referred to as (i) the bottom ground plane; (ii) a substrate integrated waveguide (SIW) with a SIW-to-coax feeding; (iii) a ground plane film; (iv) a coax-fed driven patch antenna; and (v) a chiral dielectric metasurface consisting of a 2×2 array of slotted dielectric cylinders. The ground plane films were considered to be made of lossy copper. A top-view and side-view of the main building elements of the antenna design are presented in Figure 2.

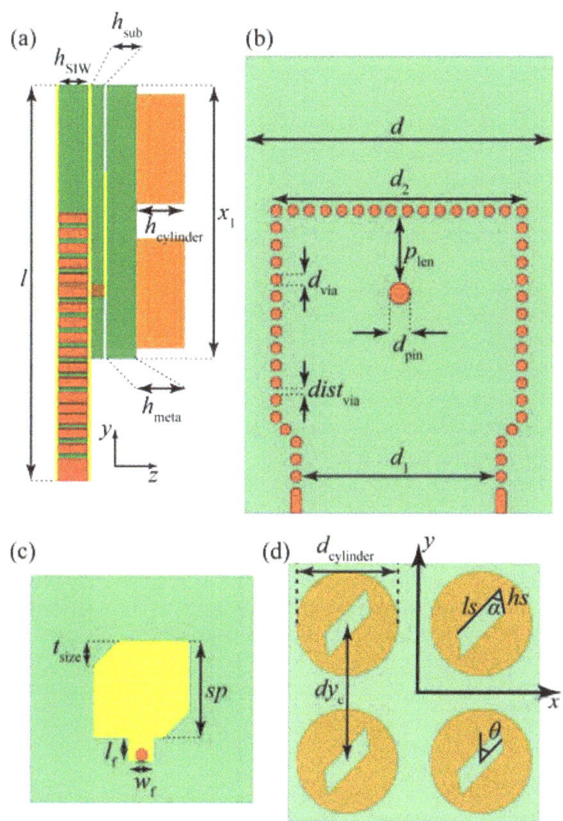

Figure 2. (**a**) Parametric side view of the antenna layers divided as: (**b**) SIW-to-coax feeding, (**c**) driven patch layer and (**d**) chiral dielectric metasurface layer. The design parameters are $x_1 = 10$ mm, $d = 10$ mm, $d_2 = 7.5$ mm, $p_{\text{len}} = 2.15$ mm, $d_{\text{via}} = 0.3$ mm, $d_{\text{pin}} = 0.4$ mm, $dist_{\text{via}} = 0.19$ mm, $d_1 = 5.71$ mm, $t_{\text{size}} = 0.86$ mm, $sp = 3.3$ mm, $l_f = 0.85$ mm, $w_f = 0.8$ mm, $ls = 2.1$ mm, $hs = 0.8$ mm, $\alpha = 60°$ and $\theta = 45°$.

The SIW-based feeding layer is etched on a slab of Rogers RT/duroid 5880 ($\epsilon_r = 2.2$, $\tan\delta = 0.0009$) , with thickness $h_{\text{SIW}} = 0.5$ mm. A ground plane, $h_{\text{sub}} = 0.035$ mm thick, was used to separate the SIW-feeding and the coax-fed driven-patch antenna. A CDM is stacked at the top of the structure, consisting of four slotted cylinders made of a ceramic dielectric material with $\epsilon_r = 16$, layered on a low-permittivity substrate of RT/duroid 5880 ($h_{\text{meta}} = 0.5$ mm). The cylinders had a height $h_{\text{cylinder}} = 0.8$ mm and diameter $d_{\text{cylinder}} = 3.6$ mm, and were separated by a center-to-center distance of $dy_c = 4.8$ mm. The slots at the center of the cylinders were parallelograms whose sides were labeled as ls and hs and the tilting angle is α. The slots are rotated by an angle θ with respect to the y-axis, as depicted. CDM is fed by a driven patch localized below the low-permittivity layer at the center of the metasurface. The patch made of lossy copper had rectangular sides sp and corner cuts t_{size}. The coax-to-patch feeding line has a width $w_f = 0.8$ mm and length $l_f = 0.85$ mm. The SIW-feeding structure is amenable to integration with planar front-end circuits. The SIW structure was designed as a rectangular waveguide input, forming a SIW-to-coax transition where the distance between the pin and the shorting wall was tuned for impedance matching [24,25]. The dimensions of the SIW-feeding and the remaining parameters are listed in the caption of Figure 2. It is worth mentioning that with the antenna size taken as $0.95\lambda_0 \times 0.95\lambda_0$ (λ_0 at the center frequency of 28.5 GHz), our concept enables the application of phase-arrayed antenna systems by designing the structure to be directly fed by a coax-wire.

2.2. Working Principle and Enhancement of ARBW

Recent approaches for enhancement of AR bandwidth in metallic antennas have employed two pair slots, as explained in Reference [21]. Alternatively, dielectric resonators with high-permittivity are known to support electric and magnetic resonances of different orders [8,10], in analogy with cylindrical nanostructures [18]. The near-field overlapping between nearby cylinders resembles the electronic-bands from well-localized atomic orbitals in the tight-binding model [26,27] in electromagnetic platforms known as metasurfaces. Here, we use dielectric cylinders with a slanted slot at the center to produce integrable, highly efficient and broad ARBW CP antennas. An analogous concept was exploited in nanophotonic platforms to produce giant enhancement of chiroptical effects [18]. The concept is based on the coupling of the electric and magnetic modes of the resonators (considered to be dipoles in a qualitative approximation) which can be expressed as

$$\tilde{\mathbf{p}} = \tilde{\alpha}\tilde{\mathbf{E}} - i\tilde{G}\tilde{\mathbf{B}}, \qquad \tilde{\mathbf{m}} = \tilde{\chi}\tilde{\mathbf{B}} + i\tilde{G}\tilde{\mathbf{E}}, \tag{1}$$

where $\tilde{\mathbf{p}}$ and $\tilde{\mathbf{m}}$ are the electric and magnetic dipoles, respectively. $\tilde{\alpha}$ and $\tilde{\chi}$ are the complex electric polarizability and magnetic susceptibility, while $\tilde{G}$ corresponds to the mixed electric-magnetic dipole polarizability. $\tilde{\mathbf{E}}$ and $\tilde{\mathbf{B}}$ are the complex electric and magnetic fields. The radiated electromagnetic energy has, therefore, a term $\sim \tilde{G}''\text{Im}\{\tilde{\mathbf{E}} \cdot \tilde{\mathbf{B}}\}$, where $\tilde{G}''$ stands for the imaginary part of $\tilde{G}$. It is just at the slanted sides hs where $\tilde{\mathbf{E}}$ and $\tilde{\mathbf{B}}$ become mixed [18], allowing the tuning of $\tilde{G}$ through the tilting angle α (see Figure 2d). The electromagnetic fields exciting the CDM are produced by a single narrow-band driven-patch antenna with truncated corners, thus being constantly fed by two orthogonal linearly polarized modes (AR $\gtrsim 5$), as will be shown later.

3. Results and Discussion

The individual contributions from the building components in the antenna design can be studied by analyzing the results in Figure 3 for S_{11}, AR and boresight gain for the patch without (W/O) metasurface (dashed lines) and with the metasurface of solid cylinders (solid lines) depicted in the inset of Figure 3a. Calculations in this figure were made using the commercial software CST Studio. Results for S_{11} are presented in Figure 3a, whereas the corresponding AR and boresight-gain are shown in Figure 3b. Poor values for the impedance-matching ($S_{11} \gtrsim -10$) and gain (< 8) are observed for the patch W/O metasurface. The antenna with the metasurface containing the solid cylinders improves the impedance-matching, AR values and boresight gains, in addition to an improved bandwidth performance. Such improvements are due to the dielectric cylindrical resonators and their corresponding near-field interactions. However, the radiation and AR performance are still low in all of these results, and additional strategies are required to design useful CP antennas.

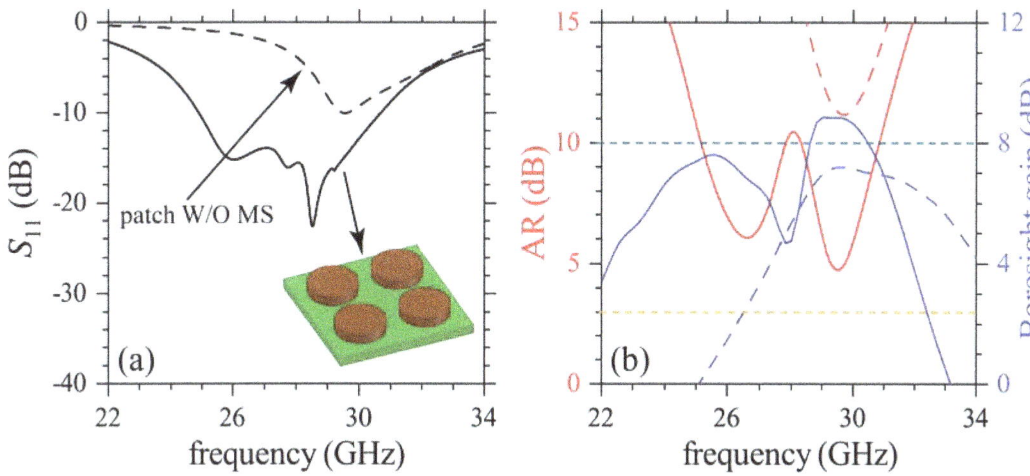

Figure 3. Numerical results for (**a**) S_{11} and (**b**) AR and boresight gain. Solid and dashed lines are for the patch antenna with and without (W/O) the DR metasurface. Simulations were carried out using the commercial software CST Studio. Horizontal dashed lines in (**b**) are eye guides for $AR = 3$ dB and boresight gain = 8 dB.

Following the reasoning of chiral dielectric metasurfaces for enhanced chiroptical effects [18], we considered slanted slots etched at the center of each cylinder, as schematized in the inset of Figure 4a. This mechanism not only improved AR but also the whole operating performance of the antenna, as it can be seen from Figure 4a,b. Calculations from CST Studio and ANSYS HFSS are presented with solid and dashed lines, respectively. Negligible differences are observed due to different convergences of the software packages. The applicability of this CDM-based CP antenna is demonstrated in Figure 4c,d with AR and gain corresponding to the system surrounded by a metal cavity (considered to be made of Al). This cavity surrounding the four-element CDM enables integration into antenna arrays by suppressing the mutual coupling between adjacent elementary cells. Although a small decrease was induced by the cavity in the impedance $\text{BW} = [25.5\text{ GHz}, 31.3\text{ GHz}]$ and 3-dB $\text{ARBW} = [27.1\text{ GHz}, 30.3\text{ GHz}]$, in relation to the impedance $\text{BW} = [25.3\text{ GHz}, 31.6\text{ GHz}]$ and 3-dB $\text{ARBW} = [26.9\text{ GHz}, 30.2\text{ GHz}]$ for the system without cavity, the corresponding gain ($\gtrsim 9$) and AR ($\text{AR} \ll 3$) were considerably improved. Moreover, a gain notch from Figure 4b was shifted out of the bandwidth range of interest when using the cavity, which certainly improves the CP radiation performance. Figure 5 shows the **E**-field distribution for the two orthogonal modes of the patch coupled to the CDM. These fields were calculated for the frequencies at which AR achieves the best results, according to Figure 4d. For illustrative purposes, results are presented for **E**-field distribution at the patch and CDM planes, and with radiation of an LCP electromagnetic field. As it can be inferred, an RCP antenna can be developed using the mirror image of CDM and the patch antenna. The evanescent behavior of the electromagnetic field in the cavity is also shown.

The numerical results from hereon were obtained with the commercial software CST Studio for the antenna without the cavity. Results of the LCP (co-polarized) and RCP (cross-polarized) radiation patterns for the CDM-based antenna are presented for 27.5 GHz and 30 GHz in Figure 6a–d, respectively. For the co-polarized radiation at the $\phi = 90°$-plane, a 3-dB power beamwidth of 66.3° (in the range from 38.3° to $-28°$) is observed for 27.5 GHz (Figure 6a) while for 30 GHz it is 58.7° (from 31° to $-27.7°$) (Figure 6c). The condition $\text{AR} \leq$ 3-dB is met when the difference between the co-polarized and cross-polarized patterns is $\geq$ 15-dB. Therefore, there is a high AR performance in the whole 3-dB power beamwidth for the 27.5 GHz mode, but not for 30 GHz which is limited to 43° (from 23° to $-20°$). In the case of $\phi = 0°$-plane, there is nearly the same 3-dB power beamwidth for both modes, 66.3° (in the range from 36.3° to $-30°$) for 27.5 GHz and 62.7° (in the range from 36° to −26.7°) for 30 GHz. However, the 3-dB AR performance for 30 GHz exhibits a very narrow beamwidth of 36° (from 24° to −12°), in contrast to 27.5 GHz that stays with at least 15-dB difference between the co-polarized and cross-polarized radiation patterns in the whole 3-dB power beamwidth. Hence, the area covered by the 27.5 GHz mode is larger than for the 30 GHz.

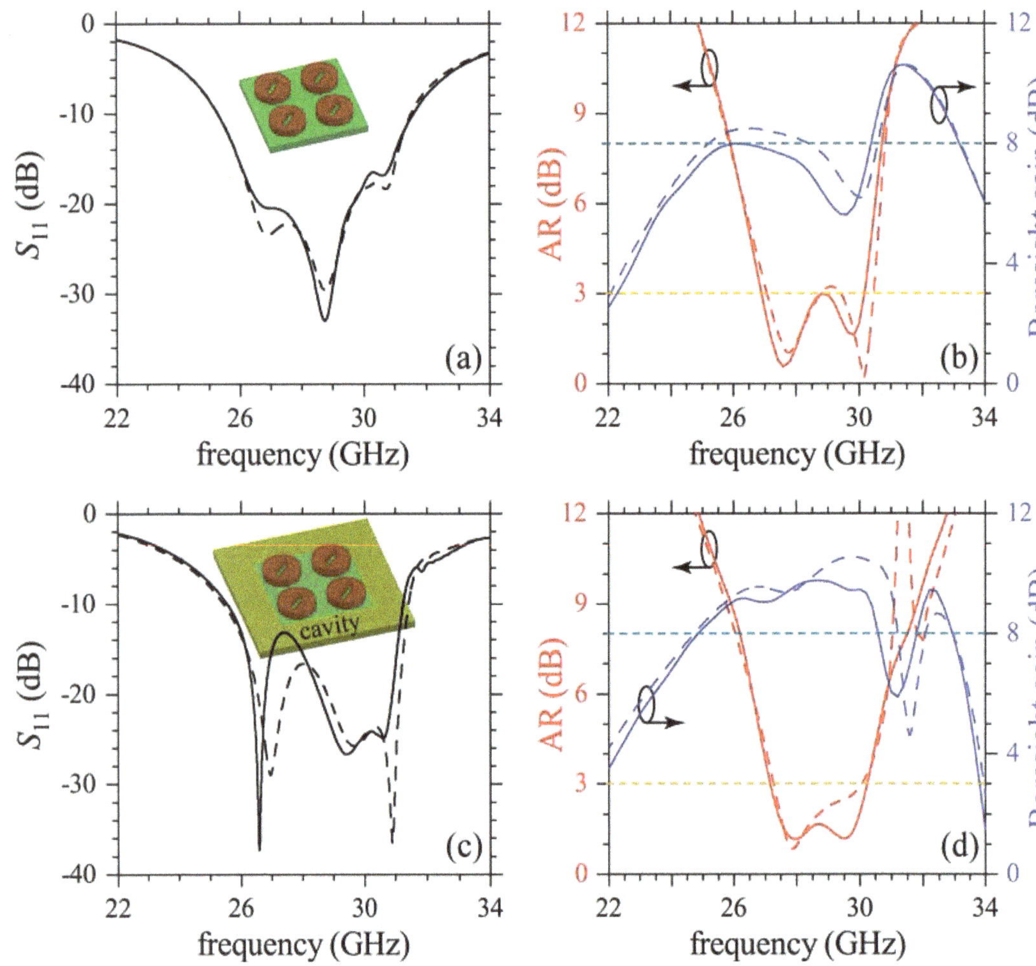

Figure 4. Numerical results for S_{11}, AR and boresight gain are presented for the CDM-based antenna (**a**,**b**) with and (**c**,**d**) without the metal cavity. Simulations from CST Studio and ANSYS HFSS are indicated by solid and dashed lines. Horizontal dashed lines in (**b**,**d**) are eye guides for $AR = 3$ dB and boresight gain = 8 dB.

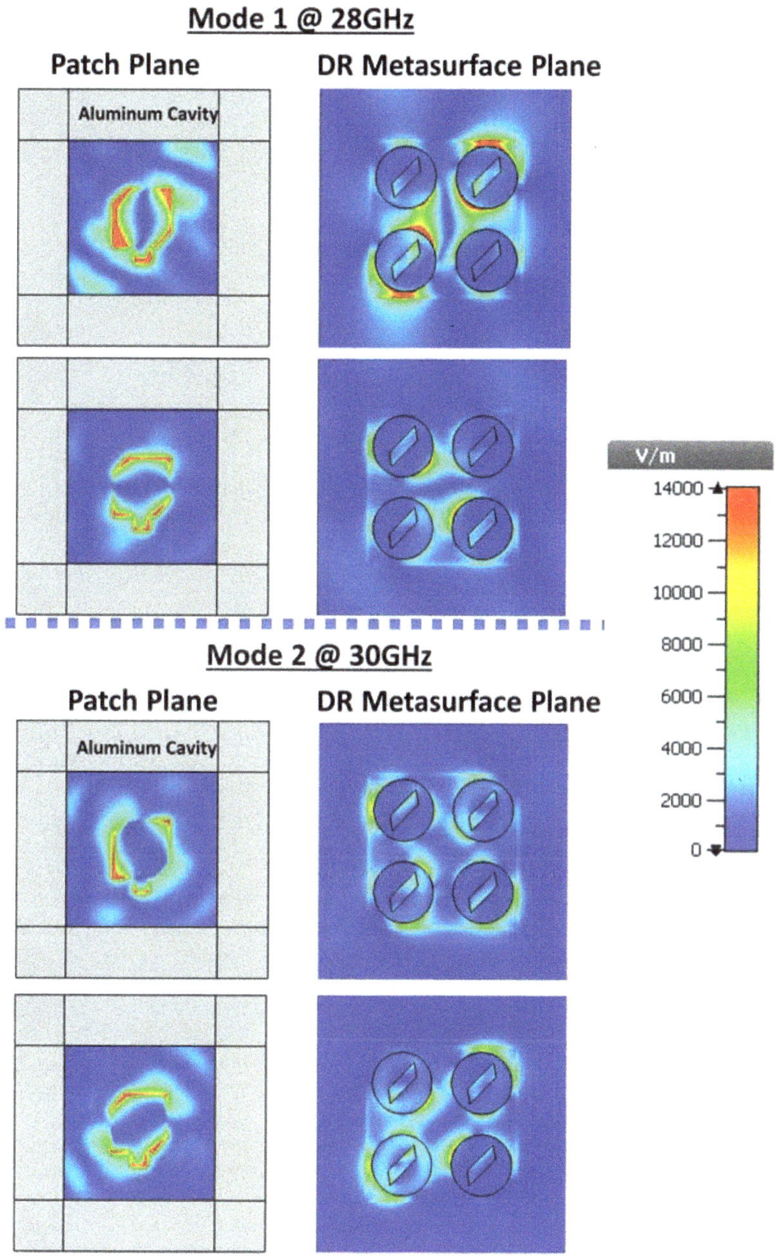

Figure 5. Simulated **E**-field of the antenna element at orthogonal phases for both CP modes and both planes: Patch and DR Metasurface Planes.

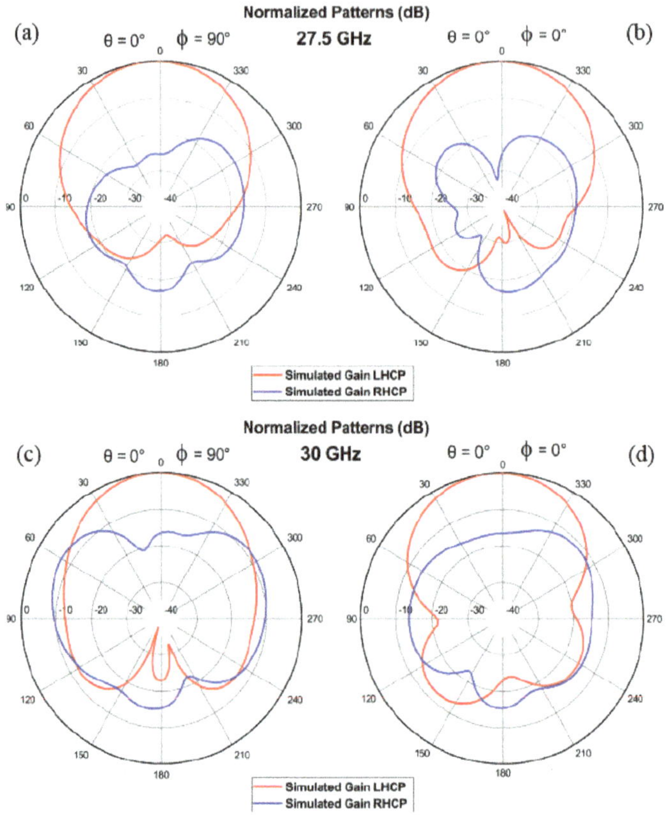

Figure 6. Simulated normalized radiation patterns at 27.5 GHz (**a**,**b**) and 30 GHz (**c**,**d**) for both cut-planes.

4. Parametric Study of the Proposed Design

To understand how different design parameters affect the functioning of the meta-antenna, we plotted S_{11} and AR for different values of θ, d_{cylinder}, ls and dy_{c} in Figures 7 and 8. There is a slight narrowing of the impedance bandwidth for increasing θ in Figure 7a, in contrast with the loss of CP radiation (AR > 3) in Figure 8a. This dependence of AR on θ highlights the importance of this parameter when designing the metasurface to obtain high-performance CP radiation. Upon increasing d_{cylinder} the modes in S_{11} are discretized with loss in the ability to generate CP radiation as indicated by the poor AR values (AR > 3) in Figure 8b. The latter behavior is explained by recalling that the CDM modes should be matched with the two orthogonal modes from the driven patch for a proper coupling and field re-radiation. Another important parameter in the design is the length ls of the slots, which can be varied to tune AR (and its bandwidth) to its optimum values with small changes of S_{11}, as noted in Figures 7c and 8c. The interaction of nearby cylinders in CDM depends on dy_{c} which then affects the quality of CP radiation. Figures 7d and 8d point to a decrease in the ability to produce CP mm-waves upon increasing dy_{c}. Excitation of several out-of-phase resonances occurs for very small dy_{c} which despite widening the impedance bandwidth, negatively affect AR.

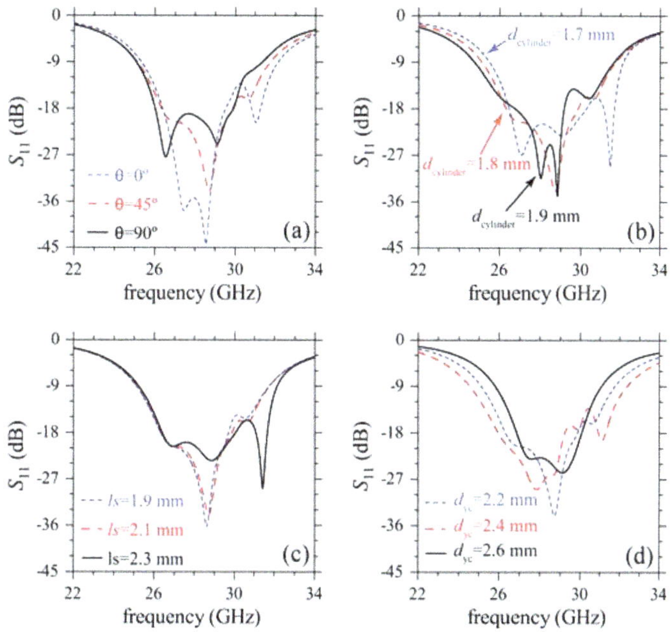

Figure 7. Simulated parametric study for the reflection coefficient of the proposed antenna. (**a**) variation of θ, (**b**) variation of $d_{cylinder}$, (**c**) variation of ls and (**d**) variation of dy_c.

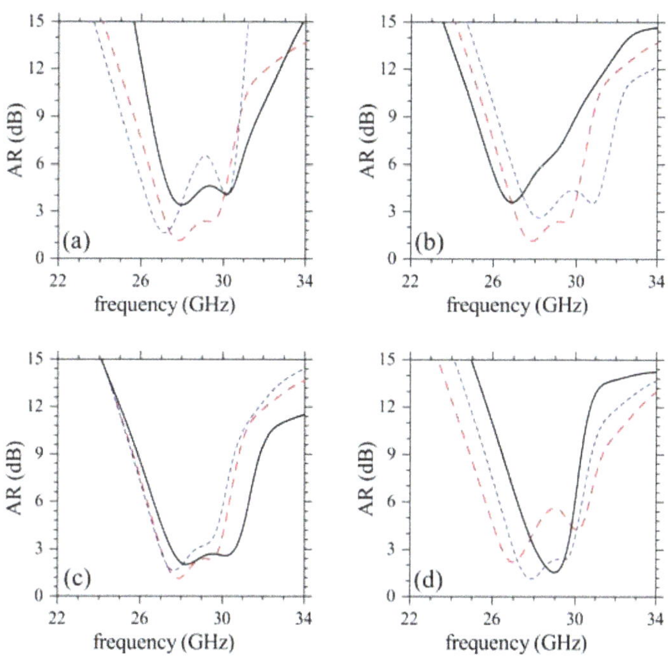

Figure 8. Simulated parametric study for AR of the proposed antenna, in relation to the structure in Figure 7. The variation of (**a**) θ, (**b**) $d_{cylinder}$, (**c**) ls and (**d**) dy_c was the same as in Figures 7a–d, respectively.

The competitiveness of our CDM-based antenna is demonstrated by comparing with recent proposals for CP antennas in Table 1, where results are given for the central frequency, impedance bandwidth (BW), ARBW, 2-dB gain bandwidth, and peak gain. We avoid comparing sizes because of the nature of the antenna and the corresponding operating frequencies. However, we emphasize that the CDM in this work has a reduced size while maintaining its efficiency. Moreover, there is a high peak gain with low gain variation in a wide bandwidth which is broader than in reports in the literature. Therefore, we expect that this work will stimulate the use of CDMs to produce CP radiation in other antenna designs.

Table 1. Comparison of performance with other CP Antennas.

Designs	Freq. (GHz)	BW (%)	3dB-ARBW (%)	2dB-Gain-BW (%)	Peak Gain (dBic)
Ref. [5]	3.5	21	8.5	19	6.5
Ref. [7]	26	12.5	26	18	7.5
Ref. [11]	5.5	25.4	22.8	34.2	7.7
Ref. [12]	11.7	27.8	24	16.5	10
this work *	28.5	22.6	11.6	34	8.5
this work **	28.5	20.4	11.2	19.4	10.4

*—W/O cavity, **—with cavity.

5. Conclusions

The introduction of a chiral metasurface on the design of low-profile CP DRAs was shown to yield a wide AR bandwidth, high-gain and low AR values. Of special relevance is the compactness of the CP antenna designed, suitable for mobile communication systems. The concepts and design were validated through numerical simulations using two commercial software packages, and no significant difference was noted in the results obtained with CST Studio and ANSYS HFSS. From these simulations one learns that the CP antennas can be tunable and integrated into front-end circuits or phase-arrayed systems (directly fed by a coax-wire). The design also included materials and conditions which are readily available for implementation, and therefore we may expect the fabrication of CP antennas for high throughput 5G and satellite communications in the *Ka*-band.

Author Contributions: Conceptualization, B.F.-G. and J.R.M.-S.; investigation B.F.-G.; supervision, J.R.M.-S.; Writing—original draft preparation, B.F.-G.; Writing—review and editing, O.N.O.J. and J.R.M.-S. All authors have read and agreed to the published version of the manuscript.

Funding: This work was partially supported by the RNP, with resources from MCTIC, Grant No. 01245.010604/2020-14, under the 6G Mobile Communications Systems project of the Radiocommunication Reference Center (Centro de Referência em Radiocomunicações—CRR) project of the National Institute of Telecommunications (Instituto Nacional de Telecomunicações—Inatel), Brazil. The authors wish also acknowledge the financial support from the Brazilian agencies FAPESP (2018/22214-6), Coordenação de Aperfeiçoamento de Pessoal de Nível Superior - Brasil (CAPES)—Finance Code 001 and the National Council for Scientific and Technological Development-CNPq (429496/2018-4, 305958/2018-6).

Institutional Review Board Statement: Not applicable.

Informed Consent Statement: Not applicable.

Conflicts of Interest: The authors declare no conflict of interest.

References

1. Wu, Z.; Li, L.; Li, Y.; Chen, X. Metasurface Superstrate Antenna with Wideband Circular Polarization for Satellite Communication Application. *IEEE Antennas Wirel. Propag. Lett.* **2015**, *15*, 374–377. [CrossRef]
2. Ta, S.X.; Park, I. Low-Profile Broadband Circularly Polarized Patch Antenna Using Metasurface. *IEEE Trans. Antennas Propag.* **2015**, *63*, 5929–5934. [CrossRef]
3. Huang, Y.; Yang, L.; Li, J.; Wang, Y.; Wen, G. Polarization conversion of metasurface for the application of wide band low-profile circular polarization slot antenna. *Appl. Phys. Lett.* **2016**, *109*, 054101. [CrossRef]

4. Niyamanon, S.; Senathong, R.; Phongcharoenpanich, C. Dual-Frequency Circularly Polarized Truncated Square Aperture Patch Antenna with Slant Strip and L-Shaped Slot for WLAN Applications. *Int. J. Antennas Propag.* **2018**, *2018*, 7684742. [CrossRef]
5. Juan, Y.; Yang, W.; Che, W. Miniaturized Low-Profile Circularly Polarized Metasurface Antenna Using Capacitive Loading. *IEEE Trans. Antennas Propag.* **2019**, *67*, 3527–3532. [CrossRef]
6. Ellis, M.S.; Effah, F.B.; Ahmed, A.-R.; Kponyo, J.J.; Nourinia, J.; Ghobadi, C.; Mohammadi, B. Asymmetric circularly polarized open-slot antenna. *Int. J. RF Microw. Comput. Aided Eng.* **2019**, *30*, e22141. [CrossRef]
7. Lu, K.; Leung, K.W. On the Circularly-Polarized Parallel-Plate Antenna. *IEEE Trans. Antennas Propag.* **2019**, *68*, 3–12. [CrossRef]
8. Mongia, R.K.; Bhartia, P. Dielectric resonator antenna—A review and general design relations for resonant frequency and bandwidth. *Int. J. RF Microw. Comput. Aided Eng.* **1994**, *4*, 230–247. [CrossRef]
9. Fakhte, S.; Oraizi, H.; Karimian, R.; Fakhte, R. A New Wideband Circularly Polarized Stair-Shaped Dielectric Resonator Antenna. *IEEE Trans. Antennas Propag.* **2015**, *63*, 1828–1832. [CrossRef]
10. Chowdhury, R.; Mishra, N.; Sani, M.M.; Chaudhary, R.K. Analysis of a wideband circularly polarized cylindrical dielectric resonator atenna with broadside radiation coupled with simple microstrip feeding. *IEEE Access* **2017**, *5*, 19478–19485. [CrossRef]
11. Sun, W.-J.; Yang, W.-W.; Chu, P.; Chen, J.-X. Design of a Wideband Circularly Polarized Stacked Dielectric Resonator Antenna. *IEEE Trans. Antennas Propag.* **2019**, *67*, 591–595. [CrossRef]
12. Yang, W.-W.; Sun, W.-J.; Tang, H.; Chen, J.-X. Design of a circularly polarized dielectric resonator antenna with wide bandwidth and low axial ration values. *IEEE Trans. Antennas Propag.* **2019**, *67*, 1963–1968. [CrossRef]
13. Gupta, S.; Sharma, A.; Das, G.; Gangwar, R.K.; Khalily, M. Wideband circularly polarized dielectric resonator antenna with polarization diversity. *IEEE Access* **2019**, *7*, 49069–49076. [CrossRef]
14. Bağmanci, M.; Karaaslan, M.; Unal, E.; Karadag, F. Microstrip Patch Antennas Covered with Chiral Metamaterial Structures. *Müh. Mim. Fak. Dergisi* **2018**, *33*, 45–254.
15. Hu, Y.-W.; Wang, Y.; Yan, Z.-M.; Zhou, H.-C. A high-gain circularly polarized Fabry-Perot antenna with chiral metamaterial-based circular polarizer. *Microw. Opt. Technol. Lett.* **2019**, *62*, 906–911. [CrossRef]
16. Hendry, E.; Mikhaylovskiy, R.V.; Barron, L.D.; Kadodwala, K.; Davis, T.J. Chiral electromagnetic fields generated by arrays of nanoslits. *Nano Lett.* **2012**, *12*, 3640–3644. [CrossRef]
17. Reyes-Gómez, F.; Mejía-Salazar, J.R.; Albella, P. All-Dielectric Chiral Metasurfaces Based on Crossed-Bowtie Nanoantennas. *ACS Omega* **2019**, *4*, 21041–21047. [CrossRef] [PubMed]
18. Reyes-Gómez, F.; Oliveira, O.N., Jr.; Albella, P.; Mejía-Salazar, J.R. Enhanced chiroptical activity with slotted high refractive index dielectric nanodisks. *Phys. Rev. B* **2020**, *101*, 155403. [CrossRef]
19. Orazbayev, B.; Kaina, N.; Fleury, R. Chiral Waveguides for Robust Waveguiding at the Deep Subwavelength Scale. *Phys. Rev. Applied* **2018**, *10*, 054069. [CrossRef]
20. Das, P.; Mandal, K.; Lalbakhsh, A. Single-layer polarization-insensitive frequency selective surface for beam reconfigurability of monopole antennas. *J. Electromagn. Waves Appl.* **2019**, *34*, 86–102. [CrossRef]
21. Ray, M.K.; Mandal, K.; Nasimuddin, N.; Lalbakhsh, A.; Raad, R.; Tubbal, F. Two-Pair Slots Inserted CP Patch Antenna for Wide Axial Ratio Beamwidth. *IEEE Access* **2020**, *8*, 223316–223324. [CrossRef]
22. Adibi, S.; Honarvar, M.A.; Lalbakhsh, A. Gain Enhancement of Wideband Circularly Polarized UWB Antenna Using FSS. *Radio Sci.* **2020**, *56*, e2020RS007098. [CrossRef]
23. Hayat, T.; Afzal, M.U.; Lalbakhsh, A.; Esselle, K.P. 3-D-Printed Phase-Rectifying Transparent Superstrate for Resonant-Cavity Antenna. *IEEE Antennas Wirel. Propag. Lett.* **2019**, *18*, 1400–1404. [CrossRef]
24. Wu, K.; Cheng, Y.-J.; Djerafi, T.; Hong, W. Substrate-integrated millimiter-wave and terahertz antenna technology. *Proc. IEEE* **2012**, *100*, 2219–2232.
25. Jiang, Z.H.; Zhang, Y.; Xu, J.; Yu, Y.; Hong, W. Integrated Broadband Circularly Polarized Multibeam Antennas Using Berry-Phase Transmit-Arrays for *Ka*-Band Applications. *IEEE Trans. Antennas Propag.* **2020**, *68*. [CrossRef]
26. Prodan, E.; Radloff, C.; Halas, N.J.; Norlander, P.A. Hybridization Model for the Plasmon Response of Complex Nanostructures. *Science* **2003**, *302*, 419–422. [CrossRef]
27. Nordlander, P.; Oubre, C.; Prodan, E.; Li, K.; Stockman, M.I. Plasmon Hybridization in Nanoparticle Dimers. *Nano Lett.* **2004**, *4*, 899–903. [CrossRef]

sensors

MDPI

Communication

Characterization of Tiled Architecture for C-Band 1-Bit Beam-Steering Transmitarray

Dmitry Kozlov [1], Irina Munina [2], Pavel Turalchuk [2], Vitalii Kirillov [2], Alexey Shitvov [3] and Dmitry Zelenchuk [4,*]

1 Nokia Bell-Labs, D15Y6NT Dublin, Ireland; dmitry.1.kozlov@nokia.com
2 Department of Microelectronics and Radio Engineering, St. Petersburg Electrotechnical University LETI, 197376 St. Petersburg, Russia; ivmunina@etu.ru (I.M.); paturalchuk@etu.ru (P.T.); vvkirillov@etu.ru (V.K.)
3 School of Physics and Astronomy, Cardiff University, The Parade, Cardiff CF24 3AA, UK; ShitvovA@cardiff.ac.uk
4 ECIT Institute, Queen's University Belfast, Queen's Road, Belfast BT3 9DT, UK
* Correspondence: d.zelenchuk@qub.ac.uk

Abstract: A new implementation of a beam-steering transmitarray is proposed based on the tiled array architecture. Each pixel of the transmitarray is manufactured as a standalone unit which can be hard-wired for specific transmission characteristics. A set of complementary units, providing reciprocal phase-shifts, can be assembled in a prescribed spatial phase-modulation pattern to perform beam steering and beam forming in a broad spatial range. A compact circuit model of the tiled unit cell is proposed and characterized with full-wave electromagnetic simulations. Waveguide measurements of a prototype unit cell have been carried out. A design example of a tiled 10 × 10-element 1-bit beam-steering transmitarray is presented and its performance benchmarked against the conventional single-panel, i.e., unibody, counterpart. Prototypes of the tiled and single-panel C-band transmitarrays have been fabricated and tested, demonstrating their close performance, good agreement with simulations and a weak effect of fabrication tolerances. The proposed transmitarray antenna configuration has great potential for fifth-generation (5G) communication systems.

Keywords: antenna array; antenna measurements; beam pattern; beam steering; equivalent circuit modelling; transmitarray

Citation: Kozlov, D.; Munina, I.; Turalchuk, P.; Kirillov, V.; Shitvov, A.; Zelenchuk, D. Characterization of Tiled Architecture for C-Band 1-Bit Beam-Steering Transmitarray. *Sensors* **2021**, *21*, 1259. https://doi.org/10.3390/s21041259

Academic Editor: Naser Ojaroudi Parchin

Received: 31 December 2020
Accepted: 6 February 2021
Published: 10 February 2021

1. Introduction

Emerging architectures of the fifth-generation (5G) new radio communication systems employ complementary use of both sub-6 GHz and beyond 24 GHz spectrum regions, whereby, in outdoor scenarios, the low-frequency bands are envisioned to provide wide uniform coverage, whereas the millimetre-wave radio would allow directed ultra-high throughput within the wide sub-6 GHz coverage area. Moreover, although millimeter-wave propagation channels exhibit many peculiar features, which may even call for the use of quasi-optical analysis and design techniques, some advanced communication principles and system architectures, primarily aimed at millimetre-wave frequencies, can be implemented and verified with the aid of low-frequency proof-of-concept prototypes.

Multiple-antenna millimeter-wave radio systems, commonly referred to as multiple-input-multiple-output (MIMO) architecture with a large number of antenna elements at the radio access nodes and user terminals enable spatial multiplexing and diversity by means of intelligent beamforming. The latter feature seems to be an indispensable attribute of the 5G communication and radar systems, alongside the exploitation of unconventional degrees of freedom in radio propagation.

Although fully digital beamforming in massive MIMO systems can, in theory, achieve optimal performance, the current state of the digital hardware makes this approach unfeasible for millimeter-wave radio, due to prohibitively high cost and as yet insufficient resolution of the analog/digital-to-digital/analog converters, [1]. On the other hand, fully analog beamforming does not provide essential flexibility in design. In the course of previous studies, it appeared that millimeter-wave channels typically have much less degrees of freedom than can be achieved with fully digital beamforming, thus making the latter redundant. Therefore, many hybrid architectures have emerged recently, aimed to efficiently exploit the sparsity of millimeter-wave channels by combining the key features of both beamforming approaches to achieve optimal performance in applications at reduced complexity and cost.

In particular, the use of refractive dielectric lenses and focusing arrays proved to be technologically advantageous and economically efficient. The use of intelligent reflecting and transmitting surfaces, [1,2], including multi-beam transmitarrays, flat and hybrid lenses, impedance-modulated holographic surfaces, programmable metasurfaces with arbitrary control of the propagated wavefronts, all of which can be realized in conventional planar multi-layer technology using either non-linear materials or surface-mount RF components, opened new avenues in the design of millimeter-wave communication and sensing systems.

Recently, the feasibility of low-bit beam-steering and phase-only beamforming has been demonstrated as a means of further cost-reduction, [3,4]. Beam-switching at the focal-plane array has also been found to be a useful feature for millimeter-wave compact small-cell architectures, [5,6]. A number of different electronically controlled transmitarray architectures have reported recently for applications from C-band to V-band, with various performance functional from merely beam collimation to wide-angle beam-steering, beamforming and complete wavefront and polarization control. A 28-GHz circularly-polarized reconfigurable transmitarray comprising 400 binary phase unit-cells of receiver-transmitter type with an integrated phase-switch network was experimentally demonstrated in [7] as an attractive solution for many applications operating in Ka-band, such as satellite communications, point-to-point links and heterogeneous wireless networks. The use of a co-designed slot-array focal source antenna enabled a significant reduction of the antenna profile. An X-band electronically reconfigurable transmitarray with enhanced transmission bandwidth and efficiency achieved by using new contactless probe-feeding of the antenna patch was demonstrated in [8], aiming at advanced communication applications. A successful attempt to extend the application of low-cost transmitarrays to V-band was experimentally demonstrated in [9], although no electronic control was available at the time for two-dimensional beam-steering. Most of the above concepts have been demonstrated using integrated transmitarrays fabricated in planar printed-circuit technology. However, fabricating large single-panel transmitarrays raises the cost of proof-of-concept prototyping and makes the technology unaffordable for teaching laboratories.

Our research is aimed at adopting the transmitarray architecture for MIMO communications in C-band. In our previous publications [10,11], we reported on a low-frequency prototype of novel 1-bit dual-polarized tiled transmitarray, whereby the required phase distribution across the array aperture was built from standalone unit cells manufactured individually and assembled in the required pattern using a rectangular latticed plastic frame, Figure 1. Some preliminary simulation and measurement results were presented, and it appeared that the tiled architecture can be a viable solution for fast prototyping and teaching experiments, without significant performance deterioration, as compared with a similar single-panel transmitarray. Moreover, the possibility of replacing and adding individual elements in the tiled array makes it both repairable and adjustable for a specific focal distance and feed type. This paper revisits previous simulations and provides new results of modelling and experimental characterization of the tiled transmitarray.

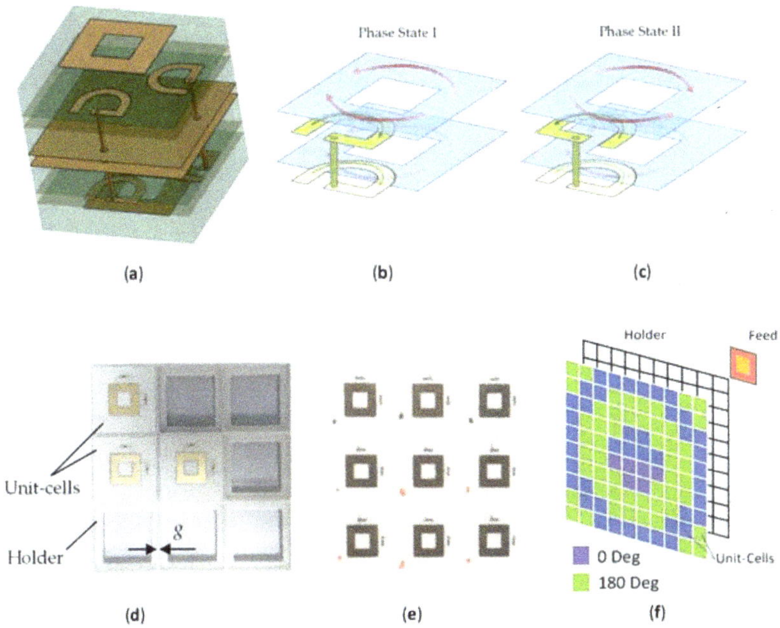

Figure 1. One-bit dual-polarized tiled transmitarray architecture: (**a**) design of the array tile (vertically exploded view), comprising two identical proximity-coupled square-ring radiators on the opposite sides of the tile connected via two U-shaped feed loops; (**b**,**c**) schematic view of the surface currents on the proximity coupled feed loops and square-ring patches in two phase states, respectively (the ground plane is not shown); (**d**) a section of the tiled transmitarray partially assembled; (**e**) a section of the integral single-panel transmitarray; (**f**) proposed device architecture, [10], including individual unit cells (different colors indicate one of the two phase states) to be mounted in the plastic grid frame and spatially fed by a focal-source patch antenna.

2. Transmitarray Model

The model of a transmitarray, first presented by the authors in [10], is given below for consistency. In a spatially phase-modulated transmitarray, the normalized wave amplitude received by a unit cell from the focal source reads:

$$a_{mn} = \frac{\lambda e^{jkR_{mn}}}{4\pi R_{mn}} \mathbf{F}_{mn}^{fs} \cdot \mathbf{F}_{mn}^{ucr} \tag{1}$$

where $m = 1, 2, \ldots, M$ and $n = 1, 2, \ldots, N$ are the row and column indexes of the array which define the position of the unit cell with respect to the reference one, k and λ are free-space wavenumber and wavelength, $\mathbf{F}_{mn}^{fs}$ is the complex vector field pattern of the focal-plane source transmitting in the direction of the unit cell defined by the corresponding polar and azimuthal angles of the local coordinate system (CS) with the origin at the focal point, $\mathbf{F}_{mn}^{ucr}$ is that of the unit cell on receive in the direction of the focal plane source defined by the respective angles of the local CS with the origin at the center of the unit cell (note that the antenna pattern on receive is conjugate of that on transmit due to the reciprocity), the dot symbol denotes the Hermitian inner product of the two complex vector patterns, and R_{mn} is the distance between the focal-source and unit-cell CSs. The unit cells are assumed to be matched to the incoming wave at all angles of incidence determined by the angular aperture of the transmitarray. The effects of the element coupling and finite aperture of the transmitarray can, in principle, be accounted for in the unit-cell antenna patterns by infinite array analysis, [12], or embedded element technique, [13].

The focal-source and unit-cell antenna patterns, in the case of linear polarization, reduce to scalar-valued functions. After sampling and retardation of the incident spherical wavefront, the complex amplitude antenna pattern of the transmitarray, $F(\theta,\phi)$, can be calculated by the pattern multiplication principle, as follows:

$$F(\theta,\phi) = \sum_{m=1}^{M}\sum_{n=1}^{N} b_{mn}F_{mn}^{uc}(\theta,\phi)e^{j\psi_{mn}^{uc}(\theta,\phi)}e^{jkd(msin\theta cos\phi+nsin\theta sin\phi)} \tag{2}$$

where $F_{mn}^{uc}(\theta,\phi)$ and $\psi_{mn}^{uc}(\theta,\phi)$ are, respectively, the unit-cell amplitude and phase patterns on transmit, θ and ϕ are azimuthal and polar angles in the spherical CS with the origin at the center of the transmitarray aperture and the polar direction aligned with the transmitarray optical axis, $b_{mn} = T_{mn}a_{mn}$-complex amplitudes of the waves radiated by each unit cell, and T_{mn}-the corresponding complex transmission coefficients. Equation (2) enables accounting for the effect of the finite array on the standalone pattern of the element, [14]. Also, the unit-cell radiation pattern is assumed to be independent of the transmission coefficient, i.e., of the specific phase shift for the phase-modulated transmitarray.

The above model can be adopted for the design of the proposed tiled transmitarray by suitably adjusting the unit-cell transmission coefficients for given focal-source and unit-cell antenna patterns. In transmitarray antennas, beam steering is achieved by spatially modulating the phase distribution of the emitted wavefront across the array aperture, as follows:

$$\arg(b_{mn}) = -k\,\boldsymbol{r}_s{\cdot}\boldsymbol{r}_{mn}, \tag{3}$$

where $\boldsymbol{r}_s(\theta_s,\phi_s)$ is the unit vector in the beam-steering direction (θ_s,ϕ_s), while the array vector $\boldsymbol{r}_{mn} = (x_{mn}, y_{mn}, 0)$ comprises the coordinates of the unit cell. For symmetrical unit cells, the required continuous local phase shift follows from (1) and (2) as:

$$\arg(T_{mn}) = \arg(b_{mn}) - \psi_{mn}^{fs} + kR_{mn} - \psi_{mn}^{uc}, \tag{4}$$

where ψ_{mn}^{fs} is the focal-source phase pattern in the direction of the unit cell (NB: typically, the phase pattern, with respect to the phase center of the antenna, is nearly flat within the angular range of the main lobe). In the proposed 1-bit transmitarray, the phase distribution (4) is discretized according to the following recipe (shown for the wrapped phase):

$$\arg\left(T_{mn}^{d}\right) = \begin{cases} 0^\circ\ \forall|\arg(T_{mn})| \leq 90^\circ \\ 180^\circ\ otherwise \end{cases}, \tag{5}$$

It is important to note that the effect of the 1-bit phase quantization on radiation characteristics was analyzed in [11,15]. It was shown that a 1-bit resolution results in the antenna gain reduction of up to 4 dB, higher sidelobe level and noticeable beam squint.

3. Unit Cell Design and Characterization

The detailed description of the unit cell design and preliminary results of the measurements inside the rectangular waveguide were reported in [10]. This unit cell structure has been employed in the current study. It is noteworthy that the proposed unit cell structure was conceived as a blank of a reconfigurable pixel of single-panel transmitarrays, using surface-mount solid-state switches to add functionality. However, in the context of the tiled transmitarray, power routing is much more challenging and thus is not addressed in this work.

The unit cell design, first reported in [10], was implemented in a stacked 6-layer structure, Figure 1a. The receiving and transmitting antennas were represented by square-ring microstrip elements with electromagnetic (proximity coupled) feeds in the form of open-ended half-wavelength semi-annular (U-shaped) microstrip loops in the layer beneath the square-ring antenna. The proximity coupling allowed a wider bandwidth when the feed loop and ring were properly aligned, [16]. The track widths of the square ring and feed loops were numerically optimized for the maximum return-loss bandwidth and low insertion loss, using CST Microwave Studio simulations with Floquet periodic boundary conditions (FPBCs) and assuming infinite ground plane. The pair of loop resonators were connected to each other by a buried via hole.

The receiving and transmitting sides of the tiled unit cell were separated by two ground plane electrodes bonded together using a 0.2 mm layer of Rogers RO4350B and protruded by the buried vias. The redundancy of the two ground planes was imposed by the manufacturing process. The metallic patterns of the square-ring radiators and feed loops were formed on and between dielectric layers of 0.51 mm thick Rogers RO4003 material (dielectric constant Dk = 3.5, dissipation factor Df = 0.0018). The layers of 0.1 mm bonding film Rogers RO4003C (Dk = 3.38) were used to stack the RO4003 layers. The lateral size of the unit cell of the single-panel transmitarray was 24 mm $\times$ 24 mm (~0.46 λ at the design frequency of 5.75 GHz) and its thickness was <0.045 λ. The tiled unit cells were trimmed by 0.5 mm around the edge in order to keep the same array period in both single-panel and tiled transmitarrays.

In the proposed unit-cell design, a 180° phase shift is implemented by switching the feed point of the U-shaped resonator on the receiving side of the transmitarray, Figure 1b,c. The state when the resonators at the receiving and transmitting sides are connected such that the currents flowing in the patches are codirectional is referred to as the phase state I (or 0° state). In the reciprocal phase state II, the resonators are connected at the opposite ends, so that the currents flow in the opposite directions thus imparting a 180° phase shift with respect to the phase state I. Two pairs of feed loops are used on each side of the structure, placed orthogonal to each other so that the unit cell can support two orthogonal linear polarizations for each phase state. The transmission and reflection coefficients measured in the waveguide were similar in both phase states and for both polarizations. The 10 dB return-loss bandwidth spanned 160 MHz from 5.67 to 5.83 GHz. The differential phase error did not exceed $\pm 6°$ across the operating band.

The transmitarray design approach adopted in our study is based upon the unit cell characterization in terms of insertion loss and differential phase shift (i.e., the phase shift in one phase state with respect to the other)-numerical with full-wave electromagnetic simulations (CST Microwave Studio), as well as experimental inside a rectangular waveguide. The tiled transmitarray has slotted dielectric substrate and ground plane, as well as additional dielectric frame 3D-printed in ABS (acrylonitrile butadiene styrene, Dk = 2.35 as measured), see Figure 1d, necessary to arrange the tiles in desired planar phase pattern. Thus, the effect of the discontinuity, i.e., the width of the gap between the adjacent unit cells, on the tile radiation performance is inherent to the design of the tiled transmitarray and we aimed to minimize its impact within the operating band.

The effect can be elucidated with the aid of the compact circuit model of the tiled unit cell shown in Figure 2a. It is noteworthy that the circuit model is loosely related to the actual geometry of the unit cell and it is derived essentially by emulating the bandpass response of the unit cell in the two phase-states. Nevertheless, the compact model provides useful insights on the interactions of different parts of the unit cell structure.

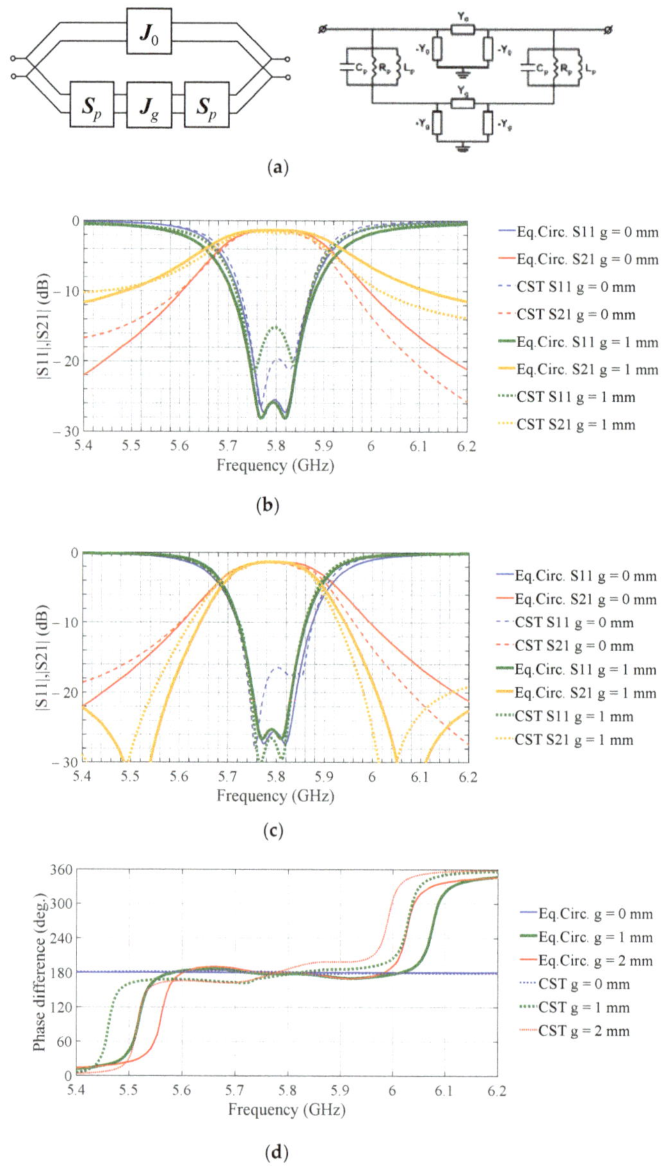

Figure 2. Equivalent-circuit modelling of the unit cells of the single-panel (g = 0 mm) and tiled (g = 1 or 2 mm) transmitarrays: (**a**) network topology (left) and compact electrical circuit model (right); (**b**) comparison of the circuit model with full-wave electromagnetic simulations (CST Microwave Studio) for the phase state I; (**c**) same as (**b**) but for the phase state II; (**d**) effect of the edge gap width, g, on the differential phase shift.

The circuit model topology constitutes a canonical parallel–parallel connection of the cascaded two-ports, Figure 2a. Being reduced to equivalent elements, the circuit comprises two parallel RLC-circuits (R_p, L_p, and C_p) associated with the receiving and emitting square-ring patches loaded by the respective U-shaped resonators and coupled via the two ideal admittance inverters, J_0 and J_g, with characteristic admittances Y_0 and

Y_g, respectively. The circuit model differs for the two phase states, due to the opposite direction of the current flowing on the receiving patch and this difference is implemented by changing the sign of the inverter admittance Y_0, with its positive value corresponding to the phase state I and negative to the phase state II. The J_0-inverter represents the primary coupling of the patches by the via connection, see Figure 1a. The effects of the edge gap are modelled by the additional inverter with a characteristic admittance Y_g, which can accurately model the out-of-band transmission zeros, see Figure 2c.

Putting R_p to zero, it can be shown that the transmission zeros appear at frequencies where the following condition fulfils:

$$Y_0 = -Y_g / \left(1 - Y_g^2 Z_p^2\right), \tag{6}$$

where Z_p is the complex impedance of the parallel LC circuit.

The resonance nulls in (6) can appear only when Y_0 and Y_g are in phase, due to the negative sign of the denominator in the vicinity of the resonant frequency of the LC circuit. This condition determines the out-of-band 180° steps of the differential phase shifts, demonstrated in Figure 2d when the differential circuit mode with the opposite direction of the currents on the receiving and emitting patches is superseded by the common mode that is driven by the floating ground plane.

The model parameters are shown in Table 1. The parameters were extracted by best-fitting to the full-wave electromagnetic simulations, as follows. Firstly, the initial 'patch' circuit parameters R_p, L_p, and C_p were fitted using the full-wave simulation of the reflection (S11) for the rectangular patch over an infinite ground plane and simplified circuit model without the inverters. In the second step, characteristic admittance of the Y_0-inverter is extracted by fitting to the full-wave simulations of transmission (S21) of the single-panel transmitarray unit cell (i.e., g = 0 mm). Finally, characteristic admittance of the J_g-inverter is obtained by fitting the model to the full-wave simulations of S21 of the two square-ring patches coupled only through the slotted ground plane, i.e., in the absence of the J_0-inverter. It appeared that the absolute value of the characteristic admittance Y_g decreases for the wider gap.

Table 1. Parameters of the compact circuit model of the transmitarray unit cell extracted by fitting to the full-wave simulations (see Figure 2b,c).

Gap Width (mm)	R_p (Ω)	L_p(nH)	C_p(pF)	$Y_0(\Omega^{-1})$	$Y_g(\Omega^{-1})$
0	300	0.038	19.84	±j0.021	N/A
1	300	0.038	19.86	±j0.021	−j0.23
2	300	0.038	19.89	±j0.021	−j0.18

Prototype unit cells emulating the structure of the tiled (1 mm gap width) and single-panel (0 mm gap width) unit cells were fabricated and measured inside the waveguide, see Figure 3a,b respectively. The results in both cases demonstrate noticeable downshift of the central frequency with respect to the design value, c.f., Figure 2b,c, as well as expected degradation of the differential phase shift for the tiled structure, c.f., Figure 2d. The observed shift of the central frequency has been attributed primarily to the specifics of the measurement setup, i.e., different boundary conditions for the unit cell in the waveguide, as compared with the FPBCs in the simulations.

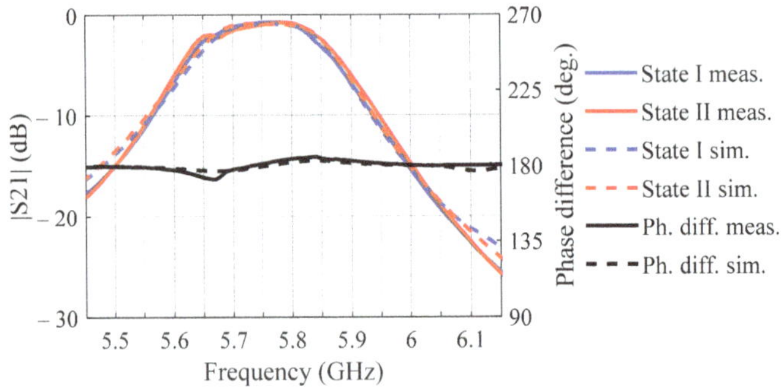

(**a**)

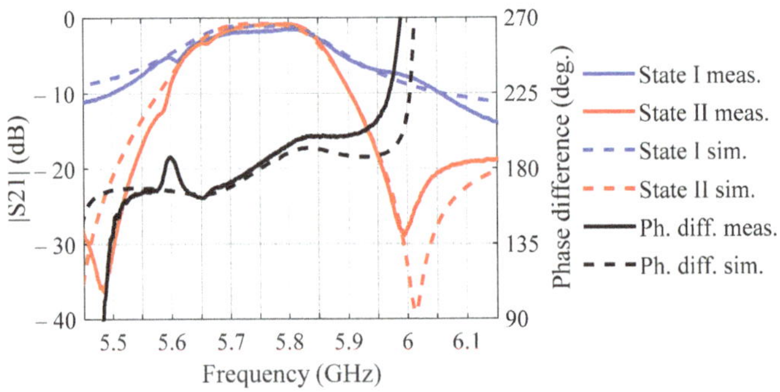

(**b**)

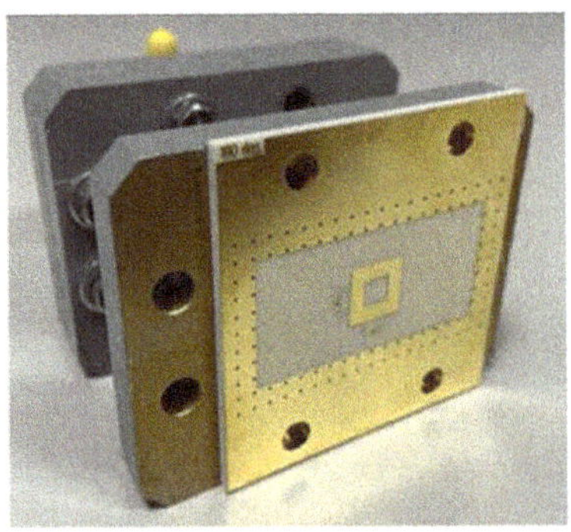

(**c**)

Figure 3. Simulated and measured transmission and differential phase shift of the single-panel without gap (**a**) and tiled with 1 mm gap (**b**) transmitarray unit cells. The results were obtained inside a rectangular waveguide (**c**).

Concluding on the results of characterization of the single-panel and tiled unit cells, it can be noticed that both structures demonstrate similar performance within the operating band. Moreover, two orthogonal polarizations demonstrated close performance, according to the full-wave simulations with FPBCs in [10]. The effects of the gap width can be modelled with a reasonably good accuracy using the compact circuit model in Figure 2a. The beam collimating and steering performances of the transmitarray with specific binary phase distributions are discussed in the next section.

4. Beam Steering by the Tiled Transmitarray

Although one may preemptively conclude from the results of the preceding section that the array performance of the tiled architecture should be commensurable with the single-panel transmitarray, there are still important factors yet unaccounted for. Here we shall apply the analytical model (2), alongside the full-wave simulations and antenna measurements, to evaluate the performance of the tiled architecture against the conventional single-panel transmitarray.

Two sets of prototype 10 × 10-element transmitarrays, viz., a set of 120 tiles hardwired for the two phase states and arranged in specific aperture pattern and a set of three single-panel transmitarrays routed for different beam scan angles (0°, 15°, and 30°), were fabricated in multi-layer printed circuit board technology by two manufacturers using similar materials, but different fabrication processes. The design of the unit cell of the single-panel transmitarrays had to be adjusted to comply with the company-specific fabrication process. That included slightly (10%) decreasing the via diameter and the width of the straight section connecting the annular track of the U-shaped feed to the via, but nevertheless, according to our simulations these changes were not expected to have a prominent effect on the transmitarray performance.

The measured and simulated boresight gains (H-plane) versus frequency of the tiled and single-panel transmitarrays illuminated by a patch-antenna feed are shown in Figure 4. The results demonstrate a 3 dB gain bandwidth of 140 MHz from 5.66 to 5.8 GHz for both transmitarrays.

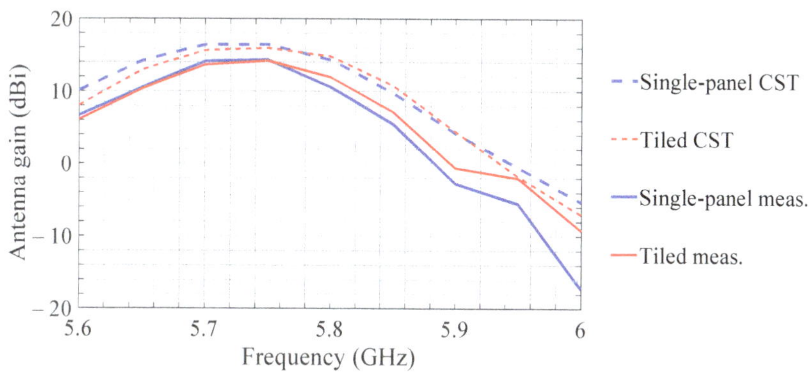

Figure 4. Measured and simulated (with the patch-antenna feed and with the plastic frame in case of the tiled array) boresight gains (H-plane) of the tiled and single-panel transmitarrays versus operating frequency.

The measured gains in Figure 4 are up to 3 dB lower than the simulated values in the operating band for both transmitarrays, which can be attributed to the simulation accuracy, particularly in estimation of the conductor and dielectric losses, as well as to the fabrication tolerances. Non-uniform amplitude distribution across the transmitarray aperture, due to slightly different transmittance of the unit cells in the two phase-states, might have been another contributing factor. This can be inferred from the measurement results in Figure 3. The measured results also indicate a noticeable (<40 MHz) upshift of the peak-gain frequency of the tiled array with respect to that of the single-panel array. Nevertheless, both arrays demonstrate adequate performance within the operating band.

The measured and simulated H-plane and E-plane beampatterns of the single-panel and tiled transmitarrays are shown in Figure 5 at the operating frequency of 5.75 GHz. It appears that the tiled array in measurements exhibits a lower gain and a higher beam-pointing error against the simulations, as compared with the single-panel transmitarray. It is noteworthy that the measured 15° beampattern of the single-panel transmitarray and simulated 15° beampattern of the tiled transmitarray feature a higher main lobe as compared with the corresponding central beampatterns. With the aid of the beampattern model (2) we have attributed this feature to the quantization error inherent to the 1-bit phase-shift design, which leads to sub-optimal radiating power combining at boresight of the transmitarray. Deviation of the differential phase shift from 180° causes decreasing peak gain of the steered beams with respect to the central beams in both transmitarrays.

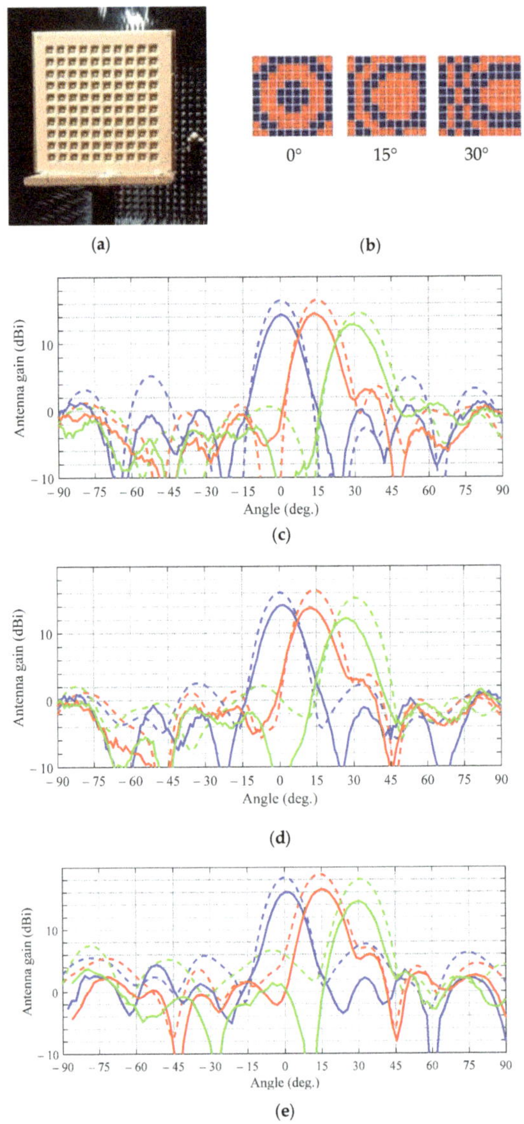

Figure 5. Measured (solid lines) and simulated with the patch antenna feed (dashed lines) beampatterns for different beam-scanning angles (viz., 0°, 15°, and 30°): (**a**) assembled transmitarray with the patch-antenna focal source visible; (**b**) binary phase distribution for different beam-scanning angles; (**c**) H-plane single-panel transmitarray beampatterns; (**d**) H-plane tiled transmitarray beampatterns; (**e**) E-plane tiled transmitarray beampatterns.

Figure 6 shows the measured beam patterns in orthogonal polarization (cross-polarization) in the principal E and H planes and diagonal D-plane. All patterns exhibit a prominent peak at boresight with respective cross-polarization ratio (CPR) ~14.5 dB. The shape of the cross-polarization beampattern is typical for dual-polarization transmitarrays, c.f., [17], and indicates polarization leakage due to coupling between the orthogonal feeds and between the patches on the opposite sides of the unit cell, as indicated by our equivalent-circuit characterisation. The measured figure agrees well with the data reported elsewhere, c.f. [17].

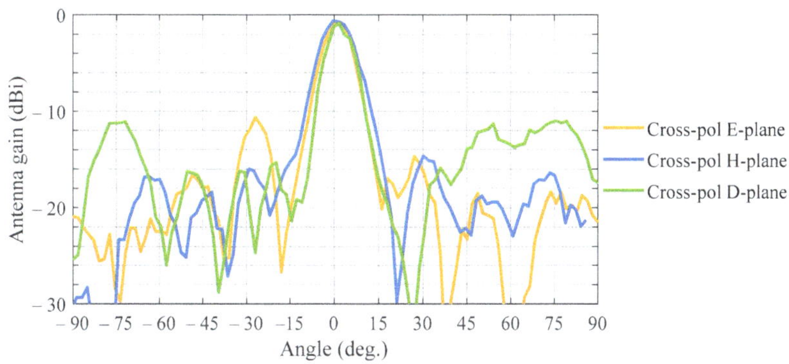

Figure 6. Measured cross-polarization patterns of the tiled transmitarray radiating at boresight plotted in the principal (E and H) and diagonal (D) planes.

Table 2 shows the performance comparison of the reference single-panel transmitarray discussed in this paper against a selection of published C-band transmitarray implementations, including the theoretical ('theor.'), measured ('meas.') and simulated ('sim.') data, [17–20]. Apart from one passive two-layer frequency-selective surface (FSS) lens, [20], the other transmitarrays adopt the conventional receiver-transmitter architecture with electronic control of the array functional (i.e., beam-steering, beam-forming or polarization conversion). As our design advances, it will integrate electronic control and provide wider bandwidth and better beam pointing accuracy.

Table 2. Comparison of some existing C-band transmitarrays.

Reference	[17]	[18]	[19]	[20]	This Work (Tiled)
Unit-cell	Two-layer stacked patches (Rx), a patch with O-slot (Tx), reflective phase shifter, vias	Patches, proximity coupled diff. feeds, balanced bridged-T phase shifters	Five stacked layers of square slot FSS and feeding networks	Passive two-layer double split-ring slot unit cells of varying size	Square-ring patch with proximity coupled U-shaped resonators, vias
Array size	8 × 8	6 × 6	5 × 5	7 × 7	10 × 10
Polarization	LP-to-LP/CP	LP	LP	LP	DLP
f0, GHz (Δf/f0)	5.4 (8.5% by AR-3 dB)	5 (10% by G0–2 dB)	5.2 (1.4% by UC S21–3 dB)	6 (15% by G0–3 dB)	5.75 GHz (2.5% by G0–3 dB)
Boresight antenna gain, dBi	17 (meas.)	20.5 (theor.) 15.0 (meas.)	18.6 (sim.) 15.6 (meas.)	16.7 (meas.)	14 (meas.)
HPBW, deg. (meas.)	13.5 (E)	20.4 (E) 18.4 (H)	14 (E) 16 (H)	12	12
SLL, dB (meas.)	−10	−21.1 (E) −14.9 (H)	−9.7	−10	−12
CPR, dB (meas)	20	35	N/A	N/A	14.5
Scan loss (scan angle), dB	0.9 (20 degree) 2.1 (30 degree)	N/A	1.4 (15 degree) 7.8 (25 degree)	N/A	2 (30 degree)
Beam pointing error (H), degree	N/A	N/A	0 (0 degree) 3 (15 degree) 8 (30 degree) 16 (45 degree)	N/A	N/A
Control	Varactors	Varactors	Varactors	None	None

5. Conclusions

A comprehensive characterization of the tiled transmitarray architecture first proposed in [10] has been carried out in this paper. A new compact circuit model has been devised to analyze the broadband transmission and differential phase shift characteristics of the tiled transmitarray unit cells.

The unit cell characterization at normal incidence has been carried out using the proposed circuit model and full-wave electromagnetic simulations. It appeared that the tiled and single-panel unit cells demonstrate commensurable performance within the operating frequency band, although the tiled unit cell exhibits a higher differential phase error.

The antenna gain and radiation patterns of the fabricated tiled and single-panel transmitarrays have been measured for different beam-scan angles, as well as compared with full-wave electromagnetic simulations. The tiled transmitarray demonstrated slightly lower gain and higher beam-pointing error as compared with the single-panel transmitarray. The measured results are in a good quantitative agreement with simulations.

In conclusion, it has been shown that the tiled transmitarrays can be effectively designed, modelled and fabricated to demonstrate the antenna performance commensurate with conventional single-panel transmitarrays. Considering the cost of manufacture and flexibility in configuring the transmitarray for various applications, the proposed tiled transmitarray architecture proves to be a feasible and economically effective solution for 5G communication systems. The future work will be carried out on advancing the analytical model by taking into account essential effects due to spillover, [21], coupling and array non-uniformity, adopting the tiled architecture for millimeter-wave applications, investigating the heterogeneous and conformal transmitarrays enabled by the tiled architecture, as well as developing hybrid approaches to beam-scanning and beam-forming by combining tiled transmitarrays with focal plane antenna arrays.

Author Contributions: Conceptualization, D.K., P.T., I.M. and A.S.; methodology, V.K.; validation, P.T., I.M., V.K. and D.K.; investigation, D.K., I.M., P.T., V.K. and D.Z.; Writing—original draft preparation, D.K. and A.S.; Writing—review and editing, D.K., A.S., I.M., P.T. and D.Z.; project administration, I.M. and D.Z.; resources, D.K., I.M., P.T. and D.Z.; funding acquisition, I.M., A.S. and D.Z. All authors have read and agreed to the published version of the manuscript.

Funding: This research was funded by the Russian Science Foundation under Project 17-79-20374 and A.S. was supported by The Royal Society under Grant IE160128 and D.Z. was supported by The Royal Society under Grant IES\R1\191236236. The APC was funded by MDPI.

Institutional Review Board Statement: Not applicable.

Informed Consent Statement: Not applicable.

Data Availability Statement: The data presented in this study are available on request from the corresponding author.

Acknowledgments: The authors would like to thank Kieran Rainey and Adrian McKernan for technical support with the antenna measurements.

Conflicts of Interest: The authors declare no conflict of interest. The funders had no role in the design of the study; in the collection, analyses, or interpretation of data; in the writing of the manuscript, or in the decision to publish the results.

References

1. Dai, L.; Wang, B.; Wang, M.; Yang, X.; Tan, J.; Bi, S.; Xu, S.; Yang, F.; Chen, Z.; Renzo, M.D.; et al. Reconfigurable Intelligent Surface-Based Wireless Communications: Antenna Design, Prototyping, and Experimental Results. *IEEE Access* **2020**, *8*, 45913–45923. [CrossRef]
2. Jamali, V.; Tulino, A.M.; Fischer, G.; Müller, R.; Schober, R. Intelligent Reflecting and Transmitting Surface Aided Millimeter Wave Massive MIMO. *arXiv* **2019**, arXiv:1902.07670.
3. Cruz, C.C.; Fernandes, C.A.; Matos, S.A.; Costa, J.R. Synthesis of Shaped-Beam Radiation Patterns at Millimeter-Waves Using Transmit Arrays. *IEEE Trans. Antennas Propag.* **2018**, *66*, 4017–4024. [CrossRef]

4. Kaouach, H.; Dussopt, L.; Sauleau, R.; Koleck, T. Design and demonstration of 1-bit and 2-bit transmit-arrays at X-band frequencies. In Proceedings of the 2009 European Microwave Conference (EuMC), Rome, Italy, 29 September–1 October 2009; pp. 918–921.
5. Moknache, A.; Dussopt, L.; Saily, J.; Lamminen, A.; Kaunisto, M.; Aurinsalo, J.; Bateman, T.; Francey, J. A switched-beam linearly-polarized transmitarray antenna for V-band backhaul applications. In Proceedings of the 2016 10th European Conference on Antennas and Propagation (EuCAP), Davos, Switzerland, 10–15 April 2016; pp. 1–5.
6. Hill, T.A.; Kelly, J.R.; Khalily, M.; Brown, T.W.C. Conformal Transmitarray for Scan Loss Mitigation with Thinned Reconfiguration. In Proceedings of the 2019 13th European Conference on Antennas and Propagation (EuCAP), Krakow, Poland, 31 March–5 April 2019; pp. 1–5.
7. Di Palma, L.; Clemente, A.; Dussopt, L.; Sauleau, R.; Potier, P.; Pouliguen, P. Circularly-polarized reconfigurable transmitarray in Ka-band with beam scanning and polarization switching capabilities. *IEEE Trans. Antennas Propag.* **2016**, *65*, 529–540. [CrossRef]
8. Wang, M.; Xu, S.; Yang, F.; Hu, N.; Xie, W.; Chen, Z. A novel 1-bit reconfigurable transmitarray antenna using a C-shaped probe-fed patch element with broadened bandwidth and enhanced efficiency. *IEEE Access* **2020**, *8*, 120124–120133. [CrossRef]
9. Frank, M.; Scheiner, B.; Lurz, F.; Weigel, R.; Koelpin, A. Low-cost transmitarray antenna designs in V-band based on unit-cells with 1 bit phase resolution. *Frequenz* **2019**, *73*, 11–12. [CrossRef]
10. Munina, I.; Turalchuk, P.; Kirillov, V.; Verevkin, A.; Zelenchuk, D.; Shitvov, A.P. A Tiled C-Band Dual-Polarized 1-Bit Transmitarray. In Proceedings of the 49th European Microwave Conference (EuMC), Paris, France, 1–3 October 2019.
11. Munina, I.; Turalchuk, P.; Kirillov, V.; Verevkin, A.; Zelenchuk, D.; Shitvov, A.P. A Study of C-Band 1-Bit Reconfigurable Dual-Polarized Transmitarray. In Proceedings of the 13th European Conference on Antennas and Propagation (EuCAP), Krakow, Poland, 31 March–5 April 2018.
12. Roscoe, A.J.; Perrott, R.A. Large finite array analysis using infinite array data. *IEEE Trans. Antennas Propag.* **1994**, *42*, 983–992. [CrossRef]
13. Kelley, D.F. Embedded Element Patterns and Mutual Impedance Matrices in the Terminated Phased Array Environment. In Proceedings of the 2005 IEEE Antennas and Propagation Society International Symposium, Washington, DC, USA, 3–8 July 2005; Volume 3A, pp. 659–662.
14. Liu, J.; Li, L.; Lv, Z.; Chen, F.; Ni, S. Impact of Element Pattern on the Performance of GNSS Power-Inversion Adaptive Arrays. *Electronics* **2019**, *8*, 1120. [CrossRef]
15. Diaby, F.; Clemente, A.; Di Palma, L.; Dussopt, L.; Pham, K.; Fourn, E.; Sauleau, R. Design of a 2-bit unit-cell for electronically reconfigurable transmitarrays at Ka-band. In Proceedings of the 2017 European Radar Conference (EURAD), Nuremberg, Germany, 11–13 October 2017; pp. 473–476.
16. Pozar, D.M.; Kaufman, B. Increasing the bandwidth of a microstrip antenna by proximity coupling. *Electron. Lett.* **1987**, *23*, 368. [CrossRef]
17. Huang, C.; Pan, W.; Ma, X.; Zhao, B.; Cui, J.; Luol, X. Using reconfigurable transmitarray to achieve beam-steering and polarization manipulation applications. *IEEE Trans. Antennas Propag.* **2015**, *63*, 4801–4810. [CrossRef]
18. Lau, J.Y.; Hum, S.V. Reconfigurable transmitarray design approaches for beamforming applications. *IEEE Trans. Antennas Propag.* **2012**, *60*, 5679–5689. [CrossRef]
19. Reis, J.R.; Caldeirinha, R.F.S.; Hammoudeh, A.; Copner, N. Electronically reconfigurable FSS-inspired transmitarray for two dimensional beamsteering. *IEEE Trans. Antennas Propag.* **2017**, *65*, 4880–4885. [CrossRef]
20. Chen, Y.; Chen, L.; Yu, J.-F.; Shi, X.-W. A C-band flat lens antenna with double-ring slot elements. *IEEE Antennas Wirel. Propag. Lett.* **2013**, *12*, 341–344. [CrossRef]
21. Plaza, E.; Leon, G.; Loredo, S.; Las-Heras, F. A Simple Model for Analyzing Transmitarray Lenses. *IEEE Antennas Propag. Mag.* **2015**, *57*, 131–144. [CrossRef]

sensors

MDPI

Article

Wideband Multiport Antennas

Mehdi Seyyedesfahlan [1,*], Abdulkadir Uzun [2,3], Anja K. Skrivervik [1] and Ibrahim Tekin [2,4,*]

1 Microwaves and Antennas Group (MAG), EPFL, CH-1015 Lausanne, Switzerland; anja.skrivervik@epfl.ch
2 Electronics Engineering, Sabanci University, 34956 Istanbul, Turkey; kadiruzun@sabanciuniv.edu
3 ASELSAN A.Ş., 34906 Istanbul, Turkey
4 Sabanci University Nanotechnology and Application Center (SUNUM), 34956 Istanbul, Turkey
* Correspondence: msesfahlan@sabanciuniv.edu (M.S.); tekin@sabanciuniv.edu (I.T.)

Received: 27 October 2020; Accepted: 3 December 2020; Published: 5 December 2020

Abstract: In this paper, a wideband four port 2–6 GHz antenna is proposed. One-, two-, and four-port antennas are implemented and characterized between 2 and 6 GHz. The isolation between the ports is improved by connecting and optimizing the ground plane sections. The results show that the antennas' reflection coefficients are better than 10 dB in the frequency band. The measured isolation between the ports is greater than 15 dB (between 2.3 and 6 GHz) and 10 dB in the whole band for two- and four-port antennas, respectively, however, it is more than 20 dB around 2.4 and 5–6 GHz for both antennas. The calculated correlation coefficient between ports is below −30 dB (>2.14 GHz) and −15 dB for the two- and four-port antennas, respectively. The measured gain and efficiency scale are 3.1–6.75 dBi and 62–98%, respectively. To the best of our knowledge, an antenna both being wideband from 2 to 6 GHz and having independent four ports is only addressed in this work. The four-port antenna can be used for MIMO systems or smartphones operating on many wireless systems simultaneously such as 3G/4G/5G Sub-6 GHz and WLAN including the next generation WiFi7 with full-duplex operation.

Keywords: wideband antenna; MIMO antenna; four-port wideband antenna

1. Introduction

Recently, there is an increasing demand for higher throughput and more reliable transceiver systems with an application on 4G wireless systems and mobile communication. Multiple Input Multiple Output (MIMO) technology could be a promising candidate for this purpose [1]. The MIMO technique is based on using multiple antennas to increase the data rate by means of uncorrelated signals.

For the MIMO system to function as expected, the mutual coupling between antenna elements should be as low as possible. A standard approach to achieve MIMO operation is to develop multiple antennas that are sufficiently separated to achieve the desired level of signal independence and port-to-port isolation. However, this will make the transceiver system bulky and result in increased assembly costs. Additionally, ease of integration and miniaturization are two major challenges ahead of MIMO antennas. Thus, the design of the MIMO antenna is the first important thing to be addressed to improve the overall system performance. Planar type antennas are preferred for MIMO applications due to ease of integration and low cost. For miniaturization purposes, there are no options but to space antenna elements closer or designing multiport, single-element antenna. Various studies have been carried out aiming to design such compact antenna systems [2,3]; they are commonly based on the planar antenna prototype [4,5]. The first approach is to decrease the spacing between antennas and keep the mutual coupling at an acceptable level by applying isolation improvement techniques. Examples of this approach can be found in various literature, such as adding a ground wall with connecting line and shorting pins [6], T-shaped ground plane [7], the corrugated ground plane with $\lambda/4$ slot [8], modified PIFA with a small local ground plane [9], techniques based on dispersion engineering called negative group delay (NGD) technique [10], use of external lumped element decoupling networks between the feed ports to allow matching of even and odd modes to a common impedance and thereby producing small cross-correlation and maximum gain over a limited frequency range [11], and other compact designs of MIMO antennas [2–5,12]. These methods can reduce the overall size of antennas and insulator regardless of the difficulty in insulator design.

The second approach takes advantage of multiport, single-element antennas to propose a more compact solution. A novel design of dual-feed, single-element antennas for 4G MIMO terminals is proposed and analyzed in [13]. The antenna consists of a radiating patch which is fed by two input ports. The idea is to use an isolated mode antenna (iMAT) [14] to reduce the antenna size and mutually couple the ports. The iMAT works based on exciting and different propagating modes of antenna for different ports. The iMAT antenna idea is also used in [15] to design a novel u-shaped single-element antenna with better performance, compared with two separate monopole antennas, in [16,17] to design a multiple compact multimode patch antenna, and in [18–20] for a multimode antenna that is not based on a patch antenna.

An important factor in MIMO systems is its bandwidth, which is determined by the bandwidth of the antenna element. Thus, a wideband single-element antenna with multiple ports could be very useful for a wideband MIMO system. In general, MIMO systems use many antennas to obtain multiport systems. Note that a multiport wideband antenna can also be used in smartphones that use many different wireless protocols at different frequency bands at the same time. There exist antennas which either are wideband or are multiport with narrow bandwidth. Nevertheless, combining multiport with wide bandwidth operation forms our antenna's novelty, which has four ports and can operate between 2 and 6 GHz. We propose a structure to increase the number of radiating element feeding/receiving ports only by rotating the main single port monopole antenna. Of course, monopole antenna is well known and there are many reports on how to make it wideband; however, increasing the number of ports while matching the ports and decreasing coupling between the ports requires many attempts. Moreover, when all the ports use common radiating elements, it needs a smart method to mitigate coupling between the ports. In this paper, we use a unit structure and bridge between the ground planes of ports to alleviate coupling between ports. We designed and optimized the antenna for frequency band between 2 and 6 GHz and achieved minimum isolation of 10 dB between the four ports. The aim is to introduce a multi-purpose (multiport and wideband) structure; however, for the desired application/band, the isolation between ports can be increased only by optimizing the ground plane and connection between ports.

The four-port antenna reported in this paper can be used for a multi-frequency system requiring many antennas. The 4×4 MIMO implemented for a WLAN on 2.4 and 5.2 GHz band is one example. The four-port antenna can also be used for the sub-6 GHz band 5G system. For a multiple radio system currently used in smartphones, let us assume there are four radios and these radios are 3G (2 GHz band), WLAN (2.4 GHz), 1–6 GHz 5G (3.6 GHz band), and WLAN (5.2 GHz). One can directly connect these four radios to the proposed four-port antenna without any switches and duplexers. RF filters can be deployed for each radio band to provide enough selectivity. However, with our antenna, all these radios can operate simultaneously. The key focus is on new mobile 5G bands including spectrum in the 3.5 GHz range that has been assigned in numerous countries. However, several countries including China and Japan plan to use spectrum in the 4.4–4.9 GHz range for 5G in addition to a growing number of countries considering the 3.5–4.2 GHz range, as well as the 2.3 and 2.5/2.6 GHz bands for 5G NR [21].

To have a wideband multiport antenna, a wideband planar structure should be selected. In this work, a printed monopole disk antenna [22] is selected for multiport use, due to its wide bandwidth operation. Figure 1 shows the monopole disk antenna with a single port. The disk monopole antenna is modified to two- and four-port versions for different frequency ranges. The geometrical symmetry of the antenna shape not only makes the design easy but also gives the versatility of adding and increasing the number of ports.

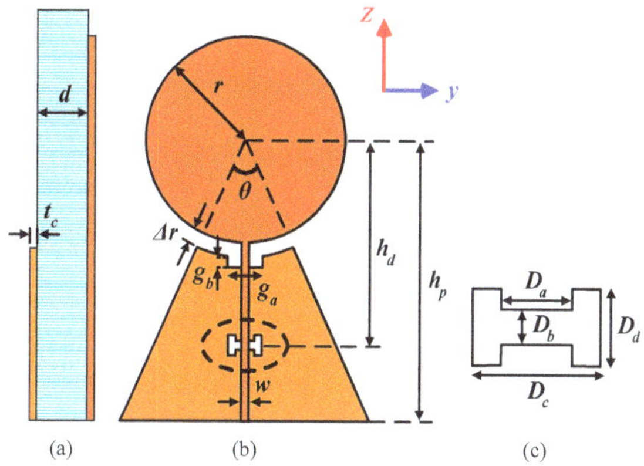

Figure 1. Schematic for the proposed single port wideband disk antenna: (**a**) cross-sectional view; (**b**) top view; and (**c**) etched ground dimensions.

In this paper, two metrics are used for the assessment of the isolation between antenna ports: the S parameter and the correlation coefficient. The correlation coefficient expresses antenna pattern independence to the S parameter, which is necessary for a MIMO antenna. This paper is organized as follows. Section 2 demonstrates the design of single-, dual-, and quad-feed disk monopole antenna with wideband operations. Section 3 presents and compares the simulated and measured results for the S parameter as well as the radiation patterns of the antennas. Section 4 summarizes and concludes the paper.

2. Multiport Antennas Design

In this section, the design process for the one-, two-, and four-port antennas are introduced. The antennas contain a radiating disk and microstrip transmission line as the antenna feed. Both two- and four-port antennas have structures similar to the single-port antenna, and the various dimensions shown in Figure 1 are optimized for each antenna to match each port to 50 Ω and decrease the mutual coupling between ports of each antenna, over the frequency band of 2–6 GHz. The scheme for increasing the number of the ports is to exploit the single-port antenna geometry (Figure 1) as the basis of n-port antennas, and then rotate/add the structure by 90° (with respect to disk center) to form the new port. The advantage of this procedure is that the ports (in multiport types) would be similar, and the design parameters in Figure 1 are optimized for all ports, simultaneously. The optimization is performed to approach the specified reflection coefficient and isolation between the ports over the desired frequency bandwidth. The antennas were simulated and prepared for fabrication on d = 0.787 mm thick (copper cladding tc = 35 μm) Rogers RT/duroid 5880 laminate with a dielectric constant of 2.2 and tangent loss of 0.0009.

2.1. Single-Port Antenna

The schematic of the single-port antenna and the parameters for which optimizations are performed are shown in Figure 1. The antenna can be divided into two major parts: the radiating disk and the transmission line that feeds the disk. The fabricated antenna with the dimensions of 6.8 cm × 4.4 cm is shown in Figure 2. In the bottom layer, an incomplete triangular shape ground plane supports the signal line in the top layer and can have coupling with the radiating disk.

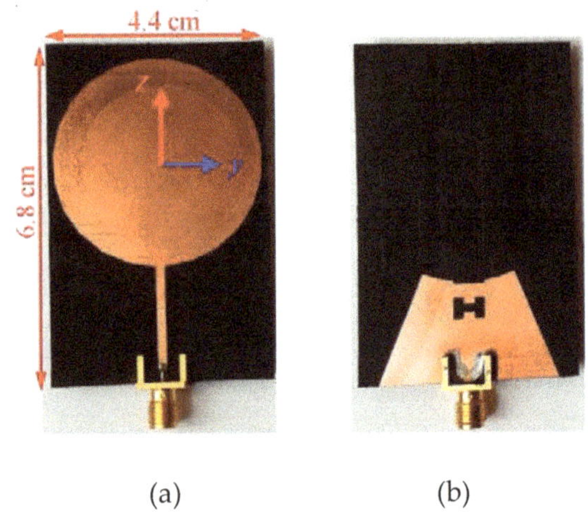

Figure 2. Fabricated single-port antenna: (**a**) top view; and (**b**) bottom view.

The dimension of the disk (r) adjusts the antenna operating frequency, while Δr is the spacing between the disk (top layer) and ground plane (bottom layer) edge. Since the antenna is similar to a monopole antenna, the disk will resonate with a quarter-wavelength diameter ($2r = \lambda/4$). The radius of the disk for resonating at 2 GHz in the free space is calculated as 18.75 mm, which is used as the initial value for r. To match the antenna to 50 Ω in the band of 2–6 GHz, other parameters (shown in Figure 1) are utilized to tune the antenna over the entire desired frequency band or in some specific frequencies. θ and h_p control the dimensions of the ground plane. The gap specified by the dimensions of g_a/g_b, as well as the location (h_d) and dimensions of the dumbbell-shaped etching, affect the antenna reflection coefficient by changing the inductance/capacitance of the transmission line and improving the feed line S_{11} magnitude.

The optimized parameters for fabricating the antenna (Figure 2) are reported in Table 1. Different parameters of the single-port antenna are swept around the optimized values to show their effect on the antenna reflection coefficient.

Table 1. Geometrical dimensions for different antennas.

	Single Port	Two-Port	Four-Port
h_p	43.7	54.7	69
r	21	20.5	20
Δr	3	2.5	3
θ	45	43	38
w	1.8	1.8	2
g_a	0.6	2.8	2.1
g_b	6	7	6
h_d	29.25	32	31
D_a	2.4	1.4	2.4
D_b	1.5	2	2
D_c	6	5	6
D_d	4	4.5	3.5
R_d, C_t	-	$R_d = 36$	$C_t = 6$
e_a	-	-	3
e_b	-	-	2

Values are in mm.

Figure 3a shows that the optimum value for a disk radius of 21 or 22 mm can give the best reflection coefficient values at less than −15 dB. As the spacing between the disk and ground plane is increased up to 3 mm, the antenna matching is improved, while greater values deteriorate the antenna performance, due to the decoupling between the microstrip line and the disk antenna, as shown in Figure 3b.

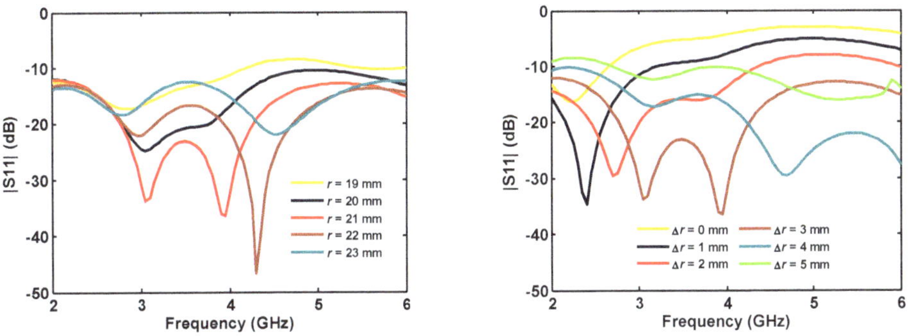

(a) (b)

Figure 3. Single-port antenna simulated reflection coefficient for different: (**a**) disk radii; and (**b**) the gap length between disk and the ground plane.

When the ground plane angle (θ) is increased, the antenna $|S_{11}|$ is improved for higher frequencies, while the impedance matching worsens in the middle of the band (Figure 4a). As shown in Figure 4b, the height variation of the truncated triangle ground causes a frequency shift in the antenna reflection coefficient.

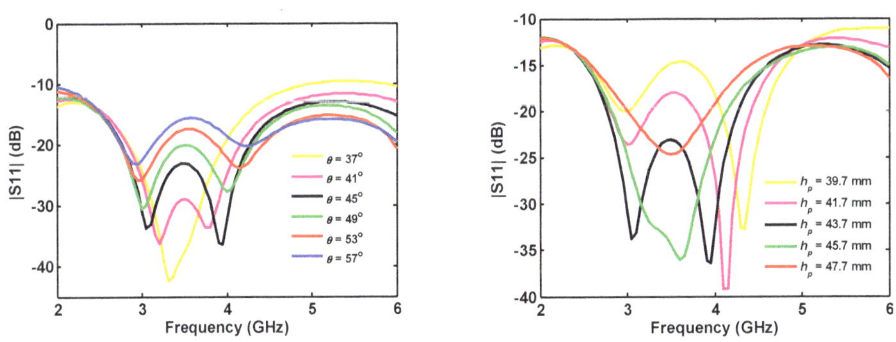

(a) (b)

Figure 4. Single-port antenna simulated reflection coefficient in terms of triangle parameters: (**a**) angle; and (**b**) height.

The effect of the dumbbell-shaped etched ground plane on matching the antenna between 2 and 6 GHz is demonstrated in Figure 5a. Note that the legend with the word "No" in Figure 5a points to the full ground (without dumbbell-shaped etching). Changing the dimensions of the etched area in Figure 5b,c shows its major effect on the antenna reflection coefficient for frequencies greater than 3 GHz. Although the effect of some parameters is not that significant in the single-port antenna, they play a drastic role in tuning the multiport antennas in the wideband operation.

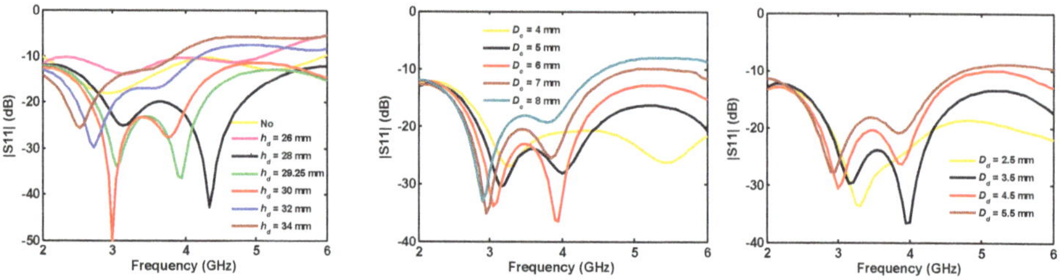

(a) (b) (c)

Figure 5. Single-port antenna simulated reflection coefficient for different (**a**) location; (**b**) width; and (**c**) height of the dumbbell-shaped etched ground.

2.2. Two-Port Antenna

The two-port antenna is obtained by rotating the single-port antenna by 90°, with respect to the center of the disk, and adding another port. As shown in Figure 6, the ground planes of the two ports are connected via a circular ring sector.

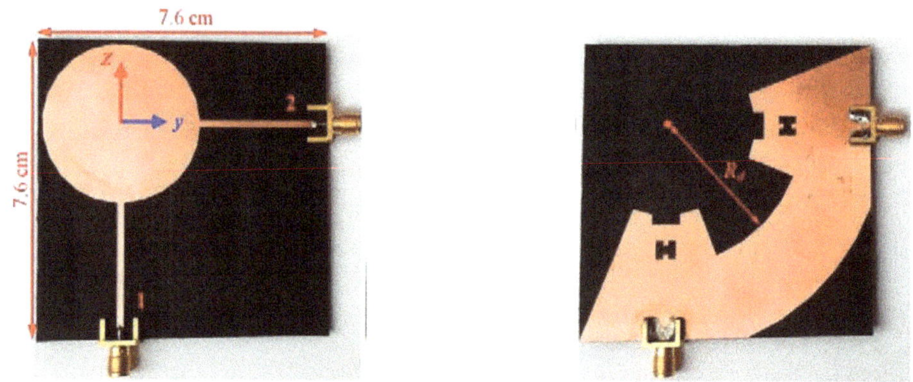

(a) (b)

Figure 6. Fabricated two-port antenna: (**a**) top view; and (**b**) bottom view.

The angle of the sector is $90°-\theta$, while its inner and outer radius are R_d and h_p, respectively. By connecting the grounds of the two ports, better isolation between the ports is obtained. When the first port of the antenna is fed, the received signal in the second port includes two components: (a) the signal that passes over the disk; and (b) the signal that flows from the connected ground of the ports. Therefore, these two components can cancel each other, if the phase difference of 180° is kept when these two components arrive at the second port. Out of phase condition between the mentioned two current trajectories improves the ports' isolation significantly and can be achieved by optimizing some of the antenna parameters. The working mechanism of the connection is shown (see Figure 7) in the simulated current distribution on the antenna at 2.4 GHz and at different phases.

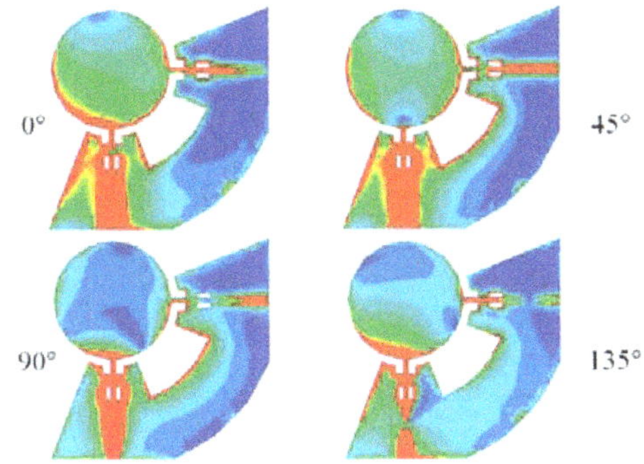

Figure 7. The simulated magnitude of current distribution on top and bottom layers at 2.4 GHz and different 0-, 45-, 90-, and 135-degree phases.

Note that the power that is dissipated in the vicinity of port two (due to cancellation) can decrease the radiation efficiency of the antenna, whereas the used PCB board is chosen to have a very small tangent loss. The dimensions of the fabricated two-port antenna are 7.6 cm × 7.6 cm and the rest of the parameters are given in Table 1.

Changing the disk radius (r) and Δr can both affect the accepted/reflected power by the first port on the disk side as well as the coupled power to the second port through the disk. The influence of this complicated process on the insertion loss between the ports, for various r and Δr, is shown in Figure 8.

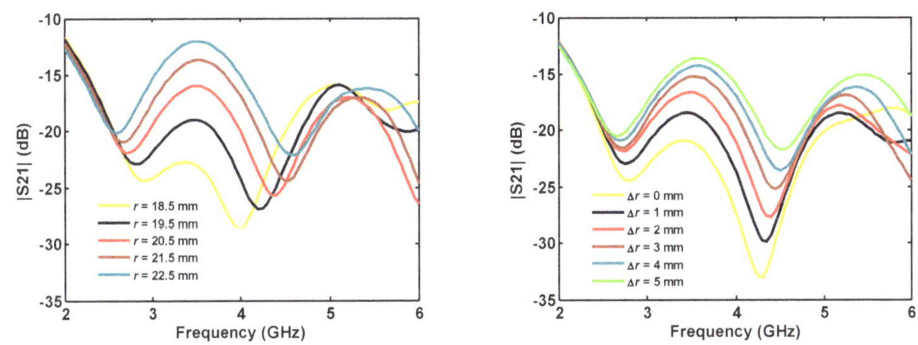

(a) (b)

Figure 8. Simulated insertion loss between the ports of two-port antenna for various: (**a**) disk radius; and (**b**) disk-ground spacing.

As shown in Figure 8, the most significant effect of the parameters r and Δr on $|S_{21}|$ is between 2.7 and 4.5 GHz. At these frequencies, insertion loss can be adjusted to be below −20 dB, by the disk size and the gap spacing. The ground plane angle also affects insertion loss between ports in a limited frequency band of 4–5 GHz (Figure 9a). Increasing the height of the ground plane shifts the $|S_{21}|$ to lower frequencies in the 3–6 GHz band (Figure 9b).

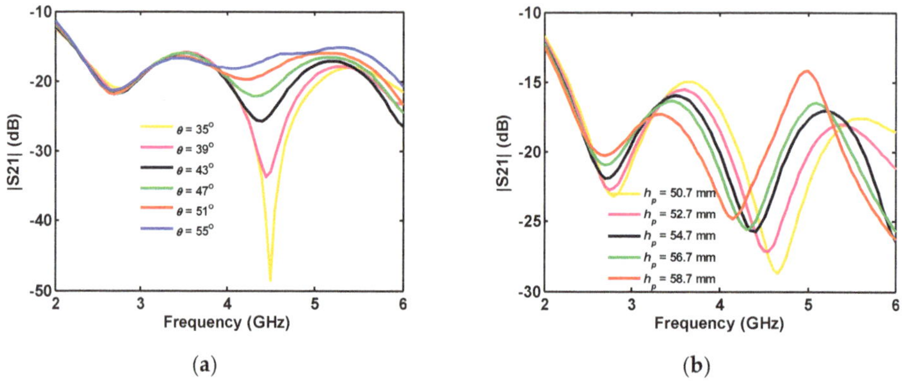

Figure 9. Simulated insertion loss between the ports of two-port antenna for various: (**a**) angle; and (**b**) height of the truncated triangular-shaped ground plane.

As discussed above, the connection between the ports' ground plays an important role in improving the insertion loss between them. The important factor R_d, which controls the dimension of the connected part, and corresponding isolation between the ports, is swept around its optimum value R_d = 36 mm, as shown in Figure 10. This results in increasing R_d, and hence decreases the thickness of the connected section and causes the S_{21} curve to shift to lower frequencies.

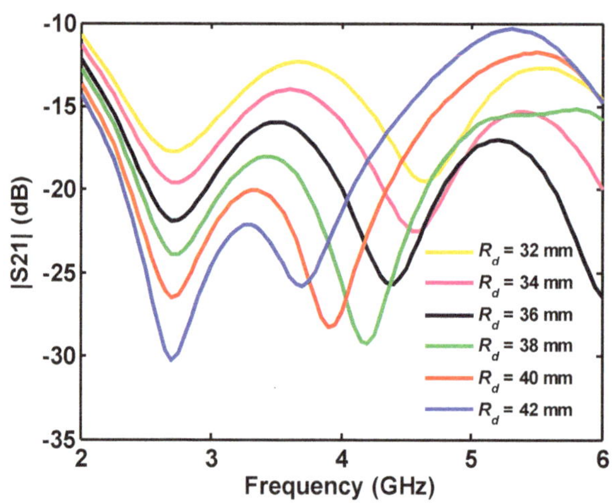

Figure 10. Effect of the inner radius of the ground plane sector on simulated insertion loss.

2.3. Four-Port Antenna

Another multiport antenna is a four-port antenna that is formed by rotating/adding a single-port antenna with respect to the disk center. In this type of antenna, the spacing between the ports is 90°, as shown in Figure 11.

(a) (b)

Figure 11. Fabricated four-port antenna: (**a**) top view; and (**b**) bottom view.

For this antenna, the basic geometry is similar to that of the single port, while the edge of the disk at the ports is etched (in top view) and the grounds are connected using a metal bar with a thickness of C_t. The etched areas on the disk are a rectangular section with dimensions of e_a and e_b (Figure 11), which alleviate the ports' reflection coefficient to be below −10 dB within the band of 2–6 GHz. The ground connection also controls the insertion loss between the ports, the same technique as used in the two-port antenna. The dimensions of the fabricated four-port antenna are 12.4 cm × 12.4 cm, with parameters given in Table 1. Note that, for an application on a mobile phone, antenna size can be made smaller by bending from the microstrip line sections. The proposed four-port antenna is designed for a wide frequency band, starting from 2 GHz. By excluding WiFi 2.4 GHz frequency, while shifting start frequency to 3 GHz, which means, if only 5G systems are chosen, the disk size and hence the overall antenna size will be smaller by a factor of 1.5 times to achieve an 8 cm × 8 cm antenna. Moreover, for mobile applications, part of the feeding network can also be placed in a different PCB layer, or a flexible board may be folded/wrapped and antenna size can be further made smaller. Moreover, the antenna can be optimized/improved by separating 5G or WiFi system and having two antennas. For example, for a 5G sub-6 GHz system, the antenna size can be optimized to 3 GHz, and for WiFi it could be around 5 GHz band.

As shown in Figure 12, etched areas on the disk (geometrical parameters e_a and e_b in Figure 11) and the ground plane (geometrical parameters g_a and g_b in Figure 1) play a very crucial role in adjusting each port's reflection coefficient below −10 dB. The legend entry "No" indicates no etching is performed on the copper.

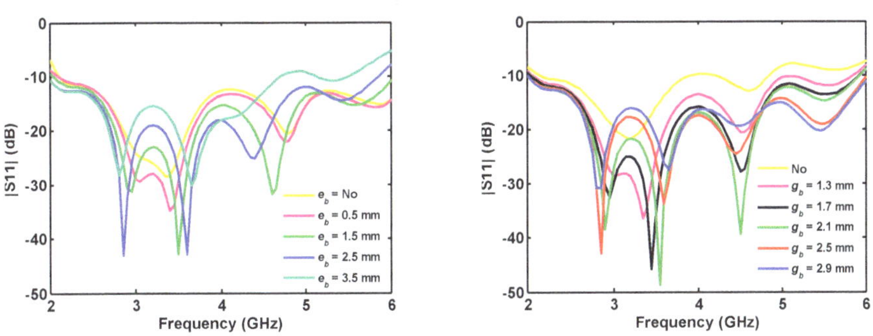

(a) (b)

Figure 12. Effect of etched area's simulated reflection coefficient of four-port antenna: (**a**) disk; and (**b**) edge of the ground plane.

Due to the wide bandwidth of 2–6 GHz and the number of ports, developing the antenna for $|S_{11}| < -10$ dB and desired isolation between the port is challenging or maybe impossible for some geometries. Consequently, it is proposed to match the ports to 50 Ω with $|S_{11}| < -10$ dB and improve ports isolation for some specific frequencies, while it exceeds 10 dB for whole the bandwidth. Therefore, isolation between the ports is optimized to target the higher values around the frequency of 2.4 GHz and bandwidth of 5–6 GHz that are used by WLAN.

Since the ports are symmetric, $S_{21} = S_{41} = S_{32} = S_{43}$, and $S_{31} = S_{42}$, only S_{21}, and S_{31} is plotted. As shown in Figure 13, as the ground plane angle (θ) is increased, the magnitudes of the S_{21} and S_{31} shift to higher frequencies.

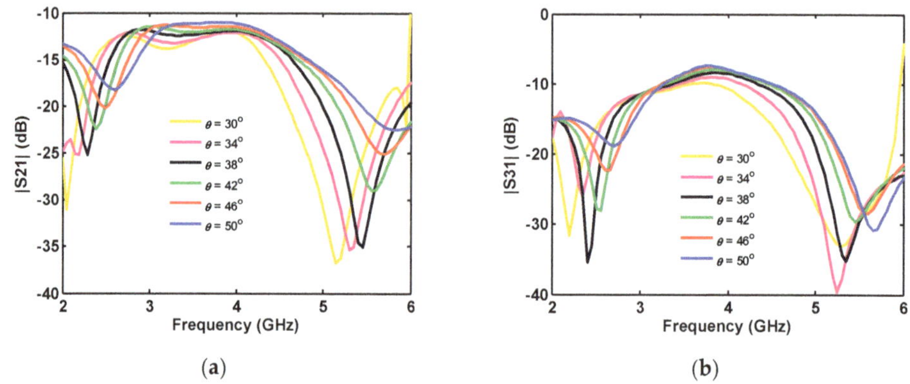

Figure 13. Effect of ground plane angle θ on the simulated isolation (a) S_{21} and (b) S_{31} between different ports of four-port antenna.

The increasing θ causes an increase in the ground plane size, and, as a result, the length of the connected part between ground planes of the ports is decreased. Moreover, increasing the height of the ground plane (h_p) increases the length of the connected part and shifts S_{21} and S_{31} to lower frequencies (see Figure 14).

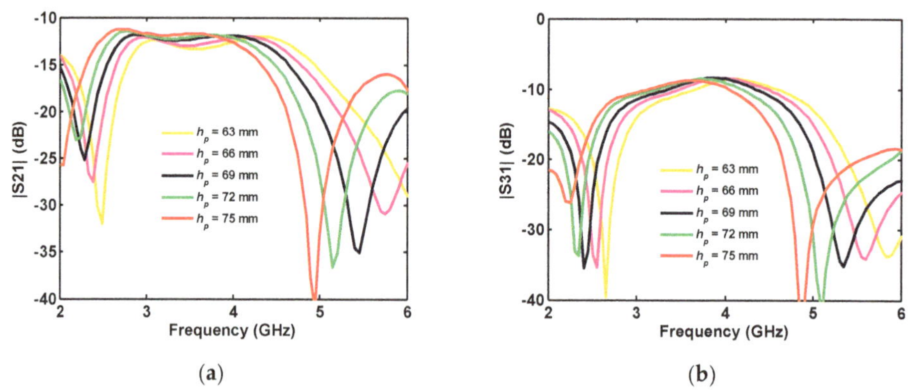

Figure 14. Effect of the ground planes height on the different ports' isolation (**a**) S_{21} and (**b**) S_{31} in a four-port antenna.

When the thickness of the connected part (C_t) is increased, as in Figure 11b, since the lower side of the connection part is limited/fixed by the triangular-shaped plane (h_p), the upper edge is extended toward the disk. Therefore, by increasing the thickness of the connected part (C_t), its average length is decreased, shifting S_{21} and S_{31} to higher frequencies (see Figure 15).

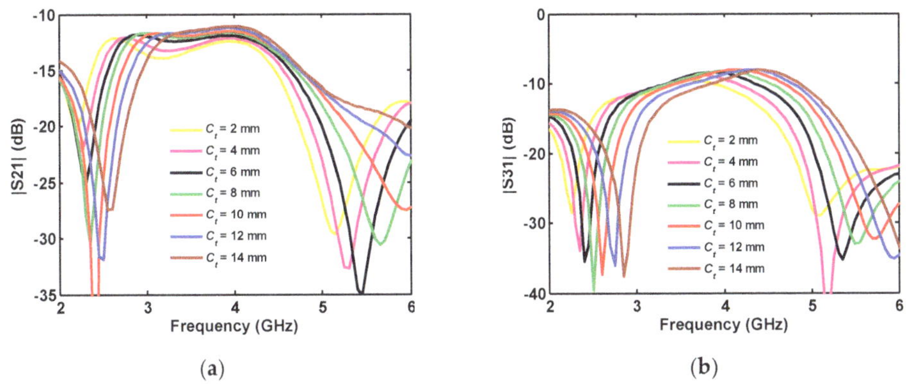

Figure 15. Influence of the ground joint thickness on the different ports simulated isolation (a) S_{21} and (b) S_{31} in four-port antenna.

3. Measurements and Simulations

The antennas were simulated, fabricated, and measured using the geometrical parameters in Table 1. The antennas were characterized for S parameters and 3D cross-polar and co-polar gain at some specific frequencies. The correlation coefficient between ports i and j of N-port antennas is calculated using the simulated and measured S parameters using (1). Equation (1) is an approximation to calculate the pattern independence between the ports using the S parameter. Its precision is increased as the radiation efficiency of the antenna is increased [23]

$$\rho_e(i,j,N) = \frac{|C(i,j,N)|^2}{\prod_{k=i,j}[1 - C(i,j,N)]}, \quad C(i,j,N) = \sum_{n=1}^{N} S^*_{i,n} S_{n,k} \tag{1}$$

The measurements were performed at Sabanci University Anechoic Chamber that is suitable for the frequency range from 700 MHz to 50 GHz and is equipped with a PNA5245A vector network analyzer (working up to 50 GHz).

3.1. Reflection Coefficient

Simulated and measured reflection coefficient results from 1 to 6 GHz of the single-port antenna are shown in Figure 16. The measured $|S_{11}|$ has some shifts for frequencies greater than 4 GHz. This shift can result from the PCB dielectric constant variation in different frequencies or the effect of the measurement setup. Although the antennas are measured inside the anechoic chamber, due to their isotropic radiation pattern (which is discussed in the next section), the absorbers, feeding cable, and setup in close distance to the antenna can affect its performance. The measurement shows that the antenna reflection coefficient is below −10 dB for the whole 1–6 GHz frequency band.

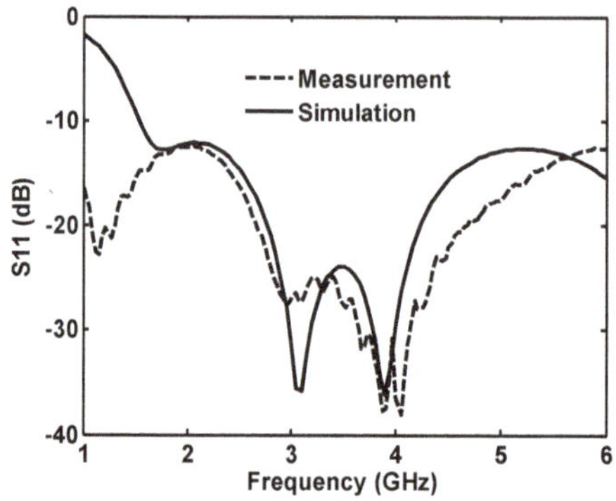

Figure 16. Measured and simulated reflection coefficient for the single-port antenna.

The measured and simulated S parameters and calculated correlation coefficient using (1) for the two-port antenna are shown in Figure 17. Some frequency shift around 200 MHz is also seen in the measured S parameters. The measured results comply with the simulated ones and the antenna is matched to 50 Ω for frequencies greater than 1.1 GHz. Isolation between the ports is better than 15 dB for higher frequencies (>2.3 GHz). After a frequency of 2.14 GHz, a correlation coefficient of better than −30 dB is obtained from the measured/calculated ρ_{21}.

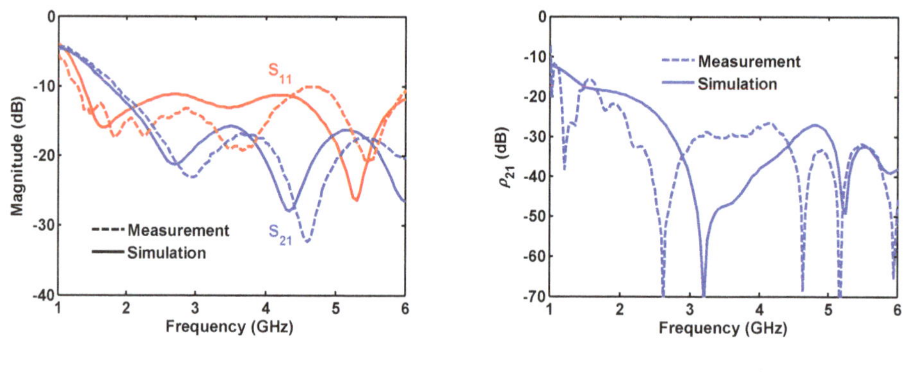

(a) (b)

Figure 17. Measured and simulated (**a**) S-parameters and (**b**) correlation coefficient for two-port antenna.

The four-port antenna is characterized for S_{11}, S_{21}, and S_{31}, as shown in Figure 18. The measured and simulated results agree well, and only some frequency shift is seen in S_{21}. As mentioned for the two-port antenna, the frequency shifts between simulation and measurement can be the effect of setup (such as cables) that reflect back the radiated field from the antenna and change the antenna performance. When cables are used for measuring two close ports such as S_{21}, their influence is more pronounced than for the other ports such as S_{31}. In addition, as the radiating disk is surrounded by the metal ground plane (the number of ports is increased), the antenna radiation on the ground plane direction is reduced. Thus, the effect of any cables, which are extended in the same plane as the ground plane, is decreased by increasing the number of ports.

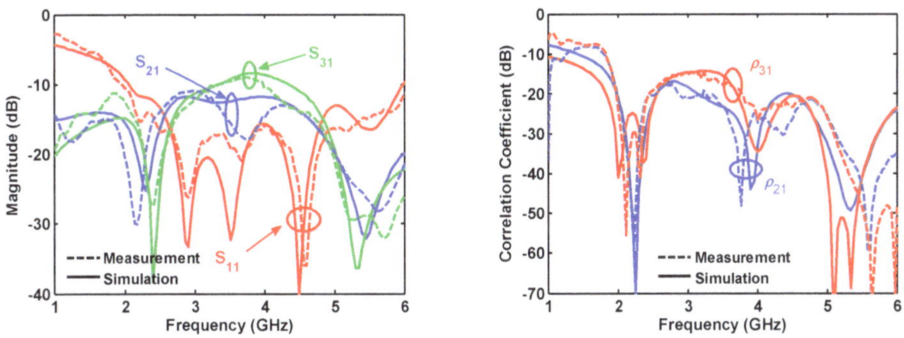

(a) (b)

Figure 18. Measured and simulated (**a**) S-parameters and (**b**) correlation coefficient for four-port antenna.

The measurements show that the S parameters are below −10 dB over the desired frequency band of 2–6 GHz. Isolation between the ports is better than 20 dB at around 2.4 GHz and between 5 and 6 GHz. In addition to the S parameters, good agreement between the measured and simulated correlation coefficient is also obtained (see Figure 18b).

For MIMO applications, although there are no specific requirements on isolation values, lower values of isolation will ease the work done by the baseband processor. Two- and four-port antennas can operate at frequencies below 6 GHz with good isolation values. For the two-port antenna, isolation is lower than 15 dB in the overall band mainly around 20 dB, which may be sufficient for MIMO applications. For the four-port antenna, isolation is lower than 10 dB in the overall band, more than 20 dB around 2.4 and 5–6 GHz, and more than 10 dB in the whole band. Further, for a 5G MIMO system, the four-port antenna can also be used for some portions of the bandwidth. The least isolation is 10 dB; however, the 5G sub-6 GHz system will not use the whole 4 GHz available, but a few hundred MHz bandwidth from the spectrum. When we consider a realizable 5G massive MIMO system with a few hundred MHz bandwidth operations, the four-port antenna may achieve more than 20 dB isolation, which may be sufficient for a MIMO system such as in 5–6 GHz band.

There is also interest in how these antennas will perform in a real environment. Most of the time, measured results are performed in a controlled environment such as an anechoic chamber. When these antennas are placed in a real environment, the isolation, as well as the return loss of the antenna, may change. However, most of these isolation and reflections are due to the antenna itself, the so-called self-interference signal. Isolation and reflection will not degrade significantly if an object is not placed in the vicinity of the antenna. Specifically, for the 5G 3 GHz frequency, the free space path loss around 3 GHz at 1 m is 42 dB, if an object is placed at 1 m from the antenna. The two-way path loss will be 84 dB lower, which will worsen the isolation and reflection. However, it will not be that significant if the isolation is around 20–30 dB range. If a very close object is placed by the antenna, this may cause a few dB change in the isolation; however, if the objects are placed far away from the antennas, similar performance should be expected. As measured in [24], the performance of high isolation antennas in a real environment will definitely change, but the normal operation of the antenna will remain stable.

3.2. 2D and 3D Gains

The 3D gain of antennas was measured at frequencies of 2, 2.4, 3.4, 4.4, 5.2, and 5.8 GHz. The origin of the Cartesian coordinate system, which describes the gain of antennas, is the disk center and is shown in Figures 2, 6 and 11 for one-, two-, and four-port antennas, respectively. In all systems, the feeding port is on the z-axis (Port 1), and the x-axis is perpendicular to the disk. Since the radiation pattern of the antennas is similar to a dipole antenna, the antennas' co- and cross-poles are indicated by

G_θ and G_ϕ components. Therefore, it is expected that the antennas co-pole gain (G_θ) will significantly dominate the cross-pole gain (G_ϕ). Furthermore, the one- and four-port antennas are symmetric with respect to the xz-plane; thus, the patterns are measured only on the hemisphere on y > 0 space.

The simulated and measured 2D gains for the one- (Figure 19), two- (Figure 20), and four-port (Figure 21) antennas were obtained at different frequencies (2.4, 4.4, and 5.8 GHz) and on xz-, yz-, and xy-planes. As shown in Figures 19–21, the measured (dashed lines) and simulated (solid lines) gains in the θ direction (G_θ in the red curve) dominate the gains in the ϕ direction (G_ϕ in the blue curve). The simulated gains for G_ϕ are smaller than the measured values, which could be due to the antennas' imprecise alignment in the measurement setup or the AUT (antenna under test) tilt during the measurement. When the antenna gain at one pole is much smaller than the other pole, some tilt in AUT can cause a significant increase in the value of the cross-pole. The effect of the feed cable is seen in the measured G_θ gain at around $\theta = 180°$ and on the xz- and yz-planes. The radiation patterns show an isotropic antenna characteristic on the xy-plane. The electric field component of G_θ gain on the xz-plane is perpendicular to disk at $\theta = 0°$ and 180° and results in null at these angles. Moreover, the fact that the antennas radiation at $\theta = 0°$ and 180° is lower (smaller gain) than at other angles shows the antennas behave very similar to a dipole antenna.

The 3D gain of the antenna at 4.4 GHz for total gain, θ polarization, and ϕ polarization are shown in Figure 22. As discussed, the level of G_θ is higher than G_ϕ for all of the antennas. For all ports of the antenna, nulls are seen around the antenna feeding ports. One can note that the antenna almost radiates in the available space, making an ideal antenna for MIMO wireless systems.

3.3. Gain and Efficiency versus Frequency

The gains of the antennas are measured in some specific frequencies including 2, 2.4, 3.4, 4.4, 5.2, and 5.8 GHz, as shown in Figure 23a. The agreement between the measured and simulated gains of the antennas is decreased as the number of the ports is reduced. Since the single-port antenna radiates in all directions, the absorbers near the antenna in the anechoic chamber change the antenna performance and specifically in the lower frequencies. Note that the ground plane around the two- and four-port antennas reduces the antenna radiation on the antenna plane (xz-plane), in the direction that the near absorbers to the AUT are positioned. Consequently, the destructive effect of these absorbers, near the two- and four-port antennas, are partially canceled and the measurements get closer to simulated results.

The antennas' measured gain varies from 3.08 dBi at 3.4 GHz for a single port to 6.74 dBi at 5.8 GHz for four-port antennas.

The radiation efficiency of the antennas is calculated using the measured (with 2° angular spacing) average 3D gain technique. The calculated efficiency plot is presented in Figure 23b. Due to the big structure of the antennas, the accuracy of the simulated radiation efficiencies is low. As the frequency is increased, each antennas' radiation efficiency is increased. The values change from 50% to 100 %.

Table 2 presents the comparison of proposed antennas with the reported sub-6 GHz MIMO antennas in [25–30]. The presented two- and four-port antennas in this work outperform previously published sub-6 GHz antennas for 5G applications in [25–30] with the larger bandwidth and single radiating element, for which the antenna size, gain, efficiency, and performance in MIMO applications are comparable to existing studies where multiple radiating elements are used.

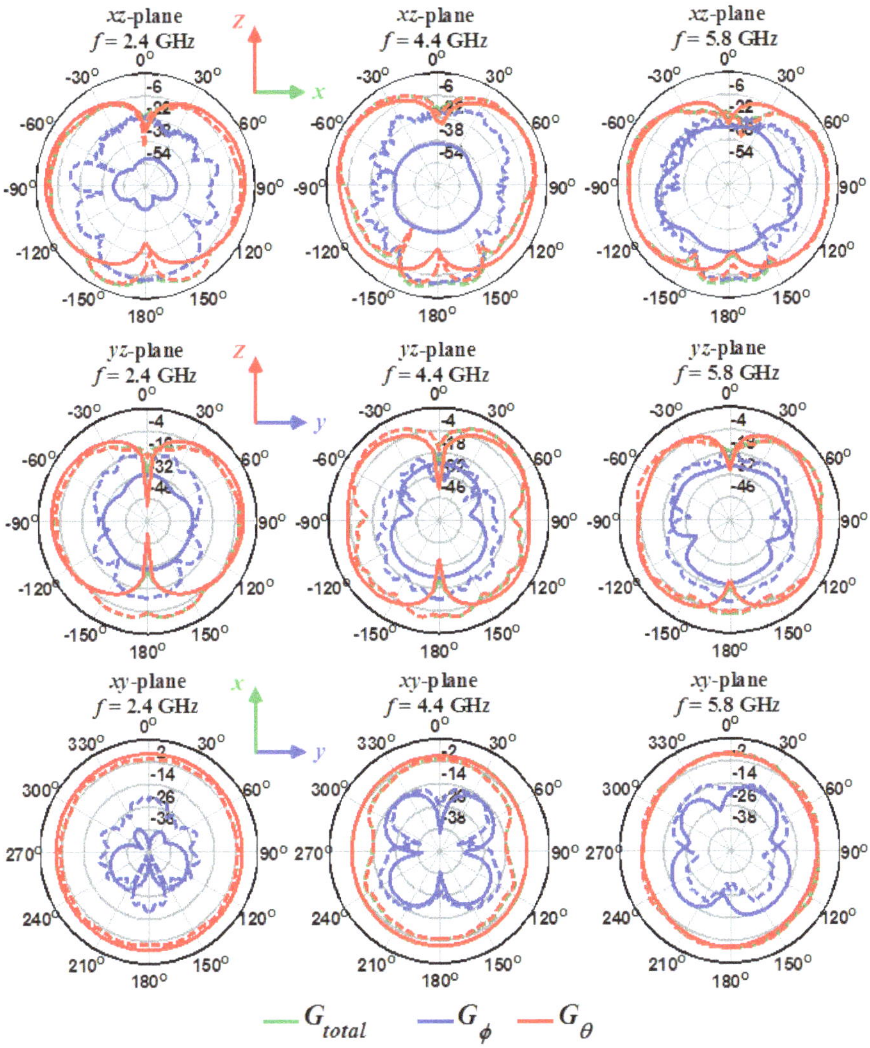

Figure 19. Measured (dashed lines) and simulated (solid lines) gain for the single-port antenna on three different planes and frequencies.

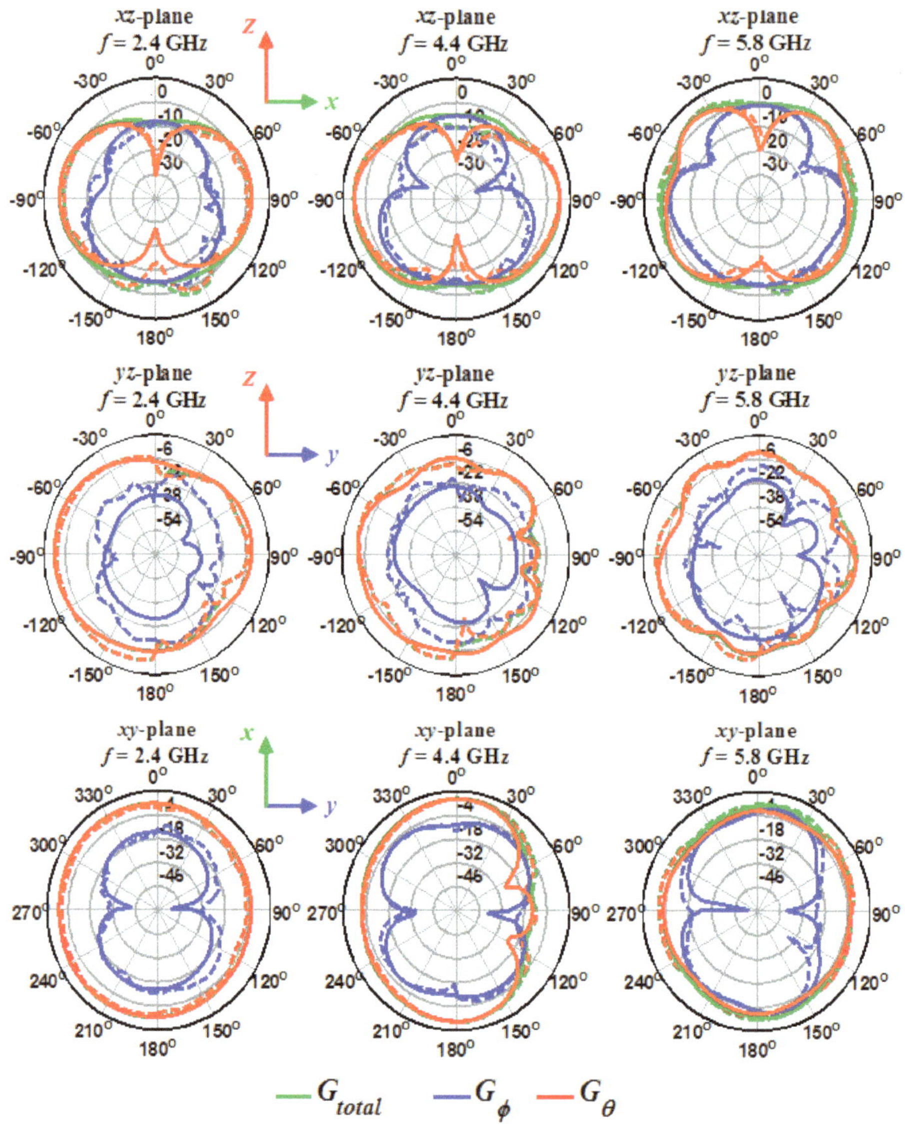

Figure 20. Measured (dashed lines) and simulated (solid lines) gain for the two-port antenna on three different planes and frequencies.

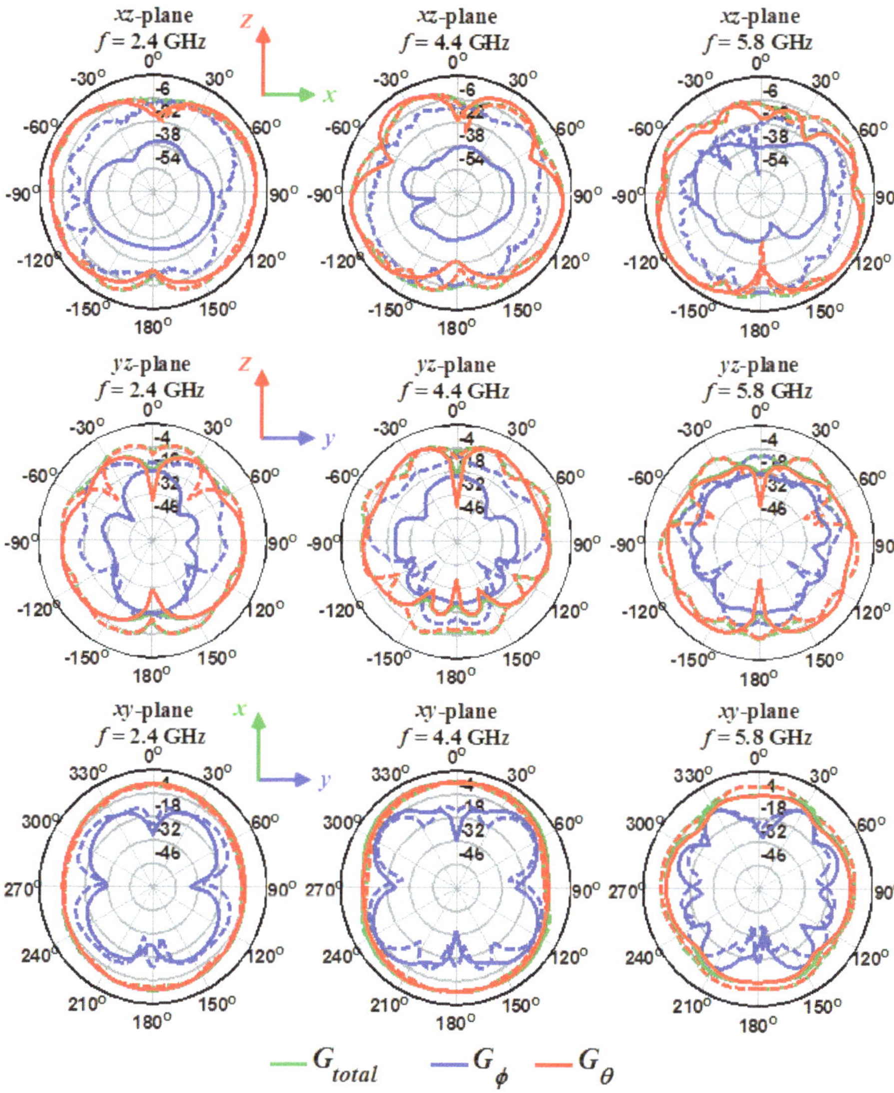

Figure 21. Measured (dashed lines) and simulated (solid lines) gain for the four-port antenna on three different planes and frequencies.

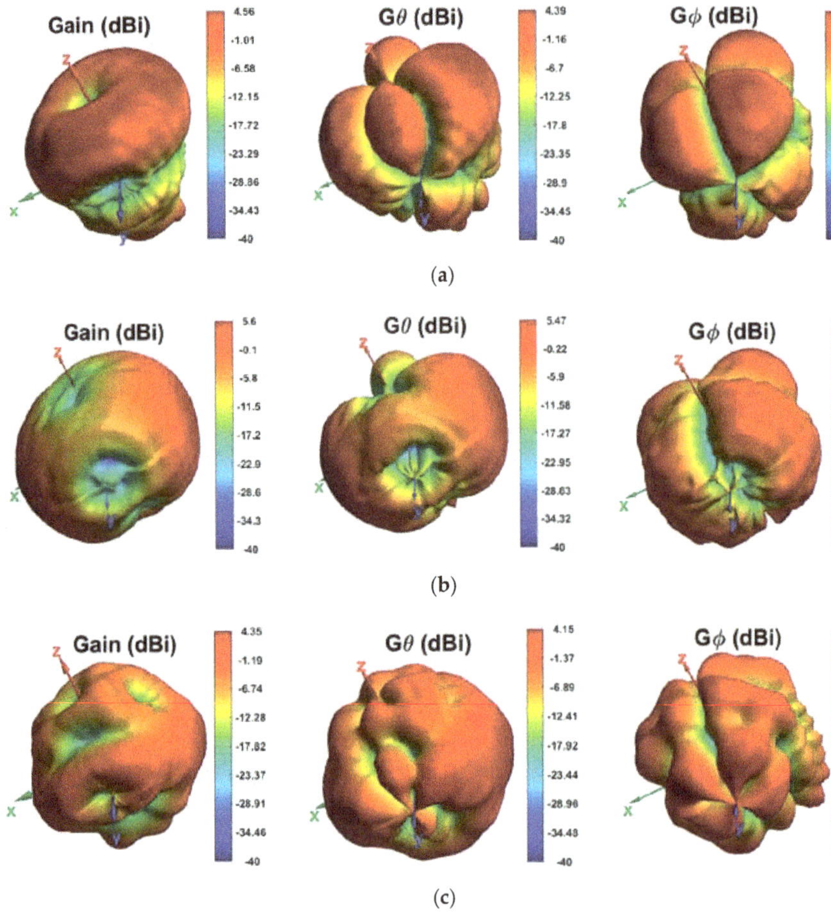

(a)

(b)

(c)

Figure 22. Measured 3D total, θ-polarized and ϕ-polarized gains at 4.4 GHz for: (**a**) single-port antenna; (**b**) two-port antenna; and (**c**) four-port antenna.

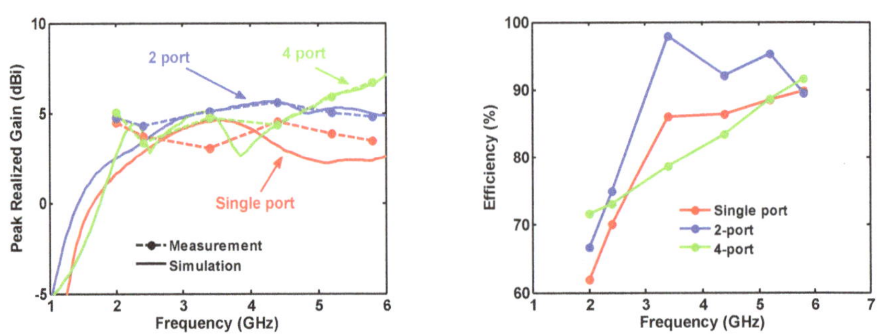

(a) (b)

Figure 23. (**a**) Measured and simulated peak gain; and (**b**) measured antenna efficiency values.

Table 2. Comparison of proposed antennas in this work and antennas in [25–30].

Ref.	Port (#)	Radiating Element	Size (cm)	Freq. (GHz)	BW Def.	BW (MHz)	Isolation (dB) >	ECC	Gain (dBi)
[25]	4	Inverted-F antennas	8.8 × 8.8	2.06–2.16	10 dB	100	22 Meas.	0.05	4.34
[25]	6	Inverted-F antennas	16 × 14	2.2–2.3	10 dB	100	24 Meas.	0.01	4.29
[26]	8	Fork-shaped dipoles, L-Shaped	13 × 10	3.6–3.83 5.15–5.925	10 dB	230 775	15 Sim.	0.070.03	2.6–3.1 2.5–4.2
[27]	4	F-shaped elements	15 × 7.5 × 0.4	3.3–4.2 4.4–5.0 5.15–5.85	8 dB	900 600 700	14 Meas.	0.05	4.39 3.66 4.62
[21]	8	Rectangular monopoles	13.6 × 6.8 × 0.62	3.6–4.2 4.4–4.9 5.15–5.925	6 dB	600 500 775	10 Meas.	0.1	-
[28]	4	Inverted L monopoles	4 × 4	2.7–4.94	10 dB	2240	11 Meas.	0.1	4
[29]	4	Elliptical ring slot antennas	12.7 × 7	3.4–3.8	10 dB	400	20 Sim.	0.01	1
[30]	2	Rhombus -shaped antennas	2 × 3.5	3.34–3.87	10 dB	530	15 Meas.	0.01	2.34
[30]	12	Rhombus -shaped antenna	20.6 × 11.5	3.4–3.8	10 dB	400	15 Sim.	-	3.2
This work	2	Single Disk	7.6 × 7.6	2.3–6.0	10 dB	3700	15 Meas.	0.01	5
This work	4	Single Disk	12.4 × 12.4	2.0–6.0	10 dB	4000	10 Meas.	0.03	3.1–6.75

Abbreviations: Ref., References; BW, Bandwidth; BW Def., Bandwidth Definition; ECC, Envelope Coefficient Coefficient.

4. Conclusions

A wideband single-port antenna, with λ/4 diameter of the radiating disk (monopole-like) and dipole-like radiation pattern, was designed and manufactured. The geometry of the single-port antenna was utilized as the prototype for the two- and four-port antennas, by rotating (90°) the single-port geometry with respect to the disk center and adding a new port. The ground planes of the ports (other than the single port) were connected to improve/increase the isolation between ports. The measured and simulated data are in good agreement. The acceptable correlation coefficient during the bandwidth makes the antenna suitable for the MIMO application for the 5G NR sub-6 GHz band. Finally, the design challenge of two or four separate antennas being near each other and any potential coupling between them can be solved by these monolithic compact antennas that contain good matching, proper isolation between the ports and omnidirectional-like radiation pattern. The antennas are not only very compatible, but their reflection coefficient/isolation between the ports can be further improved, to achieve even better values (for a limited bandwidth or single operation between 2 and 6 GHz), by optimizing the dimensions of the introduced parameters (assuming prior knowledge of their influence on Sii and Sij).

Author Contributions: Conceptualization, I.T., and M.S.; funding acquisition, I.T; software, M.S.; investigation, M.S.; writing—original draft preparation, M.S. and A.U.; writing—review and editing, I.T and A.K.S.; and project administration, I.T. All authors have read and agreed to the published version of the manuscript.

Funding: This work was supported in part by The Scientific and Technological Research Council of Turkey (TUBITAK) under Grant 114E494.

Conflicts of Interest: The authors declare no conflict of interest. The funders had no role in the design of the study; in the collection, analyses, or interpretation of data; in the writing of the manuscript, or in the decision to publish the results.

References

1. Sampath, H.; Talwar, S.; Tellado, J.; Erceg, V.; Paulraj, A. A Fourth Generation MIMO-OFDM Broadband Wireless System: Design, Performance, and Field Trial Results. *IEEE Commun. Mag.* **2002**, *40*, 143–149. [CrossRef]
2. Manteuffel, D.; Martens, R. Compact Multimode Multielement Antenna for Indoor UWB Massive MIMO. *IEEE Trans. Antennas Propag.* **2016**, *64*, 2689–2697. [CrossRef]
3. Lee, H.; Lee, B. Compact Broadband Dual-Polarized Antenna for Indoor MIMO Wireless Communication Systems. *IEEE Trans. Antennas Propag.* **2016**, *64*, 766–770. [CrossRef]
4. Bekasiewicz, A.; Koziel, S.; Dhaene, T. Optimization-Driven Design of Compact UWB MIMO Antenna. In Proceedings of the 10th European Conference on Antennas and Propagation (EuCAP 2016), Davos, Switzerland, 1–15 April 2016; pp. 1–4.
5. Sharawi, M.S.; Faouri, Y.S.; Iqbal, S.S. Design and Fabrication of a Dual Electrically Small MIMO Antenna System for 4G Terminals. In Proceedings of the German Microwave Conference, Darmstadt, Germany, 14–16 March 2011.
6. Chung, K.; Yoon, J.H. Integrated MIMO Antenna with High Isolation Characteristic. *Electron. Lett.* **2007**, *43*, 199–201. [CrossRef]
7. Wu, T.-Y.; Fang, S.-T.; Wong, K.-L. Printed Diversity Monopole Antenna for WLAN Operation. *Electron. Lett.* **2002**, *38*, 1625–1626. [CrossRef]
8. Karaboikis, M.; Soras, C.; Tsachtsiris, G.; Makios, V. Compact Dual-Printed Inverted-F Antenna Diversity Systems for Portable Wireless Devices. *IEEE Antennas Wirel. Propag. Lett.* **2004**, *3*, 9–14. [CrossRef]
9. Gao, Y.; Chiau, C.C.; Chen, X.; Parini, C.G. Modified PIFA and its Array for MIMO Terminals. *IEE Proc. Microw. Antennas Propag.* **2005**, *152*, 255–259. [CrossRef]
10. Chung, J.-Y.; Yang, T.; Lee, J.; Jeong, J. Low Correlation MIMO Antenna for LTE 700MHz Band. In Proceedings of the IEEE International Symposium on Antennas and Propagation (APSURSI), Washington, DC, USA, 3–8 July 2011; pp. 2202–2204.
11. Chaloupka, H.J.; Wang, X. Novel Approach for Diversity and MIMO Antennas at Small Mobile Platforms. In Proceedings of the 15th IEEE International Symposium on Personal, Indoor and Mobile Radio Communications (PIMRC 2004), Barcelona, Spain, 5–8 September 2004; pp. 637–642.
12. Kulkarni, A.N.; Sharma, S.K. A Compact Multiband Antenna with MIMO Implementation for USB Size 4G LTE Wireless Devices. In Proceedings of the IEEE International Symposium on Antennas and Propagation (APSURSI), Washinton, DC, USA, 3–8 July 2011; pp. 2215–2218.
13. Kiem, N.K.; Dinh, D.N.; Viet, H.T.; Dao-Ngoc, C. A Novel Design of Dual-Feed Single-Element Antenna for 4G MIMO Terminals. In Proceedings of the Progress in Electromagnetics Research Symposium, Kuala Lumpur, Malaysia, 27–30 March 2012; pp. 1827–1831.
14. Skycross, Inc. iMAT Antenna Whitepaper. 2008. Available online: http://www.skycross.com/ (accessed on 19 February 2017).
15. Caimi, F.M.; Mongomery, M. Dual Feed, Single Element Antenna for WiMAX MIMO Application. *Int. J. Antennas Propag.* **2008**, *2008*, 1–5. [CrossRef]
16. Rajo-Iglesias, E.; Quevedo-Teruel, Ó.; Sanchez-Fernández, M. Compact multimode patch antennas for MIMO applications. *IEEE Antennas Propag. Mag.* **2008**, *50*, 197–205.
17. Vaughan, R.; Andersen, J. A Multiport Patch Antenna for Mobile Communications. In Proceedings of the 14th European Microwave Conference (EuMA 1984), Liege, Belgium, 10–13 September 1984; pp. 607–612.
18. Svantesson, T. Correlation and Channel Capacity of MIMO Systems Employing Multimode Antennas. *IEEE Trans. Veh. Tech.* **2002**, *51*, 1304–1312. [CrossRef]
19. Vaughan, R. Two Port Higher Mode Circular Microstrip Antennas. *IEEE Trans. Antennas Propag.* **1981**, *36*, 309–321. [CrossRef]
20. Waldschmidt, C.; Wiesbeck, W. Compact Wide-Band Multimode Antennas for MIMO and Diversity. *IEEE Trans. Antennas Propag.* **2004**, *52*, 1963–1969. [CrossRef]
21. Sim, C.; Liu, H.; Huang, C. Wideband MIMO Antenna Array Design for Future Mobile Devices Operating in the 5G NR Frequency Bands n77/n78/n79 and LTE Band 46. *IEEE Antennas Wirel. Propag. Lett.* **2020**, *19*, 74–78. [CrossRef]

22. Agrawall, N.P.; Kumar, G.; Ray, K.P. Wide-Band Planar Monopole Antenna. *IEEE Trans. Antennas Propag.* **1998**, *46*, 294–295. [CrossRef]
23. Thaysen, J.; Jakobsen, K.B. Envelope Correlation in (N, N) MIMO Antenna Array from Scattering Parameters. *Microw. Opt. Technol. Lett.* **2006**, *48*, 832–834. [CrossRef]
24. Nawaz, H.; Tekin, I. Dual-Polarized, Differential Fed Microstrip Patch Antennas with Very High Interport Isolation for Full-Duplex Communication. *IEEE Trans. Antennas Propag.* **2017**, *65*, 7355–7360. [CrossRef]
25. Kabiri, Y.; Borja, A.L.; Kelly, J.R.; Xiao, P. A Technique for MIMO Antenna Design with Flexible Element Number and Pattern Diversity. *IEEE Access* **2019**, *7*, 86157–86167. [CrossRef]
26. Li, Y.; Zou, H.; Wang, M.; Peng, M.; Yang, G. Eight-element MIMO antenna array for 5G/Sub-6GHz indoor micro wireless access points. In Proceedings of the International Workshop on Antenna Technology (iWAT), Nanjing, China, 5–7 March 2018; pp. 1–4.
27. Ren, Z.; Wu, S.; Zhao, A. Triple Band MIMO Antenna System for 5G Mobile Terminals. In Proceedings of the International Workshop on Antenna Technology (iWAT), Miami, FL, USA, 3 March 2019; pp. 163–165.
28. Sarkar, D.; Srivastava, K.V. A Compact Four-Element MIMO/Diversity Antenna with Enhanced Bandwidth. *IEEE Antennas Wirel. Propag. Lett.* **2017**, *16*, 2469–2472. [CrossRef]
29. Chakraborty, S.; Rahman, M.A.; Hossain, M.A.; Toaha, A.; Nishiyama, E.; Toyoda, I. A 4-element MIMO antenna with orthogonal circular polarization for sub-6 GHz 5G cellular applications. *Sn Appl. Sci.* **2020**, *2*, 1180. [CrossRef]
30. Saurabh, A.K.; Meshram, M.K. Compact sub-6 GHz 5G-multiple-input-multiple-output antenna system with enhanced isolation. *Int. J. RF Microw. Comput. Aided Eng.* **2020**, *30*, e22246. [CrossRef]

Publisher's Note: MDPI stays neutral with regard to jurisdictional claims in published maps and institutional affiliations.

sensors

MDPI

Article

On the Application of K-User MIMO for 6G Enhanced Mobile Broadband †

Anil Kumar Yerrapragada * and Brian Kelley

Department of Electrical and Computer Engineering, University of Texas at San Antonio, San Antonio, TX 78249, USA; dr.brian.kelley@gmail.com

* Correspondence: anilkumar.yerrapragada@utsa.edu

† This paper is an extended version of our paper published in: Yerrapragada, A.K.; Kelley, B. Very High Throughput Internet of Things Networks with K Access Points and K Devices. In proceedings of the MILCOM 2019—2019 IEEE Military Communications Conference (MILCOM), Norfolk, VA, USA, 12–14 November 2019.

Received: 4 October 2020; Accepted: 1 November 2020; Published: 2 November 2020

Abstract: This paper presents a high-throughput wireless access framework for future 6G networks. This framework, known as K-User MIMO, facilitates all-to-all communication between K access points and K mobile devices. For such a network, we illustrate the demodulation of K^2 independent data streams through a new interference cancellation beamforming algorithm that improves spectral efficiency compared to massive MIMO. The paper derives a multi-user Shannon Capacity formula for K-User MIMO when K is greater than or equal to 3. We define an Orthogonal Frequency Division Multiplexing (OFDM) frame structure that demonstrates the allocation of time-frequency resources to pilot signals for channel estimation. The capacity formula is then refined to include realistic pilot overheads. We determine a practical upper-bound for MIMO array sizes that balances estimation overhead and throughput. Lastly, simulation results show the practical capacity in small cell geometries under Rayleigh Fading conditions, with both perfect and realistic channel estimation.

Keywords: beyond-5G; 6G; MIMO; interference alignment; K-User MIMO; OFDM

---

1. Introduction

The goal of 5G is to enable a fully connected society such that instant information is available just a touch away. 5G achieves this through three key paradigms viz., enhanced Mobile Broadband (eMBB) for gigabit data rates, Ultra Reliable and Low Latency Communications (URLLC) for latency less than 1ms, and massive Machine-Type Communications (mMTC) for 1 million connected devices/sq. km [1,2]. These paradigms are supported by a multilayer technology strategy including small cell architectures [3,4], millimeter wave communication, and massive MIMO. Millimeter wave systems facilitate communication in the high Radio Frequency (RF) bands using analog, digital, and hybrid beamforming [5,6]. Massive MIMO deploys large antenna arrays at base stations and operates in the low to mid RF bands [7–10]. To support the ubiquitous deployment of densely connected networks, this paper investigates an alternative MIMO technology, in the microwave realm, for beyond-5G or 6G networks. This technology, known as K-User MIMO, has the potential to achieve very high throughput compared to 5G Massive MIMO.

Background and Prior Research on K-User MIMO

K-User MIMO is an architecture in which there are *K* access points and *K* mobile devices, each equipped with multiple antennas, i.e., spatial dimensions. In the simplest form of *K*-User MIMO, each Access Point connects to one of the *K* mobile devices. In this paper, we consider a different form of *K*-User MIMO, known as *K*-User MIMO X, in which each of the *K* mobile devices receives signals from

K access points. This is shown in Figure 1. The all-to-all architecture achieves very high throughput whilst supporting flexibility in achieving diversity. Each of the *K* access points could send redundant information streams to maximize reliability. Alternatively, they could send unique information streams in order to maximize capacity. *K*-User MIMO X can switch between these two modes without any change in the mathematics of the algorithm. Further scenarios can be envisioned in which *K*-User MIMO X allows for adaptive allocation of power to users with more favorable Signal-to-Noise Ratios. Additional encoding of data across time and frequency could be applied so the signals could adapt to malicious behavior such as jamming and eavesdropping.

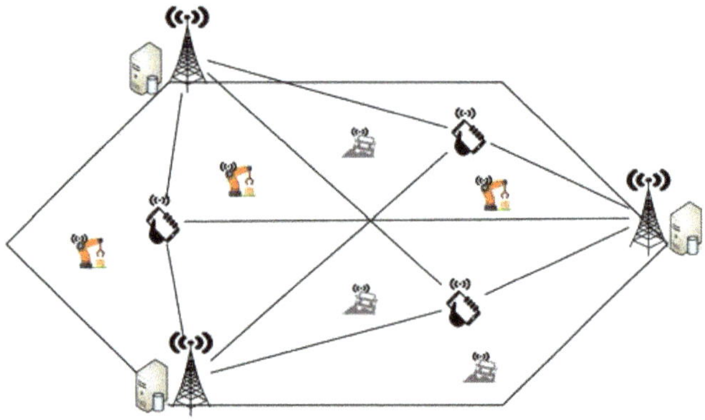

Figure 1. *K*-User MIMO X network Example for $K = 3$.

In any form of *K*-User MIMO, each mobile device receives both its desired signals and interference signals (signals meant for other mobile devices). To manage interference, *K*-User MIMO systems are often studied in the context of Interference Alignment (IA). IA is a technique that aligns interfering signal vectors in order to maximize interference-free space at each mobile device [11]. By applying suitable channel dependent precoders to the transmit signals, and beamformers to the receive signals, several interfering users can communicate simultaneously. Alignment helps confine the interference at each mobile device to a smaller dimensional subspace while projecting the desired signals into the null space of the interference.

Several works have analyzed IA on a theoretical level. A typical metric used to characterize IA is known as Degrees of Freedom (DoF). DoF is defined as the number of spatial dimensions that are free from interference [12]. The authors in [13] have provided examples showing acheivability of IA and various DoF in *K*-User interference networks with different antenna configurations. In [14], an iterative algorithm for obtaining the precoders and beamformers is presented for a Time Division Duplex (TDD) mode of operation. The precoders in this method are a function of the dual relationship between the MIMO forward and reverse channels. Another IA framework involving TDD channels is presented in [15]. In [16], interference alignment in MIMO downlink networks is investigated, where precoders are derived by eigen decomposition of the MIMO channels.

Another IA scheme for a *K*-User MIMO X network is proposed in [17]. By appropriately precoding the transmit signals, this scheme maximizes the interference-free space by limiting the interference at every mobile device to half of the received signal space. Further, by applying a zero forcing beamformer which is a function of interfering channels and precoders, interference cancellation has been achieved for $K = 3$. The algorithms in [17] are purely theoretical and in this paper, we improve upon them.

Discussion on how to demodulate symbols is not provided in [17]. Further, in order to maximize capacity, we wish to operate the *K*-User MIMO X system such that each access point is transmitting different symbols to each mobile device, whilst all being on the same frequency subcarrier. We have investigated the case where each access point is on a different subcarrier in [18] and found that the bandwidth and therefore the capacity of the K-User system is reduced by a factor of K, which is significant. This paper presents a new signal separation beamformer to regain the lost factor of K.

Interference Alignment uses precoders and beamformers that are channel dependent. Naturally, channel estimation is critical to IA [19,20]. Both [17] and our previous works [18,21,22] either consider perfect channel state information or do not consider exact channel estimation error models. Neither considers the overheads arising from transmitting pilot symbols for estimation. Park and Ko [17] assumes perfect channel state information. Yerrapragada and Kelley [18] does not explicitly estimate the channel but assumes a Cramér–Rao variance for the estimation error, which is only a lower-bound on the error. In [21], the channel is not estimated but the effects of imperfect estimation are simulated. Residual interference due to imperfect precoders and beamformers is modeled as a random variable and an expression for its distribution is provided. Yerrapragada and Kelley [22] is the first paper that introduces signal separation concepts for K-User MIMO, but it too assumes perfect channel estimation.

In this paper, we extend our previous work to provide realistic and practical capacity results for K-User MIMO X systems that account for channel estimation overheads.

2. System Model for Cellular-Based K-User MIMO

This section describes the system model for K-User MIMO X that shows transmit precoding, receive interference cancellation, and signal separation. Table 1 shows the parameters used in the key equations and derivations. Figure 2 shows the overall K-User MIMO protocol steps. Lastly, an Orthogonal Frequency Division Multiplexing (OFDM) multiple access protocol is shown. This protocol illustrates the pilot overhead resulting from serving several ($> K$) mobile devices.

Table 1. System model parameters for key derivations.

Parameter	Description	Size	For $K=3$
M	Minimum number of antennas at each access point and mobile device	$\geq K(K-1)$	≥ 6
N_D	Number of desired signals at each mobile device	K	3
N_I	Number of interference terms at each mobile device	$K(K-1)$	6
H_{ij}	Channel matrix between access point j and mobile device i	$M \times M$	6×6
v_{ij}	Precoder vector for signal between access point j and mobile device i	$M \times 1$	6×1
s_{ij}	Symbol to be transmitted between access point j and mobile device i	1×1	1×1
w_i	Additive White Gaussian Noise (AWGN) at mobile device i	$M \times 1$	6×1
P_i^I	Matrix of aligned interference column vectors at the mobile device i	$M \times \frac{N_I}{2}$	6×3
U_i	Zero forcing beamformer matrix at mobile device i	$M \times M - \frac{N_I}{2}$	6×3
P_{ij}^D	Matrix of desired column vectors at mobile device i to isolate signal from access point j	$M - \frac{N_I}{2} \times N_D - 1$	3×2
U_{ij}	Signal Separation beamformer matrix at mobile device i to isolate signal from access point j	$M - \frac{N_I}{2} \times M - \frac{N_I}{2} - (N_D - 1)$	3×1
N_s	Number of recovered copies of each symbol s_{ij}	$M - \frac{N_I}{2} - (N_D - 1)$	1

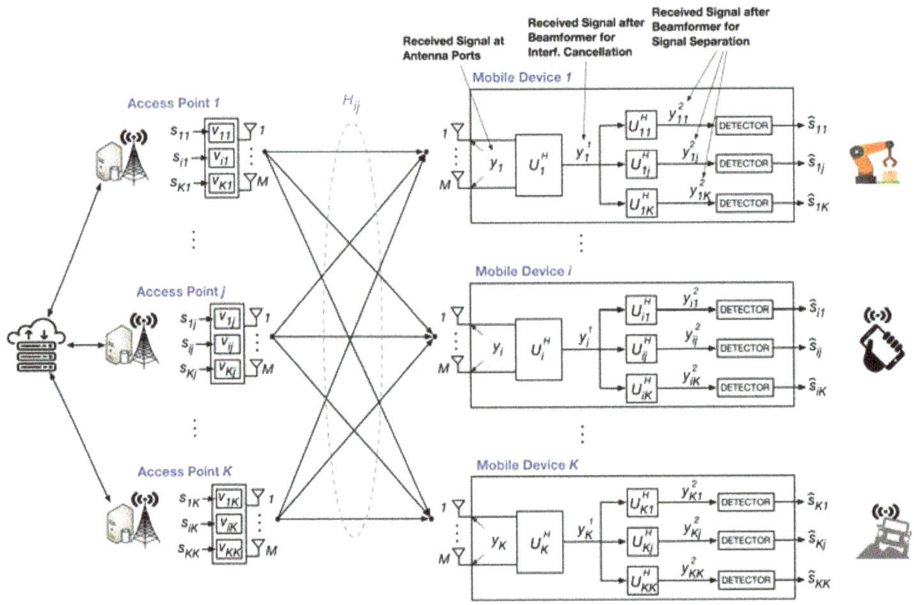

Figure 2. K-User MIMO X network showing the application of precoders at the access points as well as stage 1 and stage 2 beamformers (Equations (1)–(3)).

2.1. Received Signal at Antenna

Each of the K access points and mobile devices is equipped with M antennas. The all-to-all connectivity results in each mobile device receiving K desired signals and $K(K-1)$ interfering signals. This is shown in Figure 2 and Equation (1). Transmitted symbols s_{ij} between the jth access point and the ith mobile device are precoded by length M precoder vectors v_{ij} and transmitted over Rayleigh fading channels H_{ij}. Without loss of generality, we consider downlink transmissions. The received signal at the ith mobile device is given by

$$y_i = \underbrace{\sum_{j=1}^{K} H_{ij} v_{ij} s_{ij}}_{\text{Desired signal}} + \underbrace{\sum_{j=1}^{K} \sum_{k=1,\neq i}^{K} H_{ij} v_{kj} s_{kj}}_{\text{Total interference}} + \underbrace{w_i}_{\text{Noise}} . \tag{1}$$

This paper focuses on the maximum capacity scenario. Therefore it is assumed that each s_{ij} is unique. The precoders v_{ij} are channel dependent and are obtained by solving a system of alignment equations. The procedure for obtaining the precoder vectors is shown in Appendix A. The noise at the ith mobile device, w_i is assumed to be 0 mean Additive White Gaussian Noise (AWGN) with variance $\sigma_{w_i}^2 = E[w_i w_i^*]$, where $E[.]$ represents the expected value.

2.2. Received Signal after Beamformer for Interference Cancellation

A beamformer matrix U_i is applied to the received signal y_i to cancel the interference, as shown below:

$$y_i^1 = U_i^H \cdot y_i = U_i^H \sum_{j=1}^{K} H_{ij} v_{ij} s_{ij} + \underbrace{U_i^H \cdot \sum_{j=1}^{K} \sum_{k=1,\neq i}^{K} H_{ij} v_{kj} s_{kj}}_{I_\epsilon \approx 0} + U_i^H w_i, \tag{2}$$

where I_ϵ is the residual interference after cancellation. In the case of perfect channel state information, it is exactly equal to zero. The interference cancellation beamformer U_i is obtained by first generating a matrix, whose columns contain aligned interference vectors. The left-hand side of the Singular Value Decomposition (SVD) of this matrix contains the beamformer. The derivation of this beamformer is shown in Appendix A.

2.3. Received Signal after Beamformer for Desired Signals Separation

Symbols transmitted from access points on the same frequency subcarrier, add coherently at the mobile device. Under this model, symbol recovery is straightforward when each access point sends the same symbol to the ith mobile device. However, for maximizing capacity, each access point must be able to send different symbols to the ith mobile device. In such cases, a second beamformer operator is applied after interference cancellation, as shown in Equation (3). This second operator is applied K times at each mobile device (shown in Figure 1) and helps separate the signals sent from each of the K access points.

$$\begin{aligned} y_{ij}^2 = U_{ij}^H U_i^H y_i^1 &= U_{ij}^H U_i^H \sum_{j=1}^{K} H_{ij} v_{ij} s_{ij} + \underbrace{U_{ij}^H I_\epsilon}_{\approx 0} + U_{ij}^H U_i^H w_i \\ &= U_{ij}^H U_i^H H_{ij} v_{ij} s_{ij} + U_{ij}^H U_i^H w_i \xrightarrow{\text{To detector for symbol recovery}} \hat{s}_{ij}. \end{aligned} \tag{3}$$

The signal separation beamformer U_{ij} is obtained by first generating a matrix whose columns contain desired signal vectors from other access points that act as interference when recovering the symbol transmitted from a specific access point. The left-hand side of the SVD of this matrix contains the beamformer. The derivation of this beamformer is shown in Appendix A.

After separation, the desired signals from each of the K access points can be decoded by correcting for the effects of U_{ij}, U_i, H_{ij}, and v_{ij}. The decoding and detection process that recovers the symbols s_{ij} is shown in Appendix A.

2.4. K-User MIMO X Multiple Access Protocol

We now apply K-User MIMO X to a typical OFDM cellular scenario. Let us assume that there are nK mobile devices ($n \geq 1$) associating with K access points forming n K-User-Groups (UG). In a single UG, there are K^2M^2 Channel Impulse Responses (CIR) that need to be estimated. We propose that each of the MK transmit antennas sends pilot symbols on non-overlapping OFDM symbol times. In this paper, we assume the use of pilot signals that are configured for both synchronization as well as channel estimation. An example of such a sequence is the Zadoff–Chu sequence, commonly used in 4G Long Term Evolution (LTE). Akin to LTE pilot signals, when one transmit antenna is sending pilots, all other transmit antennas are off. This is shown in Figure 3. In the time domain, the channel needs to remain constant for at least MK symbol times. We leverage frequency domain resources to support the n User Groups. In the frequency domain, the available bandwidth B is divided into m sub-bands, where $m = \frac{B}{B_c}$ and B_c is the channel coherence bandwidth. Each of the m sub-bands is divided equally among the n K-User-Groups. As shown in Figure 3, each UG gets a chunk of bandwidth in each of the m sub-bands in which to transmit pilot signals for synchronization and channel estimation. The channels in the sub-bands not available for a certain UG can easily be obtained by interpolation. The channel estimates are conveyed by the mobile devices to a global network entity in the backhaul which makes all channels available to all access points and mobile devices through the appropriate interfaces. The overhead from this step is not considered in this paper and will make up future work.

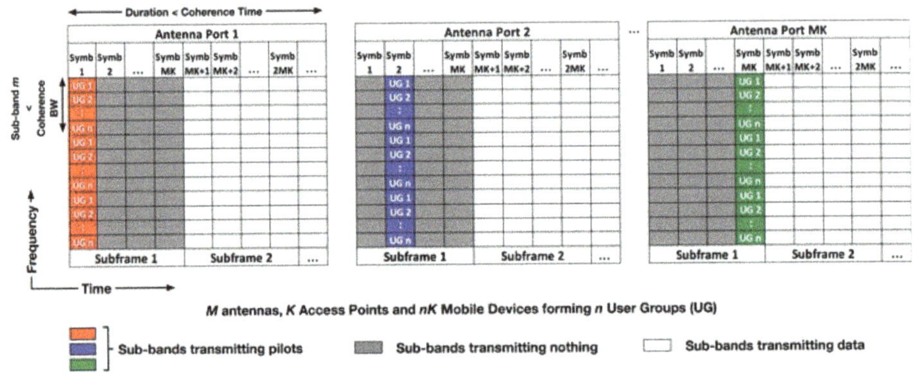

Figure 3. Example showing the channel estimation overhead for n K-User-Groups (UG). The bandwidth allocated for estimation is divided among the n K-User-Groups. Each group transmits pilot signals in its allocated band.

3. Derivation of Shannon Capacity for K-User MIMO X and Small Cell Geometric Capacity in Rayleigh Fading

This section presents an analysis of Shannon Capacity as a function of K, with and without pilot overheads. Further simulation results show the statistical distributions of capacity for $K = 3$ in Rayleigh fading and small cell geometries.

3.1. Ideal K-User MIMO X Capacity vs. K Excluding Pilot Overhead

The theorem for the upper-bound capacity for K-User MIMO X incorporates channel and beamformer gains that scale with K and antenna array size M. It is defined below.

Theorem 1. *The upper bound multi-user capacity of a K-User MIMO X system is bounded by* $C_{bits/sec} \leq BK^2 log_2(1 + \left[M - \frac{N_I}{2} - (N_D - 1)\right]\left[M - \frac{N_I}{2}\right]^2 M^2 \times SINR)$, *where B is the bandwidth, and the* $SINR = \frac{P_t d^{-\alpha}}{\sigma_w^2}$ *includes the transmit power P_t and the distance dependence based on a path loss exponent α and target distance d. The proof of Theorem 1 and verification by simulation are shown in Appendices B and C, respectively. The result of Theorem 1 gives the upper-bound multi-user capacity of the combined K^2 streams. Note that the theorem represents an unconstrained case which assumes that the entire available time-frequency resources are available only for data transmission.*

3.2. Ideal K-User MIMO X Capacity vs. K Including Pilot Overhead

Real systems are impacted by various overheads for synchronization signals, channel estimation pilot signals, and exchange of other control information. In this paper, we consider two of the most important overheads—that of synchronization and channel estimation. We propose the use of pilot signals configured for both time synchronization and channel estimation. Consequently, we define the following theorem that refines Theorem 1 to include pilot overheads.

Theorem 2. *If MK channel estimation symbols can be transmitted in less than the channel coherence time, i.e.,* $MKT_{symb} < T_c$, *the remaining time can be reserved exclusively for data transmission. The capacity equation in Theorem 1 can be modified as follows,* $C_{bits/sec} \leq \frac{T_c - MKT_{symb}}{T_c} BK^2 log_2(1 + \left[M - \frac{N_I}{2} - (N_D - 1)\right]\left[M - \frac{N_I}{2}\right]^2 M^2 \times SINR)$, *where T_c is the channel coherence time. T_{symb} is the duration of 1 OFDM symbol given* *by* $(N+CP)T_s$, *where N is the size of the Fast Fourier Transform (FFT), CP is the size of the Cyclic Prefix in samples, and* T_s *is the sampling period.*

Figure 4 shows the Shannon Capacity curves for K-User MIMO X with and without pilot overheads. The key observation from the curves is that while the unconstrained capacity from Theorem 1 continues to grow with K, that is not the case when pilot overheads are taken into account. After a certain point, it can be seen that the pilot overheads overwhelm the gains from K-User MIMO and the capacity begins to drop. In highly varying channel scenarios, it is more beneficial to operate at a lower value of K and schedule several User Groups in a multiple access framework similar to that shown in Figure 3. Figure 4 also compares the performance of our K-User MIMO X system with 5G-NR Massive MIMO [23]. It should be noted that the Massive MIMO model assumes one access point equipped with M antennas and K single-antenna mobile devices. The Shannon limit is also plotted. This limit is based on the analysis provided in [24] and assumes a system with a single access point and single mobile device each with MK antennas.

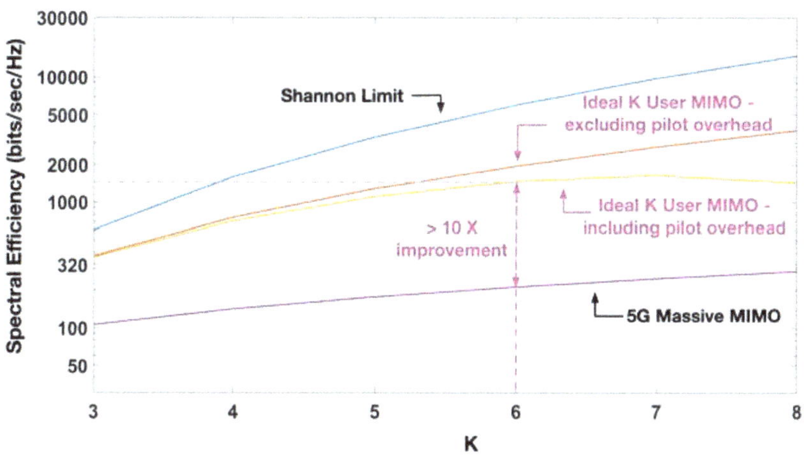

Figure 4. Comparison of our very high throughput K-User MIMO X spectral efficiencies against related technologies such as 5G-NR Massive MIMO. Spectral Efficiencies are for a 100 m cell in an Indoor A channel scenario. The curve for Massive MIMO is based on the formula shown in [23]. The number of antennas is $M = K(K-1)$.

3.3. Capacity Results for Small Cell K-User MIMO at K = 3

This section describes the simulation model, choosing $K = 3$ as an example. Simulations are done in MATLAB and the key parameters are listed in Table 2. Cell spectral efficiency performance in 500 m, 100 m, and 50 m hexagonal cells is obtained and shown in Figure 5.

Table 2. Simulation Parameters.

Channel Model	Rayleigh Fading
Channel Scenario	Indoor A [25]
Cell Radius	50 m, 100 m, 500 m
Transmit Power	16 dBW
Total Bandwidth	20 MHz
FFT Size (N)	2048
Cyclic Prefix (CP)	512 samples
Sampling frequency	30.72 MHz
Subcarrier spacing	15 kHz
Number of used sub-carriers	1320
Noise Figure	4 dB
Thermal Noise Density	−203.9 dBW/Hz
Path Loss Exponent	3

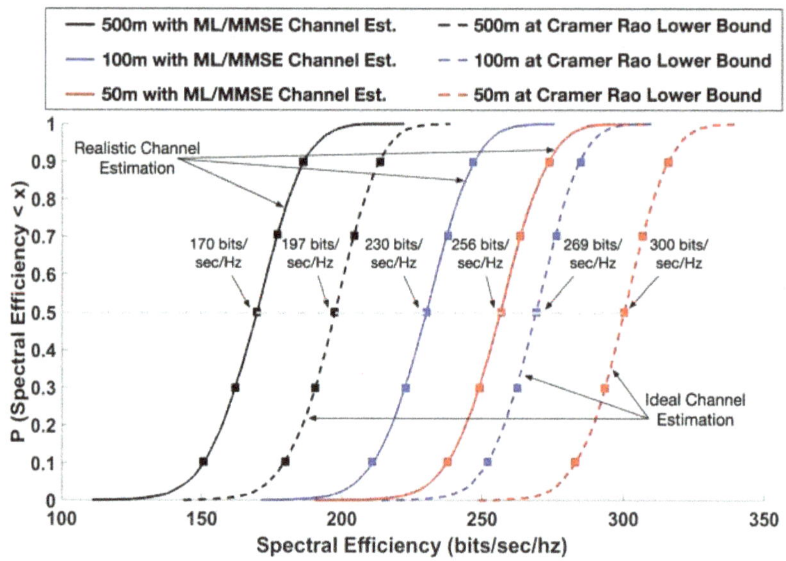

Figure 5. Cumulative distribution functions of spectral efficiencies in bits/sec/Hz for $K = 3$ in single hexagonal cells of radius 50 m, 100 m, and 500 m for Indoor A channel scenarios [25]. Results are shown both with realistic channel estimation using ML/MMSE and also at the Cramér–Rao Lower Bound.

Without loss of generality, the hexagonal geometry is chosen, for simplicity of analysis. The system can easily be translated to stochastic geometries, commonly associated with 5G systems. 3 access points are placed on alternate corners of the cell. 3 mobile devices are placed uniformly within the cell. The simulations assume an exponential path loss $L_{ij} = d_{ij}^{-\alpha}$.

While K-User MIMO X is not limited by any particular channel scenario, we note that one example of a use case is a high-throughput IoT robotic factory model. Hence, without loss of generality, a Rayleigh fading channel model with the Indoor A power delay profile [25] is used. Considering an Indoor A channel scenario with parameters such as velocity $v = 3$ kmph, carrier frequency $f_c = 1.9$ GHz, and speed of light $c = 3 \times 10^8$ m/s, the channel coherence time T_c is calculated as $T_c = \sqrt{\frac{9}{16\pi f_d^2}} \approx 80$ ms, where f_d is the Doppler shift given by $f_d = \frac{v f_c}{c}$. Let us assume that we only have a fraction of the channel coherence time, say 60 ms.

Multiple channel and location trials are run. In each trial, the channel is estimated in the first 18 symbol times (1.5 ms). This paper assumes the use of well-known estimation methods such as the Maximum Likelihood (ML) or Minimum Mean Square Error (MMSE).

For statistical analysis, it suffices to assume the presence of one K-User-Group. In such a case, the entire bandwidth is available for pilot signals. We use the Zadoff–Chu sequence for both time synchronization and channel estimation. After channel estimation, the remaining 58.5 ms is available for data transmission. The Cumulative Distribution Function (CDF) of the spectral efficiency is calculated from the multiple trials and plotted in Figure 5. These spectral efficiencies account for pilot signal overhead as well as estimation error.

In the case of $n > 1$ K-User-Groups, in the channel estimation symbols, the bandwidth can be divided as shown in Figure 3. In the data transmission symbols, the bandwidth could also be divided into sub-bands in which the different User Groups could be network scheduled using greedy or proportional fair algorithms.

Incorporation of Channel Estimation Errors

Figure 6 shows the variance of the estimation error as a function of signal-to-noise ratio for both the ML and MMSE estimates. The Cramér–Rao Lower Bound (CRLB) is also shown. As expected, the ML and MMSE variances are higher than the CRLB. The equations for the channel estimation methods and the CRLB are well researched in other works and are hence shown in Appendix D. Figure 5 also shows the spectral efficiency if the estimation error variance is at the Cramér–Rao Lower Bound (CRLB). To simulate this, we calculate the CRLB as shown in Equation (A15) in Appendix D. The CRLB is calculated for each transmit–receive antenna link, and error terms are drawn from $\mathcal{CN}(0, \sigma^2_{CRLB})$. These errors are added to the actual channels to simulate estimation error at the CRLB.

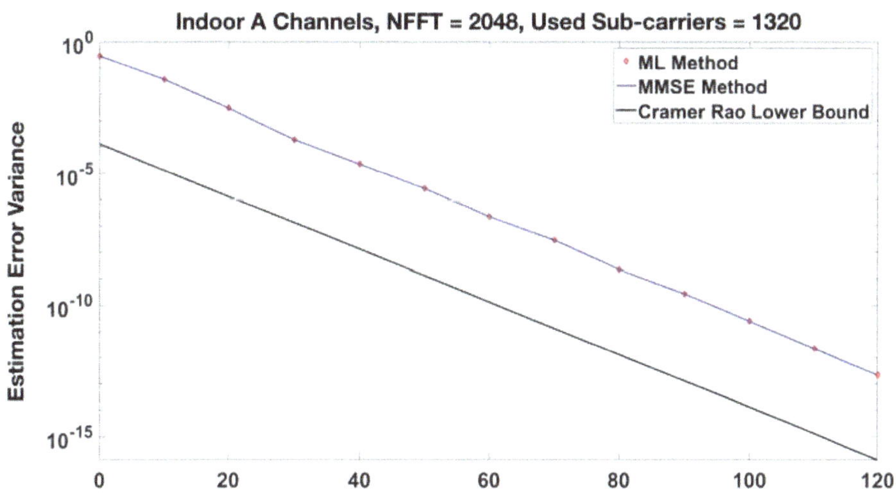

Figure 6. Comparison of channel estimation error variance between Maximum Likelihood (ML)/Minimum Mean Square Error (MMSE) and the Cramér–Rao Lower Bound (CRLB).

It can be seen from Figure 5 that the median best case spectral efficiency in 500 m, 100 m, and 50 m cells are 170, 230, and 256 bits/sec/Hz, respectively. These results are under Rayleigh fading conditions. If the channel follows a Rician distribution, it means that there is one dominant line of sight path in the Channel Impulse Response. The stronger the line-of-sight component, the rarer the occurrences of deep fades. Though this scenario is not simulated explicitly, we have investigated another scenario which is maximum capacity scheduling with Rayleigh fading. This approach schedules K-User-Groups in a particular sub-band with the best channel conditions, thus weeding out deep fade channel instances. We have found very little improvement in spectral efficiency with this type of scheduling. We believe that the same will be the case with Rician channels. The reason for this is that the non co-located nature of K-User MIMO transmitters already provides enough channel diversity to overcome the effects of deep fades.

The spectral efficiencies in Figure 5 are multi-user values and take into account the fact that channel estimation through pilot signals takes up 18 symbol times, where no data is transmitted. In a 20 MHz band, this amounts to a best-case scenario of 3.4 Gbps, 4.6 Gbps, and 5.1 Gbps, respectively; and in the case of 5 aggregated bands, data rates in excess of 17 Gbps, 23 Gbps, and 25 Gbps, respectively, can be achieved. This underlines the wide range of exciting possibilities that can be achieved in beyond-5G and 6G networks with K-User MIMO X.

The 6G extension of eMBB, defined in [26] as eMBB Plus, will serve mobile and IoT communications with data rate requirements far greater than 5G. The high throughput and spectral efficiency of K-User MIMO X lend themselves well to help support eMBB Plus. 6G will also be more machine-learning and security driven [27,28]. The all-to-all nature of K-User MIMO X, with its ability to switch between maximum capacity and maximum reliability modes, makes it particularly suitable for future machine learning integration. Machine learning algorithms could be used to come up with new encoding schemes across the K transmitters that can help adapt to spatially and temporally varying channel conditions as well as eavesdroppers and jammers. This flexibility leads perfectly into the 6G version of URLLC, known as event-defined-URLLC [28], which provides context-aware communications not thought of in 5G.

4. Conclusions

In this paper, we have reviewed and identified the massive scope for increased throughput for beyond-5G or 6G networks. Under realistic channel estimation constraints, we have provided a K-User MIMO X framework that can cancel interference, demodulate, and maximize capacity through signal separation. Further practical aspects such as OFDM multiple access for channel estimation and data transmission have been described. Lastly, cell capacity performance has been simulated and compared with related technologies.

Author Contributions: Conceptualization, A.K.Y. and B.K.; investigation, A.K.Y.; methodology, A.K.Y. and B.K.; writing—original draft, A.K.Y.; writing—review and editing, A.K.Y. and B.K. All authors have read and agreed to the published version of the manuscript.

Funding: This research was funded by US Air Force Research Laboratory and Office of the Secretary of Defense under Testing, Evaluation, and Control of Heterogeneous Large Scale systems of Autonomous Vehicles (TECHLAV) with grant number FA8750-15-2-0016.

Conflicts of Interest: The authors declare no conflict of interest.

Appendix A. K-User MIMO Interference Alignment Mathematics for $K = 3$

Appendix A.1. System of Interference Alignment Equations for Precoder Generation

Interference cancellation can be achieved subject to the following constraints.

Postulate 1 [17]: At each mobile device, interfering signals from the same access point cannot be aligned in the same direction such that $H_{ij}v_{kj} \neq H_{ij}v_{lj}$, where $i \neq k \neq l$.

Postulate 2 [17]: In a K user system, since each mobile device receives $K(K-1)$ interference components, in order to align $(K-1)$ interference signals along K dimensions, the condition to be met is $span(H_{im}v_{km}) = span(H_{in}v_{ln})$, where $k, l \neq i$.

Since there are $K(K-1)$ interference terms at each mobile device, the minimum number of antennas at each access point and mobile device has to be $M = K(K-1)$.

Table A1 shows one possible set of Interference Alignment equations for $K = 3$. These equations are formed by dividing the $K(K-1)$ interference terms at each mobile device into pairs, subject to postulates 1 and 2. Each pair represents the left- and right-hand sides of a single IA equation. Since there are $K(K-1)$ interference terms at each mobile device, there will be $\frac{K(K-1)}{2}$ IA equations. These equations will be used to solve for the precoders.

Table A1. K-User MIMO X Interference Alignment (IA) conditions, $K = 3$.

	IA conditions [17]
	$\text{span}(H_{11}v_{21}) = \text{span}(H_{12}v_{22})$
Rx 1	$\text{span}(H_{11}v_{31}) = \text{span}(H_{13}v_{33})$
	$\text{span}(H_{12}v_{32}) = \text{span}(H_{13}v_{23})$
	$\text{span}(H_{21}v_{11}) = \text{span}(H_{22}v_{12})$
Rx 2	$\text{span}(H_{21}v_{31}) = \text{span}(H_{23}v_{13})$
	$\text{span}(H_{22}v_{32}) = \text{span}(H_{23}v_{33})$
	$\text{span}(H_{31}v_{11}) = \text{span}(H_{33}v_{23})$
Rx 3	$\text{span}(H_{31}v_{21}) = \text{span}(H_{32}v_{12})$
	$\text{span}(H_{32}v_{22}) = \text{span}(H_{33}v_{13})$

Each access point will apply K precoders. From the IA equations, the precoders v_{ij} are obtained as follows:

$$\begin{aligned} v_{12} &= (H_{22})^{-1}H_{21}v_{11} \quad v_{33} = (H_{13})^{-1}H_{11}v_{31} \\ v_{21} &= (H_{31})^{-1}H_{32}v_{12} \quad v_{32} = (H_{22})^{-1}H_{23}v_{33} \\ v_{22} &= (H_{12})^{-1}H_{11}v_{21} \quad v_{23} = (H_{13})^{-1}H_{12}v_{32} \\ v_{13} &= (H_{33})^{-1}H_{32}v_{22} \quad v_{11} = (H_{31})^{-1}H_{33}v_{23} \\ v_{31} &= (H_{21})^{-1}H_{23}v_{13}. \end{aligned} \tag{A1}$$

The initial value of v_{11} is obtained by first defining a matrix E [17] as follows:

$$\begin{aligned} E = {} & (H_{31})^{-1}H_{33}(H_{13})^{-1}H_{12}(H_{22})^{-1}H_{23}(H_{13})^{-1}H_{11} \\ & \times (H_{21})^{-1}H_{23}(H_{33})^{-1}H_{32}(H_{12})^{-1}H_{11}(H_{31})^{-1}H_{32} \\ & \times (H_{22})^{-1}H_{21}. \end{aligned} \tag{A2}$$

It is to be noted that E is obtained from Equation (A1). Then, v_{11} is arbitrarily chosen to be one of the eigenvectors of E and subsequently, all the other precoders can be obtained in the order, v_{12}, v_{21}, v_{22}, v_{13}, v_{31}, v_{33}, v_{32}, and v_{23}.

Appendix A.2. Obtaining the Beamformer for Interference Cancellation

The zero forcing beamformer matrix U_i is obtained by first defining a matrix P_i^I for each mobile device and taking the Singular Value Decomposition (SVD) as follows:

$$\begin{aligned} P_1^I &= \begin{bmatrix} H_{11}v_{21} & H_{11}v_{31} & H_{12}v_{32} \end{bmatrix} \\ &= \begin{bmatrix} \bar{U}_1^{(1)} & \bar{U}_1^{(0)} \end{bmatrix} \begin{bmatrix} \bar{\Lambda}_1 \\ 0 \end{bmatrix} \begin{bmatrix} \bar{V}_1^{(1)} & \bar{V}_1^{(0)} \end{bmatrix}^H, \end{aligned} \tag{A3}$$

where P_1^I is the set of aligned interfering column vectors at the first mobile device. From Equation (A3), we can set $\bar{U}_1^{(0)} = U_1$, the zero forcing beamformer at the first mobile device. The number of columns in U_1 is equal to the number of nonzero singular values in $\bar{\Lambda}_1$. Similarly, we determine the beamformers at the second and third mobile devices by defining $P_2^I = \begin{bmatrix} H_{21}v_{11} & H_{21}v_{31} & H_{22}v_{32} \end{bmatrix}$ and $P_3^I = \begin{bmatrix} H_{31}v_{11} & H_{31}v_{21} & H_{32}v_{22} \end{bmatrix}$, where P_2^I and P_3^I are the sets of aligned interfering column vectors at the second and third mobile devices, respectively.

Appendix A.3. Obtaining the Beamformer for Signal Separation

At the ith mobile device, in order to separate the different signals sent from the K access points, we apply a second beamformer matrix. This operator is applied K times at each mobile device. This matrix is obtained by defining a matrix P_{ij}^D for each access-point–mobile-device pair and taking the SVD as follows:

$$\begin{aligned} P_{11}^D &= \begin{bmatrix} U_1^H H_{12}v_{12} & U_1^H H_{13}v_{13} \end{bmatrix} \\ &= \begin{bmatrix} \bar{U}_{11}^{(1)} & \bar{U}_{11}^{(0)} \end{bmatrix} \begin{bmatrix} \bar{\Lambda}_{11} \\ 0 \end{bmatrix} \begin{bmatrix} \bar{V}_{11}^{(1)} & \bar{V}_{11}^{(0)} \end{bmatrix}^H, \end{aligned} \tag{A4}$$

where P_{11}^D is the set of desired column vectors at mobile device 1 corresponding to access points 2 and 3. From Equation (A4), we can set $\bar{U}_{11}^{(0)} = U_{11}$. The number of columns in U_{11} is equal to the number of nonzero singular values in $\bar{\Lambda}_{11}$. The matrix U_{11}, when multiplied to the received signal after zero forcing, isolates the desired signal at mobile device 1 from access point 1. Similarly, the desired signals at mobile device 1 from access points 2 and 3 respectively can be isolated by defining $P_{12}^D = \begin{bmatrix} U_1^H H_{11}v_{11} & U_1^H H_{13}v_{13} \end{bmatrix}$ and $P_{13}^D = \begin{bmatrix} U_1^H H_{11}v_{11} & U_1^H H_{12}v_{12} \end{bmatrix}$, where P_{12}^D and P_{13}^D are the sets of desired column vectors at mobile device 1 corresponding to access points 1,3 and 1,2, respectively. The same process can be repeated at the other mobile devices.

Appendix A.4. K-User MIMO X Demodulation and Symbol Detection

The following demodulation process helps recover symbols arriving from each access point. Note that this process is applied after zero forcing and signal separation. Defining $\Gamma_{ij} = U_{ij}^H U_i^H H_{ij}$ and taking the SVD, we have

$$\Gamma_{ij} = \Phi_{ij} \cdot \Lambda_{ij} \cdot \Psi_{ij}^H, \tag{A5}$$

where Λ_{ij} is a diagonal matrix containing the singular values of Γ_{ij} such that

$$\Lambda_{ij} = \begin{bmatrix} \lambda_{ij}(1) & 0 & \dots & 0 & 0 & 0 \\ 0 & \lambda_{ij}(2) & 0 & \dots & 0 & 0 \\ \vdots & \vdots & & & \ddots & \\ 0 & 0 & \lambda_{ij}(N_s) & 0 & \dots & 0 \end{bmatrix}. \tag{A6}$$

The information symbol from access point j to mobile device i is estimated by the element wise division,

$$[\hat{s}_{ij}(1), \dots, \hat{s}_{ij}(\gamma), \dots, \hat{s}_{ij}(N_s)]^T = \frac{\phi_{ij}^H \tilde{y}_{ij}}{\psi_{ij}^H v_{ij}}. \tag{A7}$$

The demodulation process recovers N_s copies (refer to Table 1) of each symbol s_{ij} corresponding to the number of nonzero singular values in Λ_{ij}.

Due to interference cancellation, the demodulated signal obtained from Equation (A7) takes the form

$$\hat{s}_{ij} = \sum_{\gamma=1}^{N_s} \lambda_{ij}(\gamma) s_{ij}(\gamma) + \hat{w}_i, \tag{A8}$$

where $\lambda_{ij}(\gamma)$ is the γth diagonal element in Λ_{ij} and $\hat{w}_{ij}$ is the noise after demodulation.

The spectral efficiency after interference cancellation, signal separation, and demodulation can be computed as follows:

$$C_{bits/sec/Hz} = log_2\left[1 + \frac{\left|\sum_{\gamma=1}^{N_s} \lambda_{ij}(\gamma)s_{ij}(\gamma)\right|^2}{|\hat{w}_i|^2}\right]. \tag{A9}$$

Appendix B. Proof of Theorem 1

Proof. To complete the proof, we leverage two well-known results from Matrix Theory [29]. First, for any matrix A, the sum of the squares of the singular values is equal to the square of its Frobenius Norm such that $\sum_i \lambda_i^2(A) = \|A\|_F^2$, where $\lambda_i(A)$ is the ith singular value of A and $\|A\|_F$ is the Frobenius Norm of A. This follows from the decomposition of matrix A as a singular value decomposition, $A = U\Lambda V^H$. The Frobenius norm is invariant under orthogonal transformation of the left and right orthogonal matrices. Therefore, based on the fact that $\lambda_i(A)$ is the component of A along the diagonal of Λ, the sum of squares of the components of Λ equals the square of the Frobenius norm.

Second, the Frobenius Norm of the product of matrices is upper bounded by the product of the Frobenius Norms of the individual matrices such that $\|AB\|_F \leq \|A\|_F \|B\|_F$. The proof of this is based on the well-known Cauchy–Schwarz Inequality.

From Equation (A5), $\Gamma_{ij} = U_{ij}^H U_i^H H_{ij}$ and $\lambda_{ij}(1) \cdots \lambda_{ij}(N_s)$ are the singular values of Γ_{ij}. Combining the above two statements from [29],

$$\begin{aligned}\sum_{\gamma=1}^{N_s} |\lambda_{ij}(\gamma)|^2 &= \|\Gamma_{ij}\|_F^2 \\ &\leq \left\|U_{ij}^H\right\|_F^2 \left\|U_i^H\right\|_F^2 \|H_{ij}\|_F^2 \\ &\leq \sum |\lambda(U_{ij}^H)|^2 \sum |\lambda(U_i^H)|^2 \sum |\lambda(H_{ij})|^2 \\ &\leq \left[M - \frac{N_I}{2} - (N_D - 1)\right]\left[M - \frac{N_I}{2}\right] \cdot \left[M - \frac{N_I}{2}\right] M \cdot M,\end{aligned} \tag{A10}$$

where $\lambda(U_{ij}^H)$, $\lambda(U_i^H)$, and $\lambda(H_{ij})$ are the singular values of U_{ij}^H, U_i^H, and H_{ij}, respectively. It should be noted that $\sum_{\gamma=1}^{N_s} |\lambda_{ij}(\gamma)|^2$ is the gain for one stream between the jth access point and the ith mobile device. There are K^2 such streams. Therefore, the multi-user capacity is given by

$$C_{bits/sec} \leq BK^2 log_2\left(1 + \left[M - \frac{N_I}{2} - (N_D - 1)\right]\left[M - \frac{N_I}{2}\right]^2 M^2 \times SINR\right). \tag{A11}$$

□

Appendix C. Verification of Capacity Upper Bound

To verify the capacity upper bound for $K = 3$, several simulation trials were run in which Rayleigh Fading Channels with the Indoor A power delay profile were generated. It should be noted that U_{ij} and U_i are normalized to have a power of 1, whereas H_{ij} has a power of 1 only in an expected sense. The values of $\sum |\lambda(U_{ij}^H)|^2$, $\sum |\lambda(U_i^H)|^2$, and $\sum |\lambda(H_{ij})|^2$ were found to be 3, 18, and 6, which correspond to the dimensions of U_{ij}, U_i, and H_{ij}, respectively (Table 1).

The SVD in Equation (A5) is computed and a distribution of the sum of the singular value powers $\sum_{\gamma=1}^{N_s} |\lambda_{ij}(\gamma)|^2$ is plotted. This is shown in Figure A1. It can be clearly seen from the distribution that the values of the singular value power sum fall below the upper bound for $K = 3$, thus satisfying the theorem.

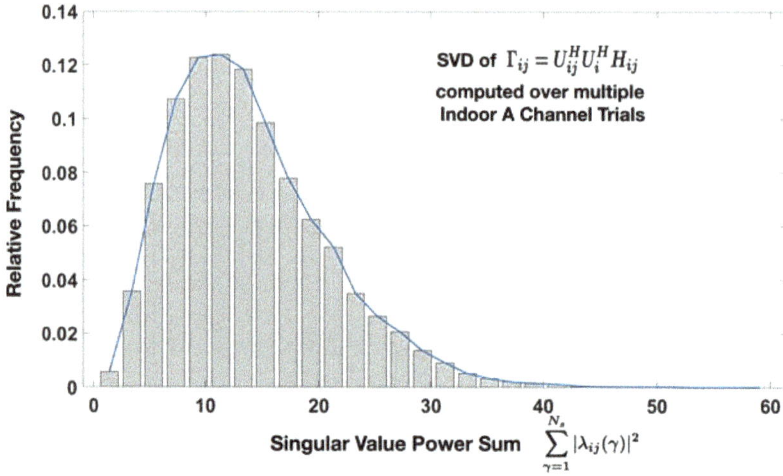

Figure A1. Relative frequency distribution of the singular value power sum for $K = 3$. The upper bound for this sum is given by Equation (A10).

Appendix D. Channel Estimation Theory and Cramér–Rao Lower Bound

In OFDM systems, for proper detection of symbols, channel estimation is performed by transmitting either known pilot symbols at certain frequency subcarriers or across the entire OFDM symbol. In the K-User MIMO X framework described above, the pilot signals are set up such that only one antenna is transmitting in any given symbol time, during the estimation phase. So, the MIMO channel estimation can be broken down into several single-antenna channel estimations. To that end, in this section, we summarize some of the well known theories on estimating a channel between a single transmit–receive antenna pair. We also present the mathematics for calculating the Cramér–Rao Lower Bound (CRLB).

Let $h = [h(0), h(1), \ldots, h(L-1)]^T$ represent the length L Channel Impulse Response (CIR). Let us denote the Channel Frequency Response (CFR) as $H = Fh$, where F is the $N \times N$ Fourier Transform twiddle factor matrix, where N is the Fourier Transform size.

Let us assume that an OFDM symbol X containing N_p pilots is transmitted, where $N_p \leq N$. The received signal Y between a single transmit–receive antenna pair in the frequency domain is given by [30]

$$Y = diag(X)Fh + W, \tag{A12}$$

where W is the noise with variance σ_w^2.

Let F_p denote the $N_p \times L$ truncated Fourier matrix. The Maximum Likelihood estimate of the channel is given by [30]

$$h_{ML} = (F_p^H F_p)^{-1} F_p^H diag(X)^H Y. \tag{A13}$$

It is to be noted that h_{ML} is in the time domain. The corresponding frequency domain estimate H_{ML} can be obtained by taking the Fourier Transform.

The Minimum Mean Square Error estimate of the channel is given by [30]

$$H_{MMSE} = R_{HH_p}(R_{HH_p} + \sigma^2(diag(X)diag(X)^H)^{-1})^{-1} H_{ML}, \tag{A14}$$

where H_p is the channel frequency response at the pilot subcarriers, R_{HH_p} is the cross-correlation between all the subcarriers and the pilot-subcarriers, and $R_{H_pH_p}$ is the autocorrelation between the pilot subcarriers.

The CRLB is referenced from [31], in which it is defined as follows:

$$\sigma^2_{CRLB} = \sigma^2 Tr\{D^{-1}\}, \tag{A15}$$

where $\frac{1}{\sigma^2}$ is the effective signal-to-noise ratio after the signal power has been normalized to 1. The operator $Tr\{.\}$ is the matrix trace. The matrix D, a function of the pilot locations is defined as $D_{n,k} = \sum_{m=0}^{N_p-1} e^{-j2\pi(n-k)i_m/N} \quad 0 \le n,k \le L-1$, where $\{i_m : 0 < m < N_p - 1\}$ represents the indices of the pilot locations.

References

1. Breivold, H.P.; Sandström, K. Internet of Things for Industrial Automation–Challenges and Technical Solutions. In Proceedings of the 2015 IEEE International Conference on Data Science and Data Intensive Systems, Sydney, Australia, 11–13 December 2015; pp. 532–539.
2. Yaqoob, I.; Ahmed, E.; Hashem, I.A.T.; Ahmed, A.I.A.; Gani, A.; Imran, M.; Guizani, M. Internet of things architecture: Recent advances, taxonomy, requirements, and open challenges. *IEEE Wirel. Commun.* **2017**, *24*, 10–16. [CrossRef]
3. Jaber, M.; Imran, M.A.; Tafazolli, R.; Tukmanov, A. 5G Backhaul Challenges and Emerging Research Directions: A Survey. *IEEE Access* **2016**, *4*, 1743–1766. [CrossRef]
4. Ishii, H.; Kishiyama, Y.; Takahashi, H. A Novel Architecture for LTE-B: C-plane/U-plane Split and Phantom Cell Concept. In Proceedings of the 2012 IEEE Globecom Workshops, Anaheim, CA, USA, 3–7 December 2012; pp. 624–630.
5. Zheng, K.; Zhao, L.; Mei, J.; Dohler, M.; Xiang, W.; Peng, Y. 10 Gb/s Hetsnets with Millimeter-Wave Communications: Access and Networking-Challenges and Protocols. *IEEE Commun. Mag.* **2015**, *53*, 222–231. [CrossRef]
6. Pi, Z.; Choi, J.; Heath, R. Millimeter-Wave Gigabit Broadband Evolution Toward 5G: Fixed Access and Backhaul. *IEEE Commun. Mag.* **2016**, *54*, 138–144. [CrossRef]
7. Hou, X.; Wang, X.; Jiang, H.; Kayama, H. Investigation of Massive MIMO in Dense Small Cell Deployment for 5G. In Proceedings of the Investigation of Massive MIMO in Dense Small Cell Deployment for 5G, Montreal, QC, Canada, 18–21 September 2016; pp. 1–6.
8. Agiwal, M.; Roy, A.; Saxena, N. Next Generation 5G Wireless Networks: A Comprehensive Survey. *IEEE Commun. Surv. Tutor.* **2016**, *18*, 1617–1655. [CrossRef]
9. Lee, S.R.; Moon, S.H.; Kong, H.B.; Lee, I. Optimal Beamforming Schemes and its Capacity Behavior for Downlink Distributed Antenna Systems. *IEEE Trans. Wirel. Commun.* **2013**, *12*, 2578–2587. [CrossRef]
10. Heath, R.; Wu, T.; Kwon, Y.; Soong, A. multi-user MIMO in Distributed Antenna Systems. In Proceedings of the 2010 Conference Record of the Forty Fourth Asilomar Conference on Signals, Systems and Computers, Pacific Grove, CA, USA, 7–10 November 2010; pp. 1202–1206.
11. Talebi, F. A Tutorial on Interference Alignment. Available online: https://www.researchgate.net/publication/255670668_A_Tutorial_on_Interference_Alignment (accessed on 5 June 2011).
12. Yetis, C.M.; Gou, T.; Jafar, S.A.; Kayran, A.H. On Feasibility of Interference Alignment in MIMO Interference Networks. *IEEE Trans. Signal Process.* **2010**, *58*, 4771–4782. [CrossRef]
13. Cadambe, V.R.; Jafar, S.A. Interference Alignment and Degrees of Freedom of the *K*-user Interference Channel. *IEEE Trans. Inf. Theory* **2008**, *54*, 3425–3441. [CrossRef]
14. Gomadam, K.; Cadambe, V.R.; Jafar, S.A. Approaching the Capacity of Wireless Networks Through Distributed Interference Alignment. In Proceedings of the IEEE GLOBECOM 2008–2008 IEEE Global Telecommunications Conference, New Orleans, LO, USA, 30 November–4 December 2008; pp. 1–6.
15. Sudheesh, P.; Magarini, M.; Muthuchidambaranathan, P. Interference Alignment for the K-user MIMO X Network Using Time Division Multiple Access. In Proceedings of the 2016 IEEE Distributed Computing, VLSI, Electrical Circuits and Robotics (DISCOVER), Mangalore, India, 13–14 August 2016; pp. 123–127.

16. Liu, W.; Sun, J.X.; Li, J.; Ma, Y. Interference Alignment for MIMO Downlink Multicell Networks. *IEEE Trans. Veh. Technol.* **2016**, *65*, 6159–6167. [CrossRef]
17. Park, S.H.; Ko, Y.C. K-user MIMO X Network System With Perfect Interference Alignment. In Proceedings of the 2011 IEEE International Conference on Communications (ICC), Kyoto, Japan, 5–9 June 2011; pp. 1–5.
18. Yerrapragada, A.K.; Kelley, B. Design of K-user massive MIMO networks. In Proceedings of the 2017 IEEE 8th Annual Ubiquitous Computing, Electronics and Mobile Communication Conference (UEMCON), New York, NY, USA, 19–21 October 2017; pp. 377–383.
19. El Ayach, O.; Lozano, A.; Heath, R.W. On the overhead of interference alignment: Training, feedback, and cooperation. *IEEE Trans. Wirel. Commun.* **2012**, *11*, 4192–4203. [CrossRef]
20. Mungara, R.K.; George, G.; Lozano, A. Overhead and spectral efficiency of pilot-assisted interference alignment in time-selective fading channels. *IEEE Trans. Wirel. Commun.* **2014**, *13*, 4884–4895. [CrossRef]
21. Yerrapragada, A.; Kelley, B. An IoT Self Organizing Network for 5G Dense Network Interference Alignment. In Proceedings of the 2017 12th System of Systems Engineering Conference (SoSE), Waikoloa, HI, USA, 18–21 June 2017; pp. 1–6.
22. Yerrapragada, A.K.; Kelley, B. Very High Throughput Internet of Things Networks with *K* access points and *K* devices. In Proceedings of the MILCOM 2019–2019 IEEE Military Communications Conference (MILCOM), Norfolk, VA, USA, 12–14 November 2019; pp. 640–645.
23. Marzetta, T.L. Massive MIMO: An introduction. *Bell Labs Tech. J.* **2015**, *20*, 11–22. [CrossRef]
24. Juntti, M. Capacity Limits of MIMO Channels. *IEEE J. Sel. Areas Commun.* **2003**, *21*,684–702.
25. Medbo, J. Channel models for HIPERLAN/2 in different indoor scenarios. *ETSI/BRAN* **1998**, *3ERI085B*, 1–8.
26. Dang, S.; Amin, O.; Shihada, B.; Alouini, M.S. What should 6G be? *Nat. Electron* **2020**, *3*, 20–29. [CrossRef]
27. Yang, P.; Xiao, Y.; Xiao, M.; Li, S. 6G wireless communications: Vision and potential techniques. *IEEE Netw.* **2019**, *33*, 70–75. [CrossRef]
28. Letaief, K.B.; Chen, W.; Shi, Y.; Zhang, J.; Zhang, Y.J.A. The roadmap to 6G: AI empowered wireless networks. *IEEE Commun. Mag.* **2019**, *57*, 84–90. [CrossRef]
29. Golub, G.H.; Van Loan, C.F. *Matrix Computations*; JHU Press: Baltimore, MD, USA, 2012; Volume 3.
30. Ozdemir, M.K.; Arslan, H. Channel estimation for wireless OFDM systems. *IEEE Commun. Surv. Tutor.* **2007**, *9*, 18–48. [CrossRef]
31. Morelli, M.; Mengali, U. A Comparison of Pilot-Aided Channel Estimation Methods for OFDM Systems. *IEEE Trans. Signal Process.* **2001**, *49*, 3065–3073. [CrossRef]

Publisher's Note: MDPI stays neutral with regard to jurisdictional claims in published maps and institutional affiliations.

sensors

MDPI

Article

Multibeam Characteristics of a Negative Refractive Index Shaped Lens

Salbiah Ab Hamid [1,*], Nurul Huda Abd Rahman [1,*], Yoshihide Yamada [2], Phan Van Hung [3] and Dinh Nguyen Quoc [3]

1 Faculty of Electrical Engineering, Universiti Teknologi MARA, Shah Alam 40450, Selangor, Malaysia
2 Malaysia-Japan International Institute of Technology, Universiti Teknologi Malaysia, Kuala Lumpur 54100, Malaysia; yoshihide@utm.my
3 Faculty of Radio Electronic Engineering, Le Quy Don Technical University, Hanoi City 100000, Vietnam; phanvanhung@tcu.edu.vn (P.V.H.); dinhnq@mta.edu.vn (D.N.Q.)
* Correspondence: 2014365251@isiswa.uitm.edu.my (S.A.H.); nurulhuda0340@uitm.edu.my (N.H.A.R.)

Received: 22 August 2020; Accepted: 2 October 2020; Published: 7 October 2020

Abstract: Narrow beam width, higher gain and multibeam characteristics are demanded in 5G technology. Array antennas that are utilized in the existing mobile base stations have many drawbacks when operating at upper 5G frequency bands. For example, due to the high frequency operation, the antenna elements become smaller and thus, in order to provide higher gain, more antenna elements and arrays are required, which will cause the feeding network design to be more complex. The lens antenna is one of the potential candidates to replace the current structure in mobile base station. Therefore, a negative refractive index shaped lens is proposed to provide high gain and narrow beamwidth using energy conservation and Abbe's sine principle. The aim of this study is to investigate the multibeam characteristics of a negative refractive index shaped lens in mobile base station applications. In this paper, the feed positions for the multibeam are selected on the circle from the center of the lens and the accuracy of the feed position is validated through Electromagnetic (EM) simulation. Based on the analysis performed in this study, a negative refractive index shaped lens with a smaller radius and slender lens than the conventional lens is designed, with the additional capability of performing wide-angle beam scanning.

Keywords: lens antenna; negative refractive index; multibeam; beam scanning

1. Introduction

In recent years, there has been a rapid development of antennas to meet the advanced mobile technology requirements. In the 5G mobile technology system, millimeter wave is used, which produces a smaller cell size and requires a multibeam radiation pattern in order to achieve massive Multiple Input Multiple Output (MIMO) operation [1,2] with the purpose to serve massive number of users, consistent interconnectivity and larger capacity. At millimeter wave, the base station size becomes less than 30cm. So, aperture antennas such as reflector antennas and dielectric lens antennas become promising candidates as compared to the present array antennas. For multibeam operation, the dielectric lens antenna can achieve very good performance by designing the lens using Abbe's sine condition [3], which is validated further in this paper.

There were several studies conducted to investigate the multibeam application using the lens antenna. The Luneburg lens antenna is one of the commonly used designs. However, major issues of the conventional Luneburg lens are large and heavy. In [4], a Luneburg lens with an operating frequency of 1–8 GHz was proposed. In the paper, multilayer Luneburg lenses were fabricated with the diameters of 0.6 and 1.0 m and weighing of 5 and 21 kg, respectively. A number of feed antennas was placed around the lens to produce multibeam with the same beam shapes. Beam scanning was achieved by switching the feed around the surface of the lens. The traditional Luneburg lenses weight has been reduced up to 8–10 times; however, in 5G base station application, this weight and size are not practical. A flat Luneburg lens antenna with a geometry of 80 × 5 mm^2 was designed to operate at Ka-band [5]. A linear antenna array with 11 E-shaped patch antenna elements is employed to feed the lens antenna. The lens is formed by rectangular shaped concentric six layered dielectric materials with a permittivity ranging from 1.74 to 10 to achieve higher aperture efficiency. This lens antenna produces a realized gain of 14.2 dBi. In total, 11 beams are produced for ±60° beam scanning. However, a lens antenna employing natural dielectric material is known to produce a thicker lens size, and thus is not suitable for base station application. In order to reduce the lens thickness, a metamaterial lens concept was proposed [6]. Lens structure became concave and reduced the lens thickness. As for metamaterial lens fabrication, a lens of refractive index of $n = -1$ was shown in [7]. The structure is obtained by using 100 unit cells of dielectric resonator structure, 0.5 of f/D with lens aperture of 5.7 λ (156 mm) at an 11 GHz operating frequency. The achieved multibeam radiation pattern is shown in [8]. To achieve a good multibeam characteristic, Abbe's sine condition design is proposed in this paper, and comparison with the energy conservation technique is analyzed.

In this paper, the application of Abbe's sine design and energy conservation design are employed to develop a metamaterial lens. Based on the lens shaping equation, the possibility of the negative refractive index design is ensured. In order to solve the design equations, a MATLAB program is developed. The design accuracies of the MATLAB program are estimated using ray tracing results and calculation of the refraction angles on the lens surfaces. For validation, the multibeam radiation pattern is calculated using a commercial electromagnetic simulator, High Frequency Structure Simulator (HFSS). Figure 1 shows the proposed base station structure.

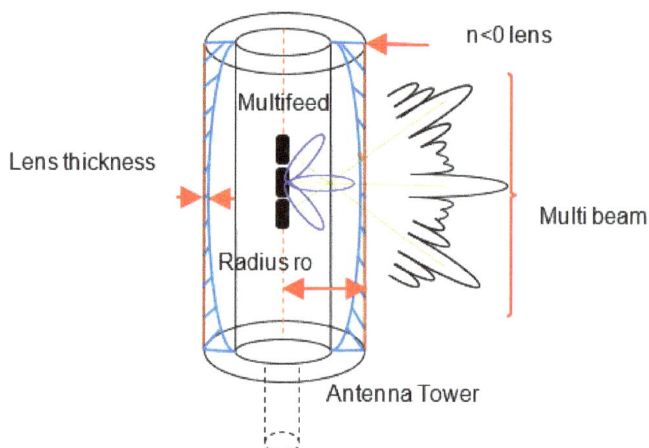

Figure 1. Proposed mobile base station structure.

2. Design Method

In designing the lens surface, the ray tracing method was employed. All rays passing through the aperture plane are designed to be parallel to the z-axis in order to achieve a flat wave front at the aperture plane.

2.1. Lens Configuration

Figure 2 shows the proposed antenna configuration and the associated radiation parameters. The feed radiator radiates signal towards the negative refractive index shaped lens, and the result is observed at the aperture. The detailed description of the parameters is shown in Table 1.

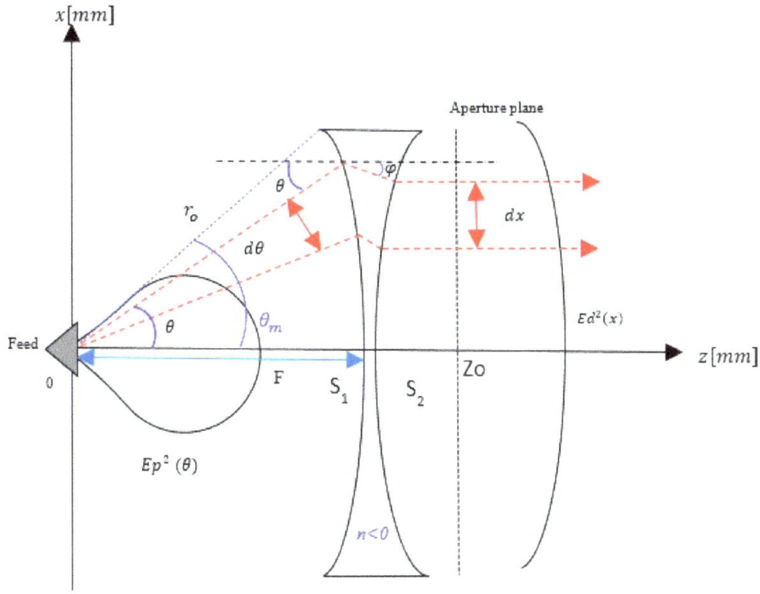

Figure 2. Lens configuration and parameters.

Table 1. Antenna parameters.

Parameter	Description
n	Refractive index
$Ep^2(\theta)$	Feed pattern
$Ed^2(x)$	Aperture Distribution
θ_m	Angle from feed to the lens edge

2.2. Lens Shape Equations

2.2.1. Aperture Distribution Designing

Based on Figure 2, at surface 1, S_1 of the lens, Snell's law is given by Equation (1) [9].

$$\frac{dr}{d\theta} = \frac{rnsin(\theta - \varphi)}{ncos(\theta - \varphi) - 1} \tag{1}$$

At the lens surface 2, the expression for the slope $\frac{dz}{dx}$ can be derived from the condition that all exit rays after refraction are parallel to the z- axis, as shown in Equation (2). The $\frac{dz}{dx}$ expression can be separated into $\frac{dz}{d\theta}$ and $\frac{dx}{d\theta}$, as shown in Equation (2) for variable change from dx to $d\theta$.

$$\frac{dz}{dx} = \frac{nsin(\varphi)}{1 - ncos(\varphi)}, \ \frac{dz}{d\theta} = \frac{nsin(\varphi)}{1 - ncos(\varphi)}\frac{dx}{d\theta} \tag{2}$$

The equal condition of ray path length can be expressed in Equation (3) where Lt indicates the total path length from the feed to the aperture plane.

$$Lt = r + n(\frac{x - rcos\varphi}{cos\varphi}) + z_o - z = \text{constant} \tag{3}$$

By using this equation, the variable φ in Equations (1) and (2) can be expressed by the variable θ. Then, through simplification, the variable of Equations (1) and (2) becomes only θ. The electric power conservation at the ray is composed of dx and $d\theta$

$$\frac{dx}{d\theta} = \frac{Ep^2(\theta)}{\int_0^{\theta m} Ep^2(\theta)d\theta} \frac{\int_0^{Xm} Ed^2(x)dx}{Ed^2(x)} \tag{4}$$

The three differential Equations (1), (2) and (4) determine the lens shape in the MATLAB program. For the electric field intensity at feed radiator, $Ep^2(\theta)$, Equation (5) is implemented. Meanwhile, Equation (6) is used for aperture illumination distribution, $Ed^2(x)$.

$$Ep^2(\theta) = cos^m(\theta) \tag{5}$$

$$Ed^2(x) = [(1 - (1 - \frac{1}{C})(\frac{x}{X_m})^2)]^P \tag{6}$$

2.2.2. Abbe's Sine Condition

In optic, a collimating lens can be designed to be a coma free for a limited scan by imposing the Abbe's sine condition. Coma refers to the aberration inherent to certain optical designs or due to imperfection in the lens or other components that results in off-axis point sources. This condition is automatically fulfilled if the inner surface of a conventional waveguide lens is spherical. The condition in the red circle in Figure 3 interprets the Abbe's sine condition. When the initial and the final ray are extended, these rays are intersecting inside the lens on a circle radius of f_e [2]. Abbe's sine law in Equations (7) and (9) are applied in the MATLAB program to design the shaped lens.

$$x = f_e sin(\theta - d\theta) \tag{7}$$

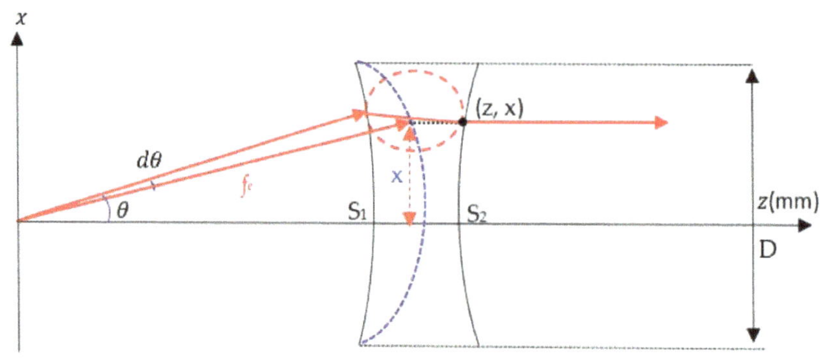

Figure 3. Abbe's sine condition.

For $d\theta << 1$, Equation (7) becomes

$$x = f_e sin\theta \tag{8}$$

$$\frac{dx}{d\theta} = f_e cos\theta \tag{9}$$

2.3. Program Flow Chart

A flow chart in Figure 4 represents the processes involved in negative refractive index lens design using MATLAB software. Program codes were developed based on the equations and formula that were previously described in Sections 2.2.1 and 2.2.2. The initial parameters (n, θ_m, r_o, d_o) determined at the lens edge are shown in Figure 5. The equations for feed radiation pattern, $Ep^2(\theta)$, and aperture distributions, $Ed^2(x)$, are given. Next, differential Equations (1), (2) and (4) or (1), (2) and (9) are solved by the MATLAB routine of "ode45".

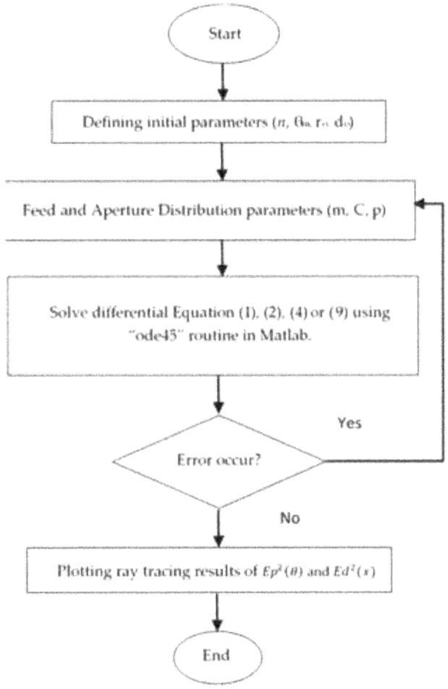

Figure 4. MATLAB Program.

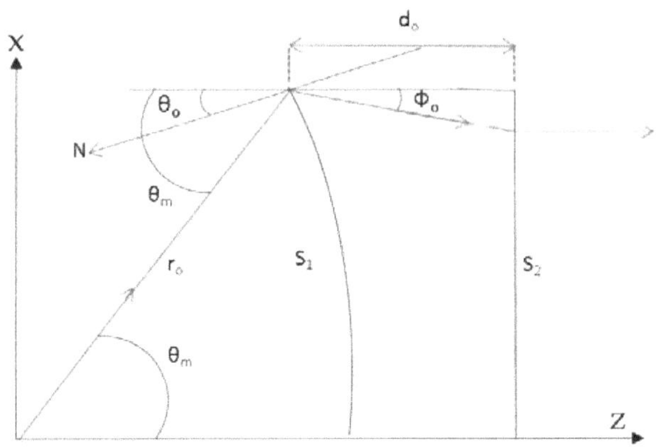

Figure 5. Initial parameters at the lens edge.

The explanation for the initial input values is shown in Figure 5. θ_0 is an important parameter that determines the lens thickness where d_0 is the initial thickness.

3. MATLAB Shape Design

3.1. Simulation of Energy Conservation Law

A negative refractive index lens is designed based on the energy conservation law by using Equations (3) and (4) in the MATLAB shape design program.

3.1.1. Aperture Distribution Design

Figure 6a shows the horn feed radiation, $Ep^2(\theta)$. The lens area is about −55° to 55° at −10 dB. It means that the area that is covered by the horn radiation beam is from −55° to 55°. The aperture distribution, $Ed^2(x)$, is shown in Figure 6b, which represents the edge level of the aperture distribution.

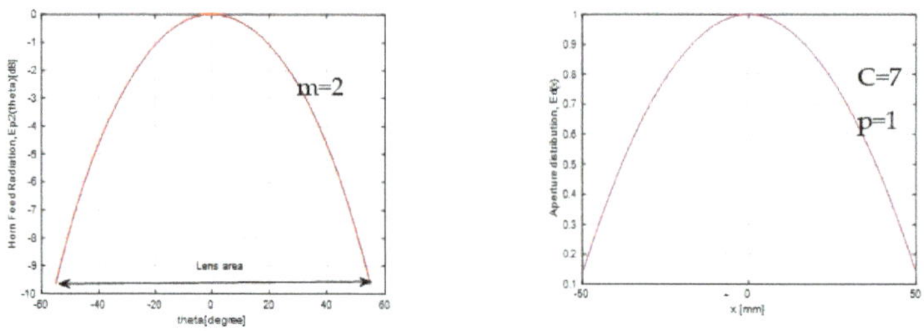

(a) Horn Radiation, $Ep^2(\theta)$ (b) Aperture Distribution, $Ed^2(x)$

Figure 6. Energy conservation design.

Aperture distribution can be expressed as in (10). When Equation (10) is compared with theoretical Equation (6) in Figure 7, it is observed that both equations have a good agreement with each other.

$$Ed^2\ (x) = \frac{d\theta}{dx}\ Ep^2\ (\theta) \tag{10}$$

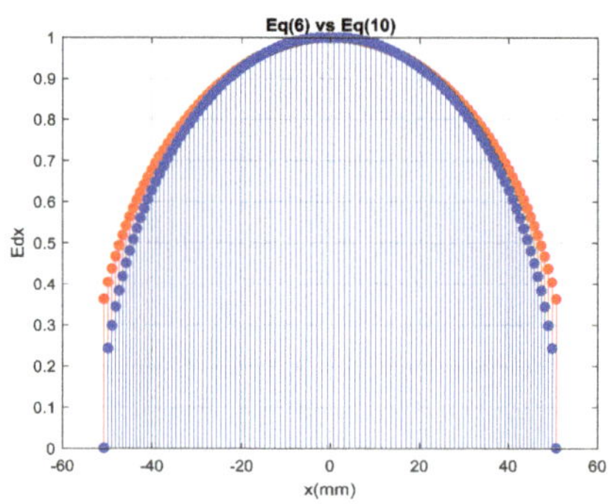

Figure 7. Comparison of aperture distribution.

Considering all the design parameters stated previously, the shaped lens was obtained as illustrated in Figure 8. For the designed lens in Figure 8, the accuracy of the MATLAB program was validated by calculating the incident angle, θ_i and refracted angle, θ_r at each surface of the lens. The refractive index value, n, was manually determined using the Snell's Law in (10). The calculated n value is expected to be similar to the n value ($n = -\sqrt{2}$), which is used in the MATLAB design program. The lens parameters are tabulated in Table 2.

$$n_1 sin\theta_i = n_2 sin\theta_r \tag{11}$$

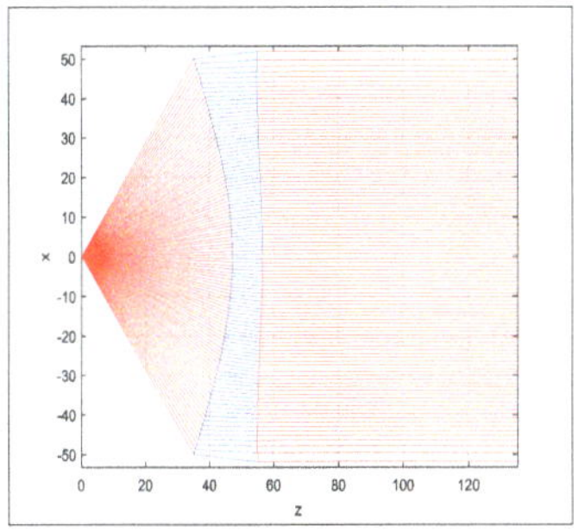

Figure 8. Shaped lens.

Table 2. Lens parameters.

Parameters	Value
Focal length (mm)	46.56
θ_m (°)	55
θ_o (°)	−24
n	$-\sqrt{2}$
C	7
d	20
Diameter (mm)	100

Table 3 represents the calculated n values considering the four first rays coming from the shaped lens in Figure 9a,b. The n_1 and n_2 values were calculated based on Equation (11) and compared to the exact n value that is used in MATLAB which is $n = -\sqrt{2}$. It is noticed that there is a small difference between the calculated values due to some errors that occurred during angle determination.

Table 3. Calculated n-value.

Ray in	θ_i (°)	θ_r (°)	n_2	Ray out	θ_i (°)	θ_r (°)	n_1
1	32	19	1.62	1	0.2	0.3	1.49
2	31	19	1.58	2	0.7	1.0	1.42
3	32	20	1.54	3	1.0	1.5	1.49
4	30	20	1.46	4	1.5	2.0	1.33

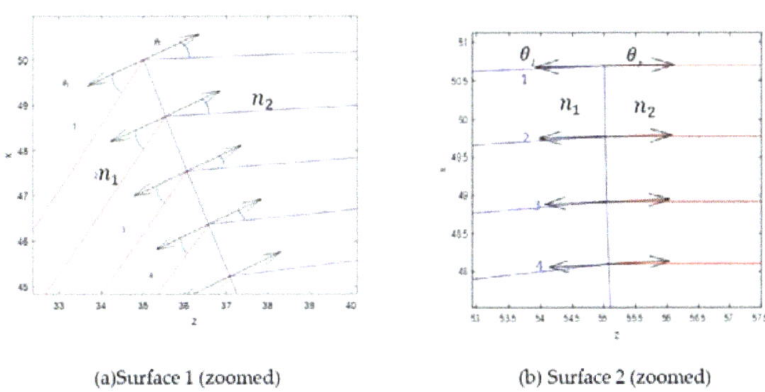

Figure 9. Rays in and rays out.

3.1.2. Abbe's Sine Condition

The same process was repeated for Abbe's sine Law by using Equations (7) and (9) in the MATLAB design program. The shaped lens is shown in Figure 10. The lens parameters are tabulated in Table 4.

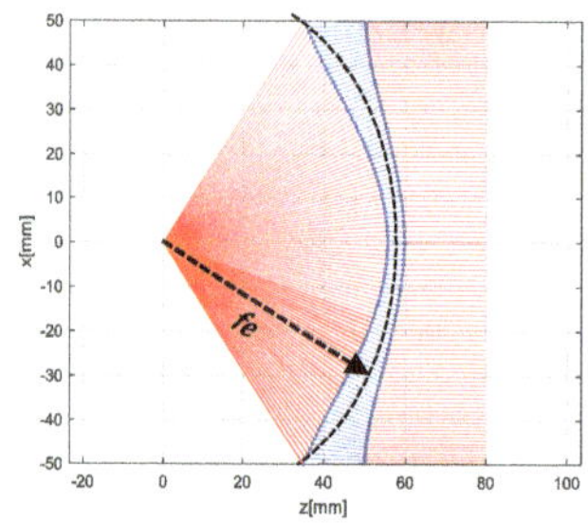

Figure 10. Abbe's sine shaped lens.

Table 4. Lens parameters.

Parameters	Value
fe (mm)	56.62
θ_m (°)	55
θ_o (°)	−24
n	$-\sqrt{2}$
C	7
d	15
Diameter (mm)	100

Table 5 represents the calculated n value considering the four first rays from the shaped lens in Figure 11a,b. A small difference between the calculated values and the exact value used in MATLAB may be due to errors during angle determination.

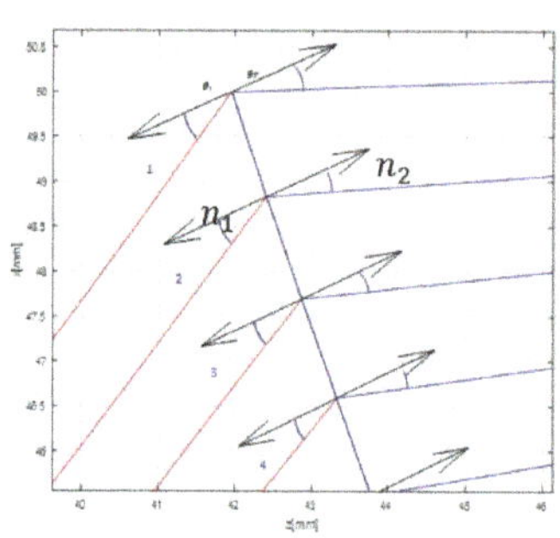

(a)Surface 1 (zoomed)

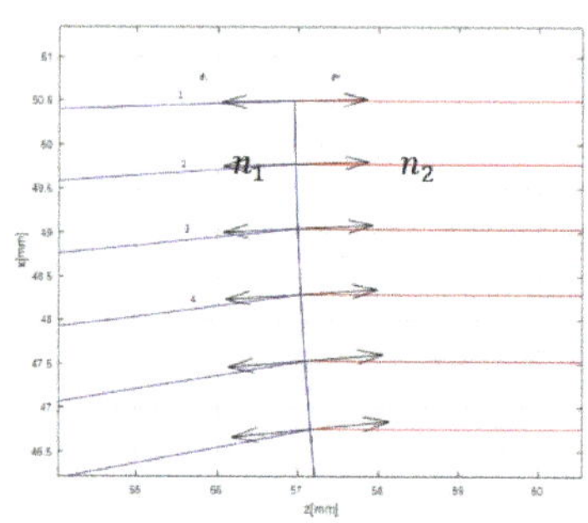

(b) Surface 2 (zoomed)

Figure 11. Rays in and rays out.

Table 5. Calculated n-value.

Ray in	θ_i (°)	θ_r (°)	n_2	Ray out	θ_i (°)	θ_r (°)	n_1
1	28	21	1.31	1	0.8	1	1.25
2	27	19	1.39	2	1.6	2	1.25
3	25	17	1.45	3	2	3	1.50
4	24	15	1.57	4	3	4	1.33

4. EM Simulations for Radiation Characteristics

4.1. Horn Feed

In order to calculate a negative refractive index material, the HFSS Simulator was used. In MATLAB, a point source was used to represent the feed, but in HFSS, a horn antenna was used to represent the feed radiator. Table 6 shows the simulation parameters using HFSS.

Table 6. Simulation parameters using High Frequency Structure Simulator (HFSS).

Parameters	Description/Value
Boundary Condition	Radiation Boundary
Refractive index,	−1.4142 ($\sqrt{-2} * \sqrt{-1}$)
Permittivity, ε_r	−2
Permeability, μ_r	−1

A conical horn antenna is designed to operate at 28 GHz as shown in Figure 12a with the radiation pattern illustrated in Figure 12b.

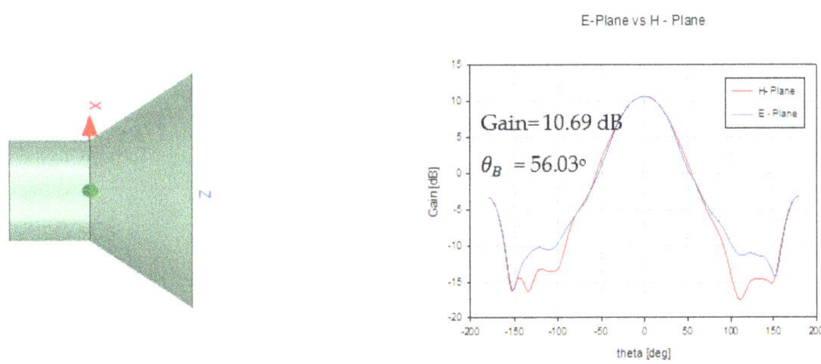

(a) Horn structure (b) Radiation pattern

Figure 12. Performance of horn antenna as a feed radiator.

4.2. *Energy Conservation Law*

4.2.1. On-Focus

The designed shaped lens structure was simulated for performance evaluation. The lens structure has a permittivity of $\varepsilon_r = -2$ and permeability $\mu_r = -1$. Figure 13a,b show the electric field distribution and the antenna gain, respectively. From Figure 13a, it is observed that the lens transformed the spherical wave into the plane wave. The obtained parameters are shown in Table 7. The theoretical value shows the uniform aperture case. This lens antenna structure produced a 27.55 dB gain with a beamwidth of 7.81°. The efficiency of the structure is about 66%.

Table 7. Lens structure simulation result.

	Theoretical	Simulation
Gain (dB)	29.35	27.55
θ_B (°)	8.01	7.81
ΔG (dB)	0	−1.80
η (%)	100	66

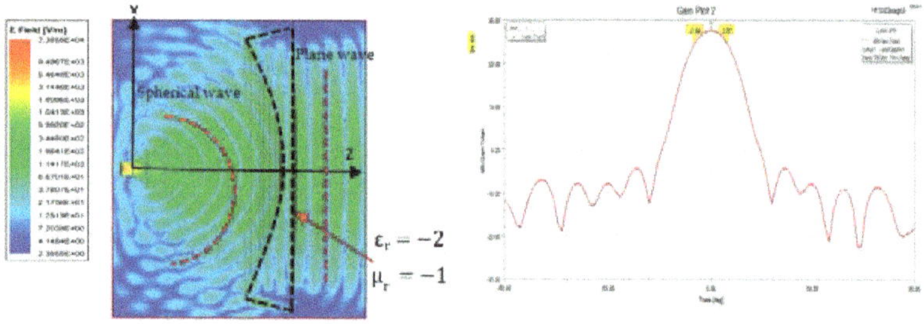

(a) Electric Field (b) Antenna Gain

Figure 13. Energy conservation shaped lens performance.

4.2.2. Off-Focus

In order to prove that the designed lens shape is suitable for wide-angle beam scanning, multibeam characteristics are investigated in this paper. The feed horn position shifts spherically by 10° using the formula of $F = fs$, where F is the distance between the feed position and the center of the lens, while fs is the focal length. The off-focus position is illustrated in Figure 14.

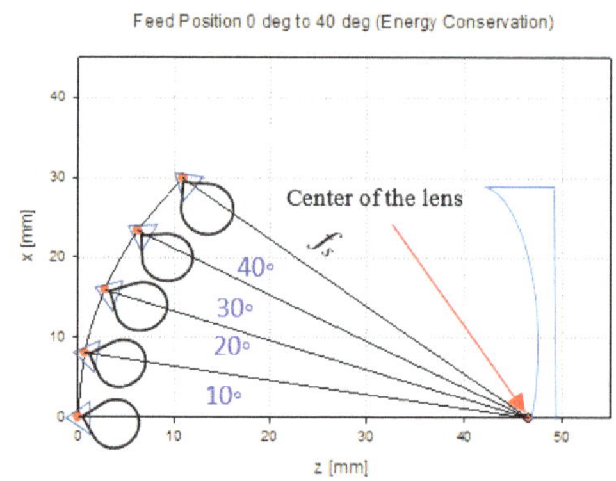

Figure 14. Feed position arrangement.

The results of antenna gain for different feed positions is illustrated in Figure 15 and Table 8. It is observed from the graph that there is no significant gain reduction for all the scanning angles. The scanning losses are all less than 3 dB, which indicates that beamwidths are consistent for all scanning angles. It can be concluded that this structure is suitable for wide-angle beam scanning.

Table 8. Simulation results of multibeam.

θ (°)	Gain (dB)	θ_B (°)
0	27.55	7.81
10	27.14	7.72
20	26.50	8.31
30	25.18	8.97
40	23.09	9.06

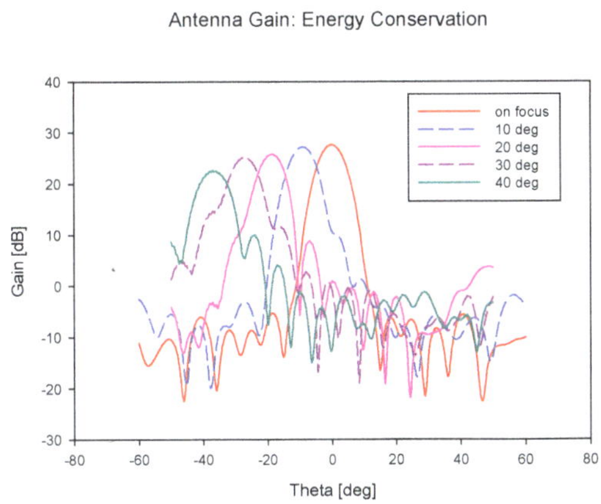

Figure 15. Antenna gain for scanning angle 0° to 40°.

Figures 16 and 17 show the results of the electric distribution in magnitude and phase, respectively, performed through near-field calculation. For on-focus condition, both intensity and phase distributions show uniform and symmetrical characteristics as shown in Figures 16a and 17a, respectively. In off-focus condition, the uniform characteristics are expected to be disturbed by some distortions due to the non-linearity in phase; thus, the behaviors are non-symmetrical, as observed in Figures 16b and 17b. Both electrical intensity and phase are slightly distorted when the feed is displaced from its original position.

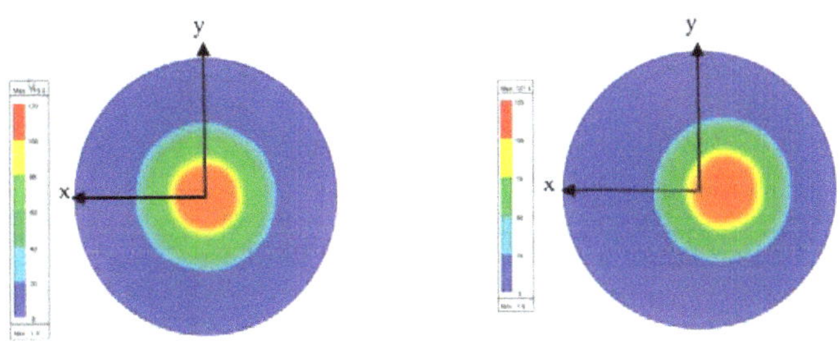

(a) On-focus (b) Off-focus

Figure 16. Electric intensity distribution.

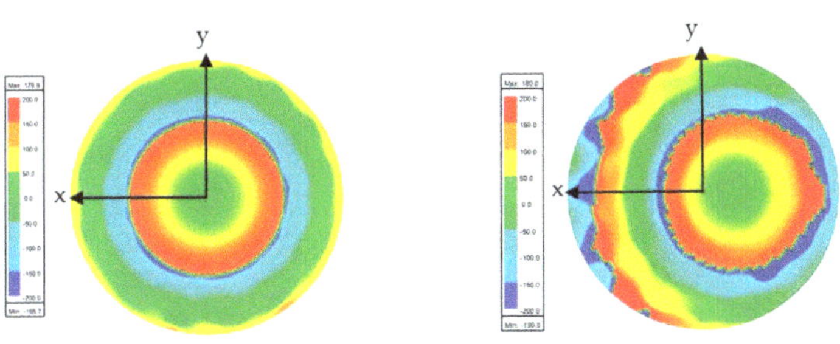

(a) On focus (b) Off-focus

Figure 17. Electric phase distribution.

4.3. Abbe's Sine Law

4.3.1. On-Focus

For Abbe's sine condition, a shaped lens structure with the permittivity value of $\varepsilon_r = -2$ and permeability value of $\mu_r = -1$ was simulated as shown in Figure 18a. The performance of the designed lens is illustrated in Figure 18b. It is observed that this Abbe's sine shaped lens produced a gain of 27.48 dB. The beamwidth is 7.64°. This lens also has 65% efficiency, as shown in Table 9.

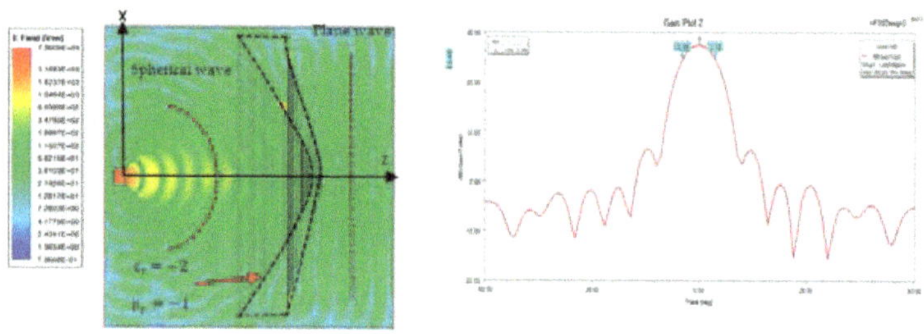

(a) Electric Field (b) Antenna Gain

Figure 18. Abbe's sine shaped lens performance.

Table 9. Simulation results.

	Theoretical	Simulation
Gain (dB)	29.35	27.48
θ_B (°)	8.01	7.64
ΔG (dB)	0	−1.87
η (%)	100	65

4.3.2. Off-Focus

For multibeam investigation, the feed position of the horn antenna is varied from 0° to 40° with 10° intervals as illustrated in Figure 19. The same condition in energy conservation ($F = fs$) was applied for Abbe's sine shaped lens. The results for all scanning angles are represented in Figure 20 and Table 10. The gain is slightly lower as compared to the energy conservation shaped lens. However, there is no significant gain reduction for every scanning angle. The beamwidths are also consistent for all scanning angles.

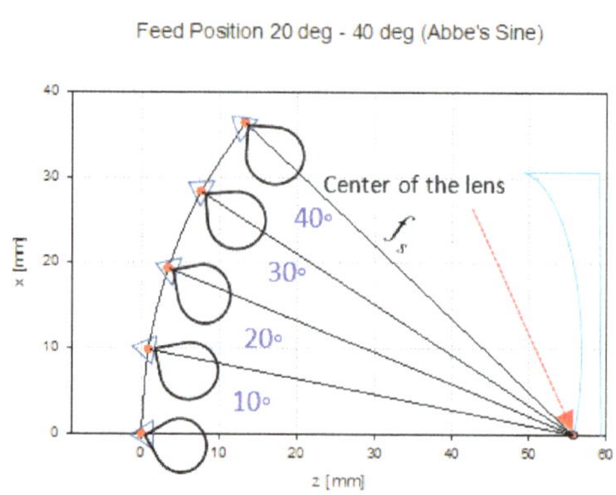

Figure 19. Feed position arrangement.

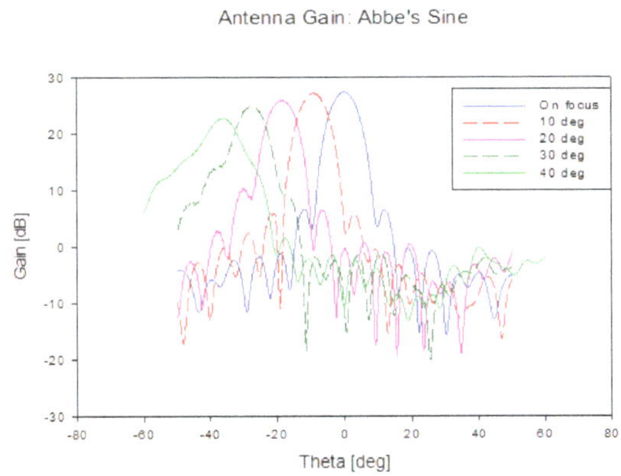

Figure 20. Antenna gain for scanning angle 0° to 40°.

Table 10. Simulation results of different feed positions.

θ (°)	Gain (dB)	θ_B (°)
0	27.48	7.64
10	27.19	7.37
20	26.73	8.08
30	25.54	8.43
40	22.92	9.43

Figure 21a,b show the results of the electric field distribution during off-focus for Energy Conservation and Abbe's sine lens, respectively.

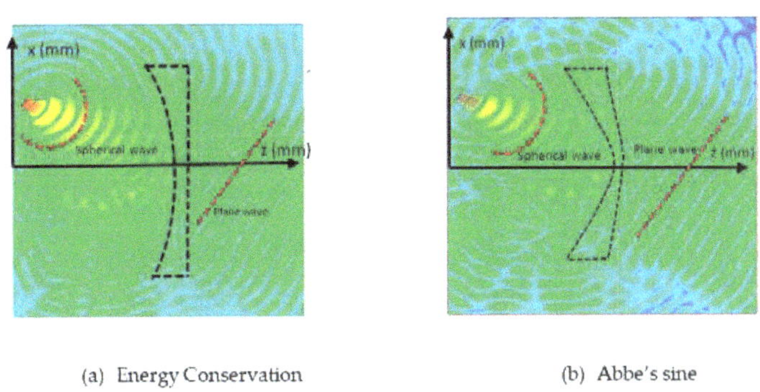

Figure 21. Electric field distribution during off-focus.

Figures 22 and 23 show the results of the electric distribution in magnitude and phase, respectively, for Abbe's sine lens. For on-focus condition, both intensity and phase distributions exhibited uniform characteristics as shown in Figures 22a and 23a, respectively. Figures 22b and 23b present off-focus condition, which shows that both electrical intensity and phase are slightly distorted. The feed displacement from its original position affects the electrical intensity and the phase distribution linearity.

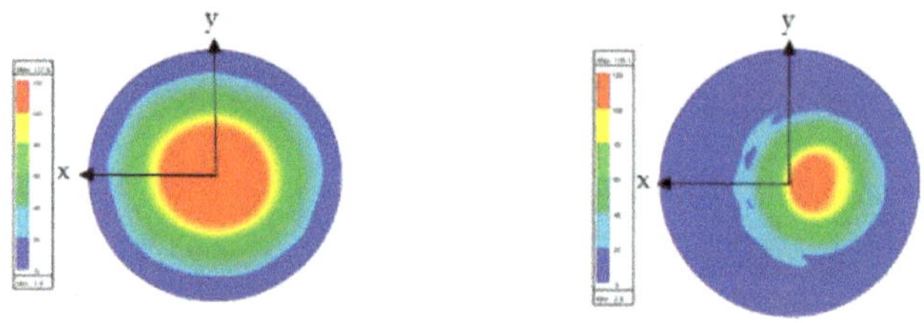

(a) On- focus (b) Off - focus

Figure 22. Electric intensity distribution.

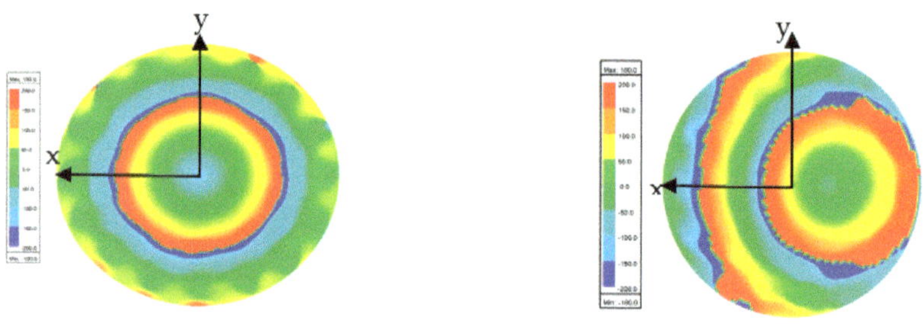

(a) On-focus (b) Off-focus

Figure 23. Electric phase distribution.

The performance for both types of lens are compared in Figure 24a,b. In terms of gain and beamwidth, the performances of both lenses are similar: there is not much difference, as shown in Figure 24a below.

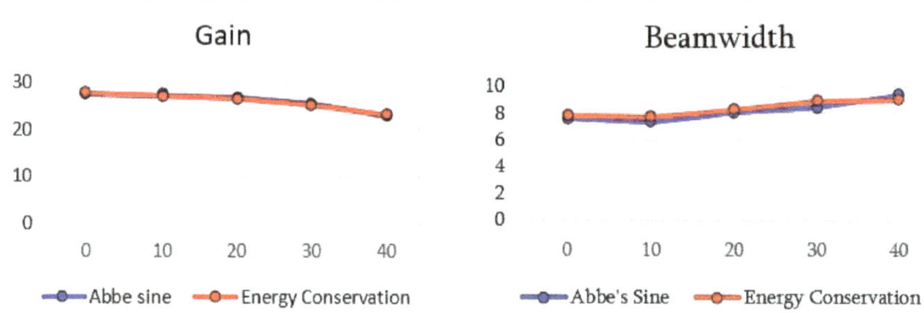

(a) Gain performance (b) Beamwidth performance

Figure 24. Performance comparison for both types of lens.

4.4. Feed Position Analysis

In this section, the performance of the feed position for scanning angles of 20° until 40° is analyzed. The performance of the designed energy conservation lens and Abbe's sine lens is evaluated when the feed is located closer and farther from the lens. The original position is $F = fs$ as shown in Figure 25. θ is the scanning angle which is varied from 20° to 40°. The original position is the calculated position based on the shaped lens. Position 2 is where the feed is placed farther from the lens, while position 3 is where the feed is placed closer to the lens.

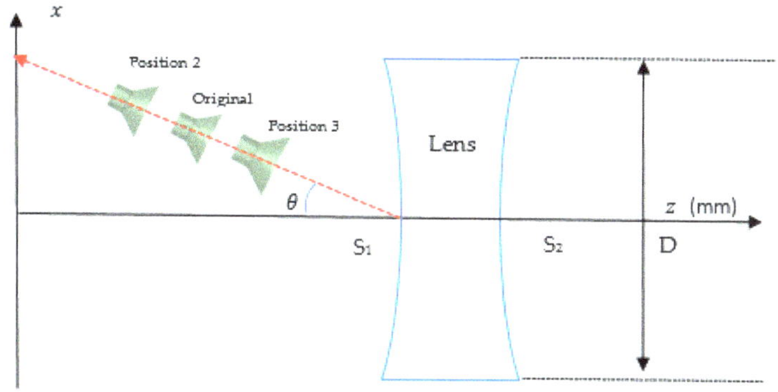

Figure 25. Feed position analysis arrangement.

Figures 26 and 27 show the performance of the designed energy conservation lens and Abbe's sine lens antenna for all feed positions from 20° to 40° scanning angle, respectively. At each scanning angle, there is no significant difference in gain, beamwidth and shift angle between the three feed positions. It can be seen that these two types of designed lens perform consistently for all the feed positions.

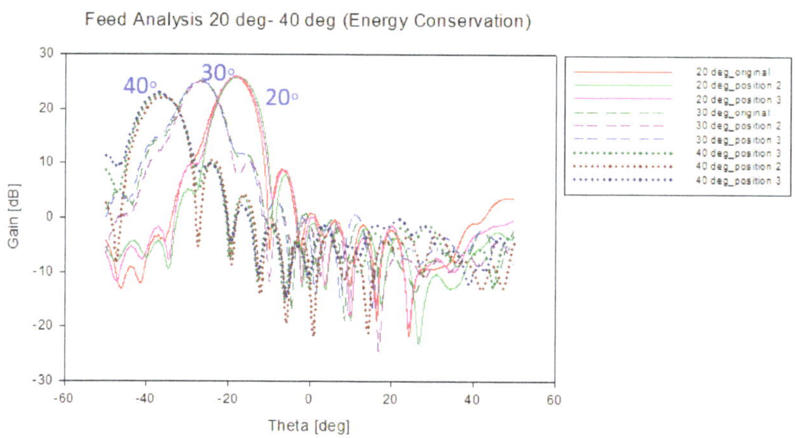

Figure 26. Antenna gain for all feed position for energy conservation lens.

Table 11 shows the overall performance for both types of lens at 20°, 30° and 40° scanning angle. From all the results shown in Figure 26, Figure 27 and Table 11, it can be seen that all the positions along the lines satisfied the $F = fs$ condition with optimum performance. It shows that the $F = fs$ condition and both lens structures are suitable for multibeam applications. Hence, this structure can be applied for 5G mobile base station.

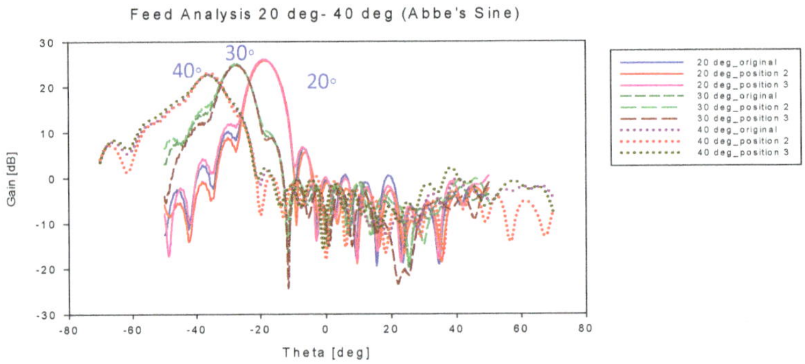

Figure 27. Antenna gain for all feed position for Abbe's sine lens.

Table 11. Performance for all positions at 20°, 30° and 40° scanning angle.

θ		Focal Length (mm)		Gain (dB)		Beam Width (o)		SLL (dB)		Shift Angle (o)	
	Position	Energy	Abbe	Energy	Abbe	Energy	Abbe	Energy	Abbe	Energy	Abbe
20	Original	46.56	56.62	26.50	26.73	8.31	8.08	−17.15	−17.95	−18	−18
	Position 2	48.68	58.74	26.12	26.55	8.36	7.95	−16.63	−17.93	−17	−18
	Position 3	44.44	54.50	26.93	26.66	7.99	8.08	−16.39	−15.93	−17	−18
30	Original	46.56	56.62	25.18	25.54	8.97	8.43	−11.65	−11.37	−25	−27
	Position 2	48.68	58.74	24.92	25.25	8.95	8.50	−11.60	−11.36	−25	−27
	Position 3	44.44	54.50	25.57	25.39	9.02	8.66	−12.24	−11.53	−26	−27
40	Original	46.56	56.62	22.61	22.78	9.48	9.34	−8.05	−7.22	−37	−36
	Position 2	48.68	58.74	22.13	23.18	9.50	9.20	−5.98	−7.94	−37	−36
	Position 3	44.44	54.50	22.92	22.73	9.61	9.41	−6.68	−7.81	−37	−36

5. Conclusions

Two types of negative refractive indexes lens were designed in MATLAB software and simulated by using the electromagnetic simulator ANSYS HFSS. Both the energy conservation lens and Abbe's sine lens offer high gain and narrow beam width characteristics with 66% and 65% efficiency, respectively. A 27.55 dB maximum gain is achieved for the energy conservation lens and 27.48 dB for Abbe's sine lens. Moreover, both shaped lenses provide optimum results for beam scanning up until 40°. High gain, narrow beam width and wide-angle beam scanning capability are the key elements for 5G application. Thus, it can be concluded that these two types of refractive index lens are good candidates for 5G mobile base station application.

Author Contributions: Conceptualization, Y.Y.; Funding acquisition, D.N.Q.; Software, P.V.H.; Supervision, N.H.A.R. and Y.Y.; Validation, Y.Y.; Writing—original draft, S.A.H.; Writing—review & editing, N.H.A.R., Y.Y. and D.N.Q. All authors have read and agreed to the published version of the manuscript.

Funding: The work by Nguyen Quoc Dinh and Phan Van Hung is funded by Vietnam National Foundation for Science and Technology Development (NAFOSTED) under grant number 102.04-2018.08. The work by Salbiah, Nurul Huda and Yoshihide Yamada is funded by Malaysia Ministry of Higher Education under grant number FRGS/1/2018/TK04/UITM/03/5.

Acknowledgments: Authors would also like to acknowledge Antenna Research Centre, Faculty of Electrical Engineering, Universiti Teknologi MARA and Malaysia-Japan International Institute of Technology, Universiti Teknologi Malaysia for supporting this project.

Conflicts of Interest: The authors declare no conflict of interest.

References

1. Stutzman, W.L.; Thiele, G.A. Aperture Antenna. In *Antenna Theory and Design*, 3rd ed.; Ashenberg, W., Ed.; John Wiley & Sons, Inc.: Hoboken, NJ, USA, 2012; Volume 1, pp. 344–428.
2. Lee, J.J. Lens Antenna. In *Antenna Theory*, 1st ed.; Lo, Y.T., Lee, S.W., Eds.; Van Nostrand Reinhold: New York, NY, USA, 1993; Volume 2, pp. 5–58.

3. Milligan, T.A. Lens antenna. In *Modern Antenna Design*, 2nd ed.; John Wiley & Sons, Inc.: Hoboken, NJ, USA, 2005; Volume 1, pp. 447–451.
4. Matytsine, L.; Lagoiski, P.; Matytsine, M.; Matitsine, S. Large Size, Lightweight, Luneburg Lenses for Multi-beam Antenna Applications. In Proceedings of the 6th European Conference on Antennas and Propagation (EUCAP), Prague, Czech Republic, 26–30 March 2012.
5. Liu, K.; Yang, S.; Qu, S.; Chen, Y. A Flat Luneberg Lens Antenna for Wide-Angle Beam Scanning. In Proceedings of the IEEE International Symposium on Antennas and Propagation and USNC-URSI Radio Science Meeting, Atlanta, GA, USA, 7–12 July 2019.
6. Christopher, C.; Tatsuo, I. *Electromagnetic Metamaterial: Transmission Line Theory and Microwave Application*; John Wiley & Son: Hoboken, NJ, USA, 2006; Chapter 7; pp. 316–322.
7. Thorton, J.; Huang, K.-C. *Modern Lens Antenna for Communications Engineering*; IEEE Press: Piscataway, NJ, USA, 2013; pp. 29–30.
8. Kamada, S.; Michishita, N.; Yamada, Y. Metamaterial lens antenna using dielectric resonator for wide angle beam scanning. In Proceedings of the IEEE Antennas and Propagation society international symposium(APSURSI), Toronto, ON, Canada, 11–17 July 2010; pp. 1–4.
9. Samuel Silver. *Microwave Antenna Theory and Design*; McGrawhill: New York, NY, USA, 1949; p. 394.

sensors

MDPI

Article

A High-Performance Transmitarray Antenna with Thin Metasurface for 5G Communication Based on PSO (Particle Swarm Optimization)

Chengtian Song [1,*,†], Lizhi Pan [1,†], Yonghui Jiao [1,†] and Jianguang Jia [2,†]

1 School of Mechatronical Engineering, Beijing Institute of Technology, Beijing 100081, China; 3220180119@bit.edu.cn (L.P.); 3220190122@bit.edu.cn (Y.J.)
2 Institute of Systems Engineering, AMS, PLA, Beijing 100091, China; jiajianguang@126.com
* Correspondence: songct@bit.edu.cn
† All authors contributed equally to this work.

Received: 2 July 2020; Accepted: 6 August 2020; Published: 10 August 2020

Abstract: A 5G metasurface (MS) transmitarray (TA) feed by compact-antenna array with the performance of high gain and side-lobe level (SLL) reduction is presented. The proposed MS has two identical metallic layers etched on both sides of the dielectric substrate and four fixed vias connecting two metallic layers that works at 28 GHz to increase the transmission phase shift range. The proposed planar TA consisting of unit cells with different dimensional information can simulate the function as an optical lens according to the Fermat's principle, so the quasi-spherical wave emitted by the compact Potter horn antenna at the virtual focal point will transform to the quasi-plane wave by the phase-adjustments. Then, the particle swarm optimization (PSO) is introduced to optimize the phase distribution on the TA to decrease the SLL further. It is found that the optimized TA could achieve 27 dB gain at 28 GHz, 11.8% 3 dB gain bandwidth, −30 dB SLL, and aperture efficiency of 23% at the operating bandwidth of 27.5–29.5 GHz, which performs better than the nonoptimized one. The advanced particularities of this optimized TA including low cost, low profile, and easy to configure make it great potential in paving the way to 5G communication and radar system.

Keywords: high gain; transmitarray (TA) antenna; metasurface (MS); PSO; side-lobe level (SLL) reduction

1. Introduction

Followed by the increasing demands of high-speed data switching rate, the fourth-generation wireless transmission protocol is not enough for the demanding data usage in many areas as more and more devices are connected in the narrow channels exactly like the traffic jam. To handle this problem, the next-generation system, 5G, is developing with a better performance in capacity, information transmitted, and energy efficiency, and showing great prospects for applications [1–4]. Allocated by the Third Generation Partnership Project (3GPP), the spectrum available for 5G network is subdivided into lower frequency part and higher frequency part which is made 6 GHz as the separation [1,2]. Although compared with the narrow channels and the higher loss in sub-6 GHz bands, the higher frequency band such as 66 and 28 GHz, with the general properties of lower path loss caused by the atmospheric absorption and less occupied channels, will be a competitive candidate for the development of 5G communication [3,4], and the research on the array antenna are getting more and more attention in recent years to satisfy the stricter requirements of 5G.

Historically, in the field of the international communication industry, the array antenna plays an irreplaceable role since it was invented. Owning to the performance of high gain, low side lobe level (SLL), and wide bandwidth, the array antenna always requires different forms to satisfy the growing corresponding needs [5–14]. The phased-array (PA) antenna performs various functions in the communication system [5–8]. It has been demonstrated that the PA has several advantages, such as compact array and high gain [5–8]. To circumvent the complexity of feeding networks, single microstrip antennas for high-directivity have been proposed [9]. PA, however, suffers from the complicated feeding networks and complex design of the excited phase at different ports.

To simplify the design and further enhance the gain, thin planar array with separated feed antenna, such as the reflectarray (RA) antenna and transmitarray (TA) antenna, are proposed in the areas of long-distance communication and radar system [10,14–16]. As for the RA, the metal ground usually covers one side of the array to reflect the incident wave, so only the single layer with almost 0 dB (close to 1) return loss is required, and only the phase distribution is needed to be considered in the design without the worry of the amplitude distribution. The radiation mechanism and a general design method of the RA is discussed in detail in Ref. [10]. In Ref. [11], the patches with identical phase-delay lines but different element rotation angles were used to regulate the reflected phase distribution to obtain high gain and beam scanning. Regardless of the advantages of low return loss and single layer, RA also faces some constraints, like the patch feed and the horn feed have the blocking effect to the reflected wave, which will enhance the backward radiation of RA [12,13]. In Ref. [17], ultrawideband low-profile TA feed by Vivaldi array, which has relatively large size, is discussed.

To overcome these shortages, the TA is developed to overcome the shortcoming of the RA and has greater tolerance to the surface errors [18]. Unlike the incident wave and reflected wave at the same side of the RA, the incident wave illuminates on the TA and passes through the array becoming the transmitted wave, so the blocking effect could be avoided largely. In general, there are mainly two ways to classify the TA. One is according to the number of layers of the dielectrics; the four-layer structures are presented in [15,19–22] and three-layer structure is proposed in the Ref. [23], while the two-layer structures are exploited, respectively. The other is based on the various working principles, TA could be further categorized into four different types, namely, frequency selective surfaces (FSS) [15,18–20,22,24], metasurface (MS) [21,23,25,26], the spatial bandpass-filter based structure [27], receive/transmit-mode structure, and so on [16,28,29]. Owing to the transmission, amplitude and phase should be taken into consideration at the same time, how to simply the TA while keeping the specific properties of lower SLL as well as greater surface tolerance for the application in 5G communication has become an urgent research hotpot. The method of optimization can be divided into two categories: local and global search algorithms [30]. Literally, the local optimization is worked in a limited space with designed initial values, whereas the global optimization is forced on the entire solution hyperspace and just the random initial values are needed and the best result will be found in the whole search space [31]. As for the optimization of phase distribution, there are always hundreds of elements that need to be optimized and it is difficult to set up suitable initial values for all the elements befittingly, so the global search algorithms are used here for the optimization of phase distribution. Considering the flexible and independent properties of the elements on the array, heuristic method is widely adopted in the optimization of array antenna among which, genetic algorithm (GA) and particle swarm optimization (PSO) show great superiority at the aspect of efficiency and operability [31–36]. Though PSO and GA both belong to the population-based stochastic optimization, PSO has only two lines of kernel codes compared to the recurrent process of crossover, breeding, and selection in GA. PSO also benefits from less computational bookkeeping with less computational resource [33,34].

In this paper, we propose a two-layer MS TA antenna feed by the optimized horn antenna to actualize the reduced SLL and high gain at the frequency range around 28 GHz for the 5G communication. This thin TA comprises a single substrate and two circular patches on both sides of the substrate, and four metal vias are inserted into the substrate at the centrally symmetrical positions for each element, which is based on structure presented in the Ref. [28] working at 20 GHz but with more simple and efficient design. The positions of vias are fixed at the changeless positions on the array in this paper, while the positions of vias changed with the varying patch size to match the requirements of phase distribution in the published paper. Only two sizes of circular patches are used, and the length of slots on the patches will control the transmitted phase. To the author's knowledge, it is the first time when the transmission coefficients are controlled using this structure. Hence, a TA structure with lower cost and easier fabrication is presented in this paper. Then, an optimized Potter horn antenna is exploited as feed structure of the TA, which has good symmetry in principal plane patterns as the corrugated horns, and the smooth wall of the Potter horn greatly reduces the difficulty of manufacturing. The TA and the Potter horn antenna make the whole prototype as a 5G TA antenna, and a 27 dBi maximum gain in the main direction is achieved, while the 3-dB gain bandwidth and aperture efficiency are maintained at 11.8% and 28%, respectively. The SLL is maintained at −20 dB level and after the optimization, the SLL can achieve the level of −30 dB, which can be used as the high-gain array antenna for 5G directional communication. Our contributions are listed as follows: in Section 2, first, a horn antenna is optimized by $\cos^q(\theta)$-mode method to get the approximate uniform radiation in azimuthal direction; second, a transmission unit cell with single substrate layer was designed in an easier way without using multisubstrate layer structure or air gaps; and third, particle swarm optimization is used to optimize the SLL of TA antenna radiation pattern. In Section 3, the optimized results are compared with the published paper and show better performance in the SLL decreasing.

2. Transmitarray Antenna Design

In the design of thin planar array antenna with separated feed, there are always five steps [37]. The first step is to find a suitable feed structure. The feed at the optimized position is an important factor related to the gain enhancement and the SLL reduction. The following step is the analysis of unit cell. Periodical boundary condition with plane wave incidence are used to get series of transmission coefficients by sweeping the related dimension. The third step will be the design of TA according to the transmission coefficients at 28 GHz, which are extracted from the previous step. In addition, the fourth step will be the joint simulation of TA and feed structure to achieve the desired performance, furthermore, the final step is to optimize the transmission coefficients of elements at different positions on the array to reduce the SLL, which will be discussed in the next section.

2.1. Feed Antenna Design

Typically, patch antenna and a horn antenna are often used as the feed source of TA and RA [12–15,18–23,26–28,38]. The substrate integrated waveguide (SIW) antenna is often used as the feed of MIMO TA antenna for the beam forming, and the Vivaldi antenna can also be used as the feed for the beam steering [25,39]. The patch antenna was chosen because of its easy manufacturing and small transmission direction size, which shows the inherent advantage of the low-profile property, but the patch antenna is often limited by the narrow bandwidth, in general, the −10 dB bandwidth of patch antenna is less than 10%. Another shortcoming of the patch antenna is that it is difficult to keep the symmetric radiation pattern in the azimuthal direction because the asymmetrical radiation will bring the distortion to the radiation when the TA or RA is feed by off-set source or in the situation of beam steering [18,19].

The horn antenna itself has the advantages of higher gain and wider bandwidth, so it can bring a higher gain enhancement and broader bandwidth for the TA antenna. Though the horn antenna often suffers from the bulky volume and larger size in the direction of wave propagation, it is easy to achieve the symmetric radiation pattern in the azimuthal direction. However, the conventional pyramid and conical horn antenna fail to perform the symmetric pattern due to the fundamental mode excitation, like the Potter horn or corrugated horn, higher-order mode is introduced into the cavity [15,19,20,27,28,38,40]. In this paper, we utilize the advantages of Potter horn antenna to optimize a feed source at the frequency range of 26–30 GHz with symmetrical radiation pattern for the SLL reduction in full azimuthal directions.

Figure 1 shows the cross-sectional view of the Potter horn antenna operating at 28 GHz. The feed structure is 21.16 mm × 21.16 mm × 27.96 mm as the effective volume, which consists of a circular waveguide, two conjunctions, and a flared section. The horn is made of copper with a thickness of 1 mm. The gain is 14.6 dBi at 28 GHz, and half power beam width (HPBW) is 36° at E-plane and H-plane in order to simplify the further optimization about the radiation and the aperture efficiency. In Ref. [37], the horn antenna can be molded as the function of the $\cos^q(\theta)$ in the scalar expression of far-field radiation pattern for a planar antenna array, and it is reasonable that q = 6.5. The curve of $\cos^q(\theta)$ and the radiation pattern of the Potter horn antenna in E-plane and H-plane are shown in Figure 2. After optimization, it is obvious that the radiations not only in E-plane but also in H-plane match well with the theoretical curve of $\cos^q(\theta)$ in the theta range from −45° to 45°.

The values of the final optimized parameters are: L_w = 3.2 mm, L_1 = 4.05 mm, L_2 = 16.14 mm, L_c = 8.9 mm, R_w = 3.39 mm, R_1 = 5.99 mm, R_2 = 8 mm, and R_c = 10.58 mm. All the results are simulated by CST Microwave Studio. L_w is the length of the circular waveguide and the waveguide excites the dominant TE_{11} mode and generate a directive beam with asymmetric radiation pattern. L_1 and L_2 are the length of the two abrupt changes of the waveguide in which the first junction keeps the power in the TE_{11} mode, the second junction excites the TM_{11} mode, and meanwhile, the flared horn is worked to avoid the propagation of TM_{12} mode. The simulation result of return loss (S11) with a −10 dB bandwidth of the feed source is 14%, which is from 26.39 to 30.37 GHz. The side lobe levels are −26.2 dB and the Half Power Beam Width (HPBW) are about 36° in both the E-plane and the H-plane. As shown in Figure 3, the electric fields at the cross sections of the optimized Potter horn are presented. It is obvious that the symmetric aperture fields are generated at the horn's cross section which lead to the symmetric radiation patterns regardless of the azimuthal angles.

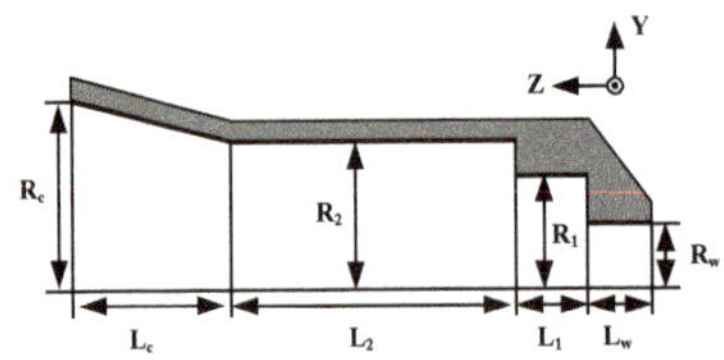

Figure 1. The cross-sectional view of the Potter horn antenna as the feed source.

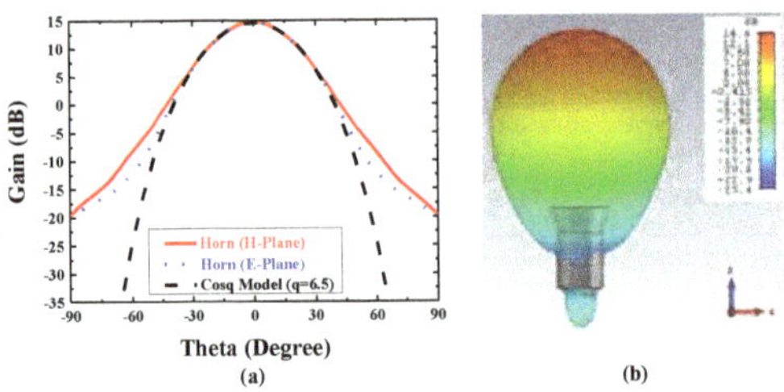

Figure 2. The radiation pattern of the proposed antenna. (**a**) Comparison of $\cos^q(\theta)$ curve and the radiation pattern of the horn in E-plane and H-plane. (**b**) The 3D radiation pattern at 28 GHz.

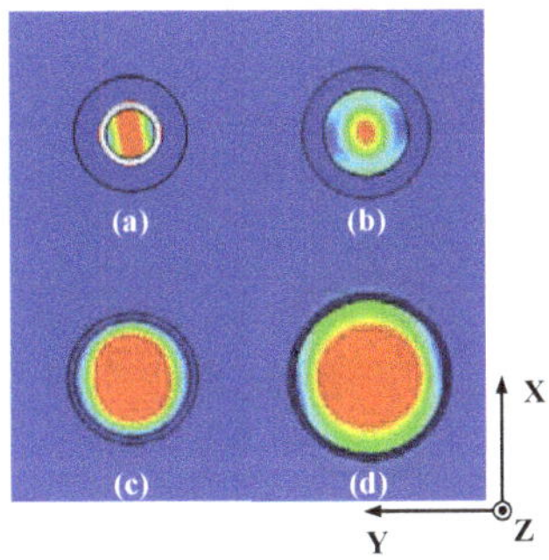

Figure 3. The electric field magnitude at the cross sections inside the optimized horn antenna: (**a**) inside the circular waveguide, (**b**) at the first abrupt change part, (**c**) at the second abrupt change part, and (**d**) at the flared aperture.

2.2. Design of Unit Cell of MS

The proposed unit cells with two identical conductive layers printed on the both sides of the substrate and the slots control the changing of transmission coefficients; the whole structure is presented in Figure 4. The material of substrate is Rogers 4003C, which has the dielectric constant (ε_r) of 3.38 and loss tangent of 0.0027. To get the phase and amplitude of transmission coefficient (S_{21}), CST Microwave Studio is used to simulate the unit cell structure with period boundary condition in X and Y direction and the excitation of the normal plane wave in Z direction. In Figure 4c, the side length of square substrate (*P*) is 5.3 mm, the thickness (*T*) of substrate is 1.524 mm, and four vias are inserted in the substrate to connect the upper and bottom metallic layers.

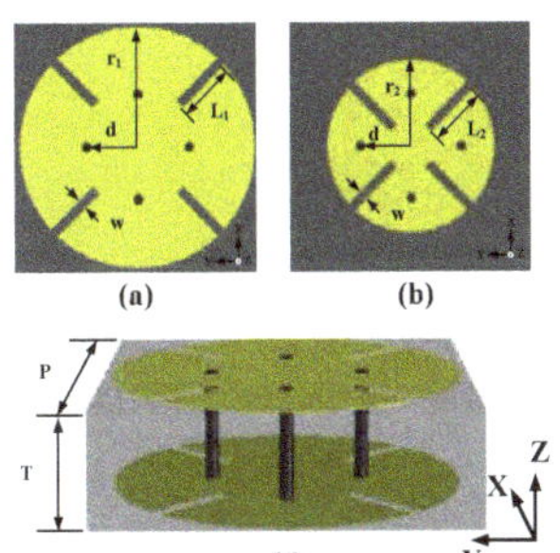

Figure 4. The structure of unit cell: (**a**) Unit Cell A, (**b**) Unit Cell B, and (**c**) side view of unit cell.

The unit cells can be classified into two group: Unit Cell A as shown in Figure 4a and Unit Cell B as shown in Figure 4b. For Unit Cell A, r_1 = 2.55 mm, L_1 changes from 1. 28 to 1.85 mm, and a phase shift range from 360° to 156° is covered. For Unit Cell B, r_2 = 1.83 mm, L_2 varies from 0.61 to 1.29 mm, while the phase shifts from 156° to 0°. The residual parameters like the position of vias (*d*) and the width of slots (*w*) are kept the same in both Unit Cell A and Unit Cell B: *d* = 1.1 mm and *w* = 0.2 mm. Each unit cell has four vias at the symmetric positions away from the center. Only two different sizes of circular metallic patches are used to build the TA, and meanwhile, the length of slot is changed to get different transmission phase, which will be used to focus the beam. Unlike in the Ref. [28], the 304° phase shift is achieved by changing the side length of square patch and the position of via is varied with the patch size proportionally; the TA in this paper can cover the 360° phase shift with sweeping the length of slots on Unit Cell A and B, correspondingly. What's more, the positions of via do not change with the dimension of patches, which means the positions of via are fixed on the array, and those will reduce the cost and the difficulty in the procedures of design and fabrication.

Figure 5 shows that the transmission coefficients change with the length of slots while other parameters remain unchanged in this process. Apparently, in Figure 5a, there is a seamless connection between the phase shift of Unit Cell A and Unit Cell B and the full 360° can be covered with a gentle slope to realize the requirement of focusing effect of transmission wave. The transmission magnitudes are all over −2 dB, which is proposed in Figure 5b, and for the high gain of TA design, design with less transmission loss is desirable; it is also be helpful in the optimization of SLL reduction in the following research of TA antenna.

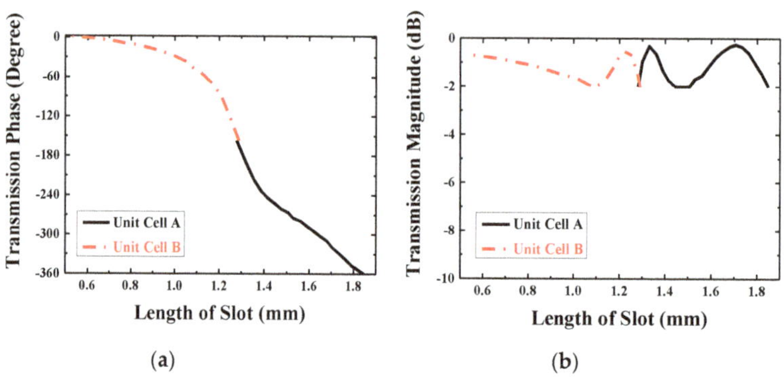

Figure 5. Transmission coefficients: (**a**) transmission phase and (**b**) transmission magnitude.

The length of slots is varied from 1.28 to 1.85 mm with a 0.03 mm step in Unit Cell A, whereas in Unit Cell B, it is varied from 0.61 to 1.29 mm with a step of 0.02 mm, thus the relationship between the length of slots and the transmitted phase will be built for Unit Cell A and Unit Cell B, separately. In this way, it will be easier to confirm the specific length of slot at different positions on the array according to the phase distribution, which will be discussed in the next part.

2.3. Array of MS

The whole circular TA shown in Figure 6a is established by 660 elements with a diameter of 160 mm (D = 160 mm), and from the plot, we can figure out the approximate distribution of Unit Cell A and B. This is because that Unit Cell A covers the phase shift from 0° to 156° and Unit Cell B covers from 156° to 360°, and the related phase at the center of TA is 0°, so all elements are near the center based on Unit Cell B and away from the center of array; the elements based on Unit Cell A and those based on Unit Cell B will appear one after another. After confirming whether Unit Cell A or Unit Cell B is to be used for every element on the array, the corresponding relationship in Figure 5a is used to determine the length of slot for each element, so the final TA will be presented as the structure in Figure 6a. In this process, the required phase φ_{xy} at the position of xy^{th} element on the array is based on (1) [37]:

$$\varphi_{xy} = k(R_{xy} - \hat{r_0} \cdot \vec{r_{xy}}) + \varphi_0 \quad (1)$$

where k is the wave number, R_{xy} is the distance between the feed position and the center position of xyth element on the array, $\hat{r_0}$ is the main beam direction, $\vec{r_{xy}}$ is the position vector of xyth element, and φ_0 represents the constant phase at the central point of the TA. Equation (1) is worked to calculate all the detailed phase values on the array as showed in Figure 6b. With the approximate functions of the element's excitation, the radiation pattern can be simplified into the scalar form [24]:

$$RP(\theta,\varphi) = \sum_{x=1}^{X}\sum_{y=1}^{Y} \frac{\cos^{q}(\theta_f(x,y))}{\left|\vec{r_{xy}} - \vec{r_f}\right|} \cdot e^{jk(\vec{r_{xy}}\cdot\hat{\mu} - |\vec{r_{xy}} - \vec{r_f}|)} \cdot e^{j\varphi_{xy}} \quad (2)$$

$$RP_n(\theta,\varphi) = 20 \times \log_{10}\left(\frac{|RP(\theta,\varphi)|}{\max(|RP(\theta,\varphi)|)}\right) \quad (3)$$

where θ is in the range of −90° to 90° and φ is in the range of 0–360°, $\theta_f(x, y)$ is the spherical angle in the feed's coordinate system, q is the feed pattern factor, which is used in the design of horn antenna with the value of 6.5, $\vec{r_f}$ is the position vector of the feed, and $\hat{\mu}$ is the observation direction. Based on (2), the normalized radiation pattern $RP_n(\theta,\varphi)$ can be achieved and the main beam has a beam width from −10° to 10° with the SLL of −19 dB.

The element at the center of TA is selected carefully. It is true that φ_0 can be an theoretically arbitrary value because the focusing effect of the TA is based on the related phase distribution, which means that the related phase at the center being 0° will make the calculation of phase distribution easier so the phase at other positions will change based on the related 0° to imitate the function of optical lens. However, in this process, the transmission efficiency is regarded as 1, which means no loss is taken into consideration. However, in the actual situation, the design of the TA has to face the problem of transmission loss inevitably and this is always a research direction for the TA. What's more, in the design of TA, there are mainly three parts of approximations needed to be paid attention. First, the transmitted phase and amplitude values obtained from the cell analysis in Figure 5 are in the case of normal incident plane wave, but in fact, elements in the array except the central one are illuminated by different degrees of oblique incident, so there is some deviation. Second, full 360° phase shifting is needed to achieve the focusing effect of the TA, as ideally, the theory requests continuous varying of phase. However, the elements on the TA are all block structures with the same volume, and we just regard the elements as volumetric free and put the center-point coordinate of every element on the array into the (1) to calculate the required phase and use those phases to build the model, so the phase values are presented in the form of cubes in Figure 6b. As a result of which, the phase shifting will be discontinuous, some phase errors will be caused inevitably. Third, the adjacent elements on the array are different from those around them and there is a feed source in the TA antenna so the mutual coupling are complex. However, in the analysis of unit cell, the data are acquired under the situation of periodic boundary condition with the plane wave excited. In general, we ignore the impacts of oblique incidence, discontinuity, and actual mutual coupling to design the TA.

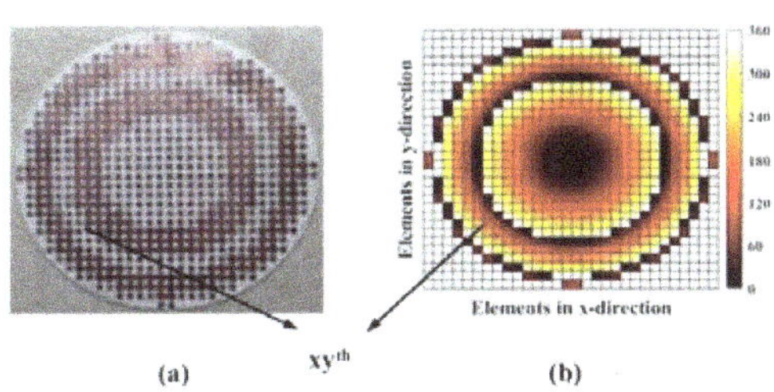

Figure 6. (**a**) The top view of the fabricated transmitarray (TA), and (**b**) the phase distribution of TA calculated by Equation (1).

After the optimization of unit cell, we can keep all the transmission coefficient in the level of −2 dB, which is shown in Figure 5b. From that plot, it is obvious that the most part in Unit Cell B has a higher and smoother tendency changing with the length of slots than the magnitude tendency of Unit

Cell A. Additionally, according to the first and second approximation discussed before, the farther from the center, the greater the design error may occur. Hence, we should put the elements with less transmission loss around the center as much as possible. In this situation, the Unit Cell B with the length of 1.29 mm (L_2 = 1.29 mm) is selected as the four identical elements in the center because of the symmetrical distribution of elements. Other elements are determined based on the relationship between the length of slots and the related phase in Figure 5a; therefore, the distribution of elements with different dimensional information is presented in Figure 6a. Figure 7 presents the positional relation between TA and the feed horn. As described above, D is the diameter of TA, which has the value of 160 mm, and f is the distance from the center of source antenna to the center of TA, and the ratio between f and D (f/D) is an important criterion to judge the SLL and low-profile of a TA antenna. The smaller value of f/D, the better low-profile performance of a TA antenna and the worse of the backward radiation, so there is a tradeoff needed to be made.

For the working principle, the TA is aimed to imitate the function of lens and the feed source is placed at the virtual focal point. When the quasi-spherical wave emitted by the feed crosses over the TA, it will compensate the phases and form a uniform wave front similar to a plane wave. Reversely, when the TA is illuminated by the plane wave, the wave will gather at one point theoretically just like the optical reversibility in the Fermat's principle. In this way of analysis, a plane wave is placed at one transmission side of the TA like shown in Figure 8a, and the power density on the E-plane and H-plane of the other TA side will be shown as in Figure 9. Obviously, the energy is focused at an area, and this area is the approximate position to place the feed. The normalized energy distribution along the intersection line of E-plane and H-plane is proposed in Figure 8b. The peak in Figure 8b is the theoretical optimum of the feed position (f) which has a value of 104.1 mm, so the theoretical ratio of f/D will be around 0.65, and slight changing near 0.65 will be needed in the simulation to find the best position of feed. The simulation result is shown in Figure 10, in which the far-field gain of TA antenna is 27 dB at 28 GHz, SLL in E-plane is −19.6 dB and in H-plane is −21.7 dB, and compared with the radiation pattern of horn, the focusing effect of TA is significant in two cut planes.

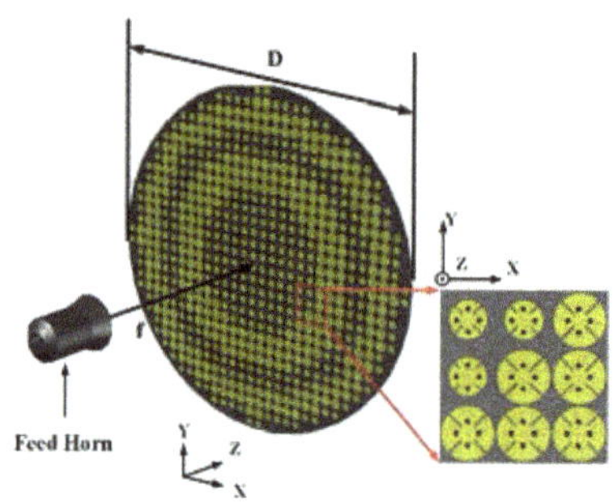

Figure 7. The simulation model of final TA and source antenna.

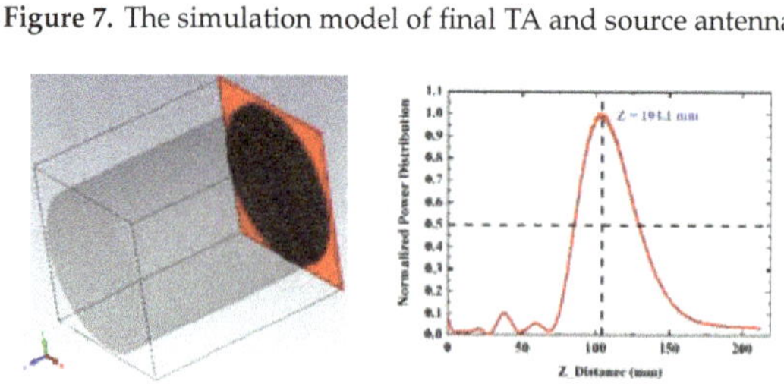

(**a**) (**b**)

Figure 8. (**a**) TA with plane wave and (**b**) the normalized power distribution along intersection line of E-plane and H-plane.

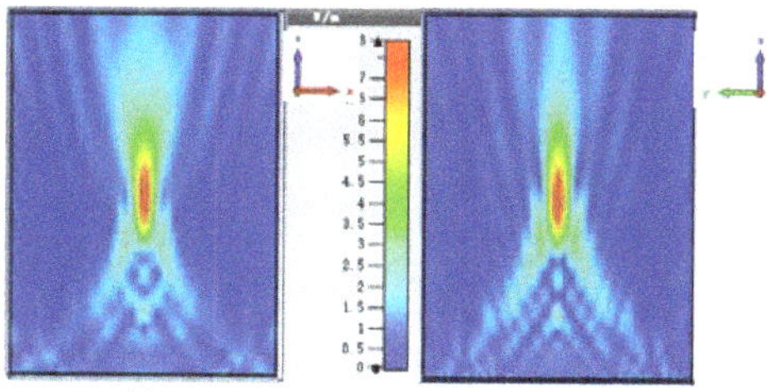

(a) (b)

Figure 9. Electric field (Ex) amplitude distribution: (**a**) E-plane and (**b**) H-plane.

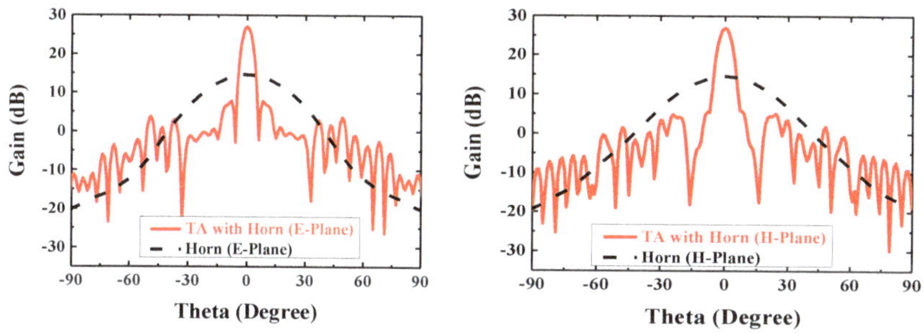

(a) (b)

Figure 10. Simulated far-field radiation pattern of feed source and TA antenna in two cut planes at 28 GHz: (**a**) E-plane comparison and (**b**) H-plane comparison.

2.4. Optimization Result

As the simulation of TA antenna established before, it has a gain of 27 dB with the SLL of −20 dB at 28 GHz, and this result is based on the optimized *f/D* of 0.65, but the position of feed is not the only part that can be optimized in a system of TA antenna. In order to reduce SLL or form multibeams in the desired direction, the phase and magnitude distribution of array can also be optimized [36].

In the Refs. [38,39], various parameters and conditions about the PSO were discussed and compared. In this paper, a 660-dimensional hyperspace is chosen as one possible solution to this phase optimization with 660 elements on the array. In this hyperspace, there are 15 particles with 5000 iterations flying around to find the best position separately (p_{best}), and meanwhile, they share the information with others dynamically (g_{best}), hence they have the tendency to be in the best position, which is the best one for all the particles. This swarm behavior is based on the Equations (4) and (5) of velocity updating and position updating:

$$v_{n+1} = \omega \times v_n + c_1 \times rand() \times (P_{best,n} - x_n) + c_2 \times rand() \times (g_{best,n} - x_n) \quad (4)$$

$$x_{n+1} = x_n + v_n \times t \quad (5)$$

where v_n and x_n are the particle's velocity and the coordinate in the nth dimension, respective; ω is the inertial factor that relates to the previous direction of velocity, and it decreases from 0.9 to 0.4 linearly, which means that it has weaker and weaker relationship with the previous position and turns to the local best position at that moment. C_1 and C_2 are the leaning factor in which C_1 is the self-learning factor and C_2 is the global learning factors, and they are set as the same value of 2.0 to balance autonomy and group learning ability [38,39]. The function of random number rand (1) give a stochastic value between 0.0 and 1.0 in every iteration so that the particle can explore as largest possible hyperspace. Equation (5) shows how the particle move to the next position with the common time step of 1.0. The aim of this PSO is to optimize the phase distribution on the array so that the TA will have a SLL lower than −30 dB. The detailed process of PSO is presented below. First, the radiation pattern in the optimization process is regarded as (6) and (7):

$$RP_1(\theta,\varphi) = \sum_{x=1}^{X}\sum_{y=1}^{Y} \frac{\cos^q(\theta_f(x,y))}{\left|\vec{r_{xy}} - \vec{r_f}\right|} \cdot e^{jk(\vec{r_{xy}}\cdot\hat{\mu} - |\vec{r_{xy}} - \vec{r_f}|)} \cdot e^{j\varphi_{opt}} \tag{6}$$

$$\varphi_{opt} = \varphi_{xy} + \varphi_{fluc} \tag{7}$$

where φ_{opt} is the phase distribution, which needs to optimize. It contains two parts, one is the normal phase distribution based on (1), whereas another is the fluctuations added at the corresponding phase, and their sum are used in (6) to get the optimized radiation pattern in every iteration for all particles. The normalized radiation pattern in the optimization is Equation (8):

$$RP_{1n}(\theta,\varphi) = 20 \times \log_{10}\left(\frac{|RP_1(\theta,\varphi)|}{\max(|RP_1(\theta,\varphi)|)}\right) \tag{8}$$

The ideal goal of the optimization has not related to φ and shown in Equation (9):

$$RP_{goal}(\theta,\varphi) = \begin{cases} 0 & -10^\circ \le \theta \le 10^\circ \\ -30dB & 10^\circ \le |\theta| \le 90^\circ \end{cases} \tag{9}$$

Then, the fitness function can be built based on the difference between RP_{1n} and RP_{goal}, which is shown in Equation (10):

$$fitness = \begin{cases} \sum\limits_{-90^\circ}^{90^\circ}\sum\limits_{0^\circ}^{360^\circ} (RP_{1n} - RP_{goal}) & RP_{1n} > RP_{goal} \\ 0 & else \end{cases} \tag{10}$$

The value of this fitness function is called as the Cost of the optimization, and the lower cost, the closer to the goal of −30 dB SLL. Theoretically, when the Cost equals to zero in the iterations, the optimization finishes completely.

But in fact, in most cases of optimization, when the Cost is small as an acceptable value and the Cost no longer changes in the process of iterations, the PSO is regarded as finished. In this paper, for 15 particles with 5000 times of iterations for each of them, the final Cost has a value of 101.2 and the Cost has hardly changed at the last 1000 iterations. Generally, the values of φ_{fluc} is the optimized phase fluctuations, and the TA will be established based on the phase distribution of φ_{opt}. The final simulation will use this optimized structure to achieve high gain and low SLL, which will be presented in the next section.

3. Result and Discussion

The simulated far-field radiation pattern in E-plane and H-plane of the feed antenna and the TA before optimization at 28 GHz is shown in Figure 10, the gain increases from 14.6 to 27.1 dB. As shown in Figure 10a, the HPBW reduces from 36° to 5.0° in E-plane and from 36° to 5.4° in H-plane as shown in Figure 10b.

Figure 11 demonstrates the 3D simulated radiation pattern of TA antenna at 28 GHz, which indicates that a pencil-shaped beam is achieved when the quasi-spherical wave emitted by the feed horn passes through the optimized TA with a low SLL. A good performance of focusing effect is visible, and the beam points clearly to the z positive direction. The experimental setup of the proposed antenna can be seen in Figure 12. The whole TA occupies an area of 160 mm × 160 mm × 140.1 mm with a *f/D* value of 0.65. Because of the symmetry of the structure, the parameters of unit cell of the TA can be obtained by one radial row. The optical parameters for two kinds of unit cells of first line along horizontal direction from center to edge are L_{11} = 1.27 mm, L_{12} = 1.29 mm, L_{13} = 1.26 mm, L_{14} = 1.27 mm, L_{15} = 1.16 mm, L_{16} = 1.02 mm, L_{21} = 1.81 mm, L_{22} = 1.70 mm, L_{23} = 1.36 mm, L_{17} = 1.27 mm, L_{18} = 0.95 mm, L_{24} = 1.80 mm, L_{25} = 1.47 mm, L_{26} = 1.27 mm, and L_{19} = 1.12 mm, and $L_i(i = 1, 2)$ refers to Unit Cell A and Unit Cell B, respectively.

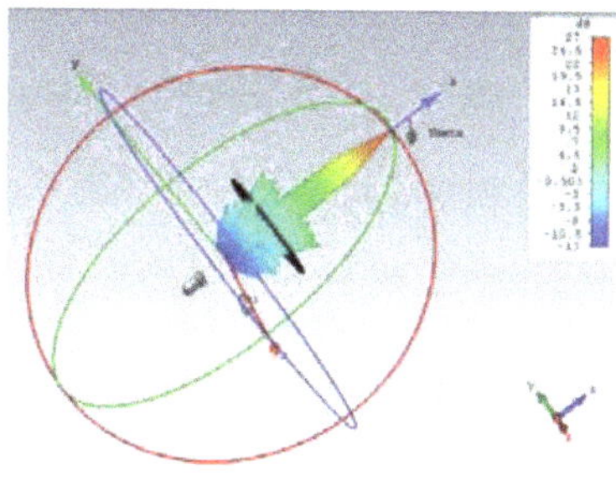

Figure 11. A 3D radiation pattern of the optimized TA antenna at 28 GHz.

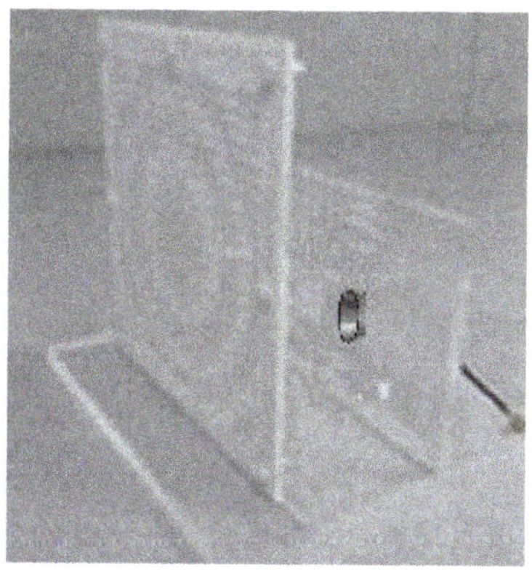

Figure 12. Photograph of experimental setup of the proposed antenna.

The reflection coefficient of the proposed antenna is shown in Figure 13. The feed antenna has wide passband, and the Gain the whole antenna is limited by TA. Figure 14 indicates the compared radiation pattern of measured results with the simulated results before the optimization in two cut planes, and it is easy to find that the SLL in E plane changes from −19.6 to −30 dB and in H plane from −21.7 to −30 dB, and all the side lobes, except the first side lobe, are kept below −35 dB. In summary, the SLL of this TA antenna achieves the level of −30 dB from the level of −20 dB by using the PSO. Figure 15 illuminates the measured gains within the operating frequency band. The PSO is good for the design of directional array antenna, which has great potentials in the design of beam steering antenna and MIMO (Multi Input Multi Output, MIMO) array antennas. Besides the analysis of gain and SLL, the property of aperture efficiency is also an aspect to evaluate the performance of TA antenna. The method to enhance the aperture efficiency is also a research direction, which is related to the f/D, spillover radiation, and edge scattering. Generally, the aperture efficiency is based on (11) and (12).

$$e_{ap} = \frac{10^{G/10}}{D_{\max}} \tag{11}$$

$$D_{\max} = \frac{4\pi \cdot A}{\lambda_0^2} \tag{12}$$

where e_{ap} is the aperture efficiency, $D_{\max}$ is the directivity in the main beam (unit: dimensionless), G is the maximum gain (unit: dB), the physical area of TA is marked as A (unit: m^2), λ_0 is the wavelength at the desired central frequency point (unit: m^2), which is 28 GHz in this paper. Then (12) will be substituted into (11), so we can get the aperture efficiency of the optimized TA antenna at *Ka* band, which shows 23% in simulation. Finally, it is obvious that the TA antenna can achieve a performance of high gain, low SLL and high aperture efficiency with the optimized feed horn and the optimized phase distribution using PSO. As shown in Table 1, the TA antenna has higher gain, wider bandwidth and higher aperture efficiency at the central frequency of 28 GHz.

Table 1. Comparison of the optimized TA antenna with some published array antenna.

Ref.	Freq. (GHz)	TA Thickness (λ_0)	Array Size (λ_0^2)	SLL (dB)	Number of Elements	Bandwidth	$\text{Gain}_{\max}$	Aperture Efficiency (%)
[25]	28	0.18	89.5	−18	484	11.%	24.2	24.5
[28]	20	10.5	368.5	−22.5	2032	10.2%	33.0	40
[38]	13.5	0.76	164.7	−22	621	9.8%	29.95	50
This paper	28	0.14	173.3	−30	660	11.2%	27	28

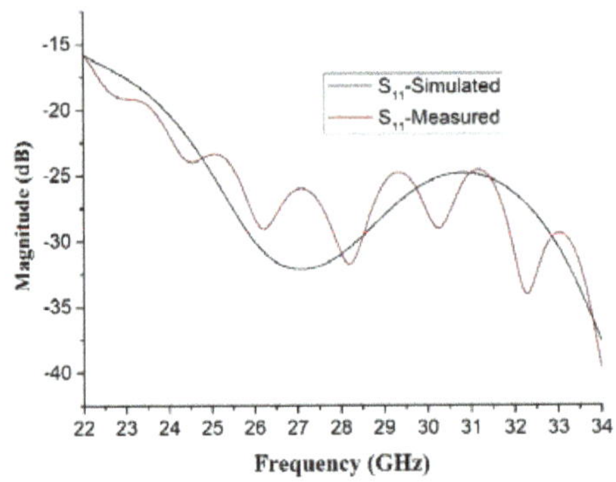

Figure 13. Simulated and measured S11 of the proposed antenna.

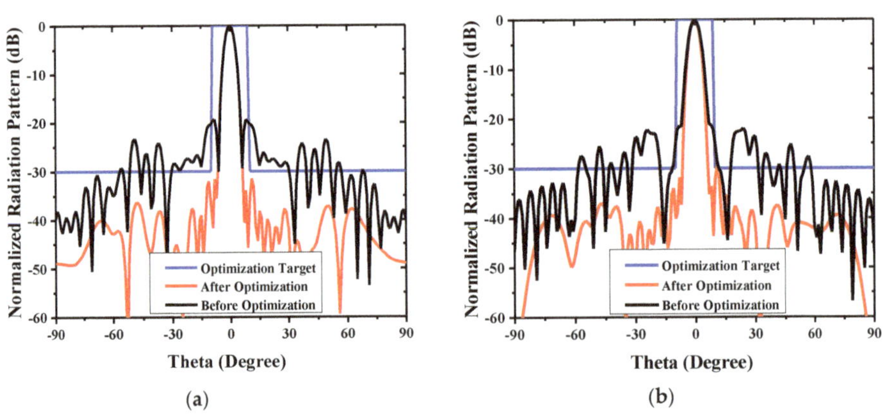

Figure 14. Comparison of the radiation pattern in two cut planes at 28 GHz: (**a**) E-plane comparison and (**b**) H-plane comparison.

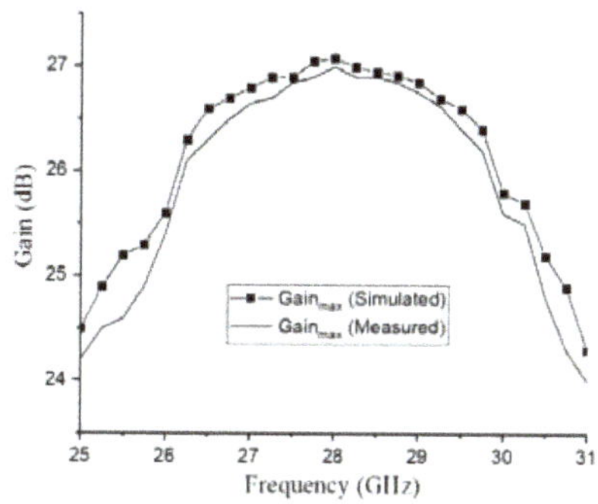

Figure 15. Simulated and measured maximum gains within the operating frequency band.

4. Conclusions

This paper presents a SLL reduced TA antenna based on the two-layered transmitarray (TA), and the whole TA occupies an area of 160 mm × 160 mm × 140.1 mm with a *f/D* value of 0.65. The two-layer structure uses the mutual coupling of vias to accumulate the transmission phase without additional substrate layers. The optimized horn antenna with a $\cos^q(\theta)$ mode simplifies the radiation-pattern calculation of the TA. Finally, the PSO is applied to optimize the SLL from −20 to −30 dB, which achieves a peak gain of 27 dBi at 28 GHz, 11.8% 3 dB gain bandwidth, and 23% aperture efficiency. This optimized TA could also be excited by the offset feed or MIMO antenna as the function of beam forming or beam steering in the field of 5G communication and radar system. The proposed TA antenna will play great prospects in the further research, especially when the high gain, low SLL and wideband is needed.

Author Contributions: All authors contributed equally to this work. Supervision, C.S.; visualization, J.J.; writing—original draft, L.P.; writing—review and editing, Y.J. All authors have read and agreed to the published version of the manuscript.

Funding: This work was supported by National Natural Science Foundation of China (Grant No. 61973038).

Conflicts of Interest: The authors declare no conflicts of interest.

References

1. Rappaport, T.S.; Sun, S.; Mayzus, R.; Zhao, H.; Azar, Y.; Wang, K.; Wong, G.N.; Schulz, J.K.; Samimi, M.; Gutierrez, F. Millimeter wave mobile communications for 5G cellular: It will work! *IEEE Access* **2013**, *1*, 335–349. [CrossRef]
2. Akpakwu, G.A.; Silva, B.J.; Hancke, G.P.; Abu-Mahfouz, A.M. A survey on 5G networks for the Internet of Things: Communication technologies and challenges. *IEEE Access* **2017**, *6*, 3619–3647. [CrossRef]
3. Azar, Y.; Wong, G.N.; Wang, K.; Mayzus, R.; Schulz, J.K.; Zhao, H.; Gutierrez, F.; Hwang, D.; Rappaport, T.S. 28 GHz propagation measurements for outdoor cellular communications using steerable beam antennas in New York city. *ICC Int. Conf. Commun.* **2013**, 5143–5147. [CrossRef]
4. Roh, W.; Seol, J.Y.; Park, J.; Lee, B.; Lee, J.; Kim, Y.; Cho, J.; Cheun, K.; Aryanfar, F. Millimeter-wave beamforming as an enabling technology for 5G cellular communications: Theoretical feasibility and prototype results. *IEEE Commun. Mag.* **2014**, *52*, 106–113. [CrossRef]
5. Syrytsin, I.; Zhang, S.; Pedersen, G.F.; Morris, A.S. Compact quad-mode planar phased array with wideband for 5G mobile terminals. *IEEE Trans. Antennas Propag.* **2018**, *66*, 4648–4657. [CrossRef]
6. Khalily, M.; Tafazolli, R.; Rahman, T.A.; Kamarudin, M.R. Design of phased arrays of series-fed patch antennas with reduced number of the controllers for 28-GHz mm-wave applications. *IEEE Antennas Wirel. Propag. Lett.* **2015**, *15*, 1305–1308. [CrossRef]

7. Pal, A.; Mehta, A.; Mirshekar-Syahkal, D.; Nakano, H. $2\times2 $ Phased Array Consisting of Square Loop Antennas for High Gain Wide Angle Scanning with Low Grating Lobes. *IEEE Trans. Antennas Propag.* **2016**, *65*, 576–583. [CrossRef]
8. Wang, Y.; Zhu, L.; Wang, H.; Luo, Y.; Yang, G. A compact, scanning tightly coupled dipole array with parasitic strips for next-generation wireless applications. *IEEE Antennas Wirel. Propag. Lett.* **2018**, *17*, 534–537. [CrossRef]
9. Anguera, J.; Andújar, A.; Jayasinghe, J. High Directivity Microstrip Patch Antennas Perturbing TModd-0 modes. *IEEE Antennas Wirel. Propag. Lett.* **2020**, *19*, 439–443. [CrossRef]
10. Rajagopalan, H.; Xu, S.; Rahmat-Samii, Y. On understanding the radiation mechanism of reflectarray antennas: An insightful and illustrative approach. *IEEE Antennas Propag. Mag.* **2012**, *54*, 14–38. [CrossRef]
11. Huang, J.; Pogorzelski, R.J. A Ka-band microstrip reflectarray with elements having variable rotation angles. *IEEE Trans. Antennas Propag.* **1998**, *46*, 650–656. [CrossRef]
12. Yao, W.; Yang, H.; Huang, X.; Tian, Y.; Guo, L. An X-band parabolic antenna based on gradient metasurface. *AIP Adv.* **2016**, *6*, 75–83. [CrossRef]
13. Shafiee, M.A.S.M.; Yusoff, M.F.M.; Johari, Z.; Ismail, M.K.H.; Rahim, M.K.A. Reflectarray antenna performances using combination of rectangular and Jerusalem unit cells at 5.2 GHz. In Proceedings of the 2016 IEEE Asia-Pacific Conference on Applied Electromagnetics (APACE), Kedah, Malaysia, 11 December 2016; pp. 163–166.
14. An, W.; Xu, S.; Yang, F. A metal-only reflectarray antenna using slot-type elements. *IEEE Antennas Wirel. Propag. Lett.* **2014**, *13*, 1553–1556. [CrossRef]
15. Ryan, C.G.; Chaharmir, M.R.; Shaker, J.R.B.J.; Bray, J.R.; Antar, Y.M.; Ittipiboon, A. A wideband transmitarray using dual-resonant double square rings. *IEEE Trans. Antennas Propag.* **2010**, *58*, 1486–1493. [CrossRef]
16. De la Torre, P.P.; Sierra-Castaner, M. Design of a 12 GHz transmit-array. *IEEE Antennas Propag. Soc. Int. Symp.* **2007**, *2007*, 2152–2155.
17. Feng, P.Y.; Qu, S.W.; Yang, S. Ultrawideband Low-Profile Transmitarray with Vivaldi Array Feed. *IEEE Trans. Antennas Propag.* **2020**, *55*, 3265–3270. [CrossRef]
18. Plaza, E.; Leon, G.; Loredo, S.; Las-Heras, F. A simple model for analyzing transmitarray lenses. *IEEE Antennas Propag. Mag.* **2015**, *57*, 131–144. [CrossRef]
19. Nematollahi, H.; Laurin, J.J.; Page, J.E.; Encinar, J.A. Design of broadband transmitarray unit cells with comparative study of different numbers of layers. *IEEE Trans. Antennas Propag.* **2015**, *63*, 1473–1481. [CrossRef]
20. Abdelrahman, A.H.; Elsherbeni, A.Z.; Yang, F. Transmitarray antenna design using cross-slot elements with no dielectric substrate. *IEEE Antennas Wirel. Propag. Lett.* **2014**, *13*, 177–180. [CrossRef]
21. Li, H.; Wang, G.; Xu, H.X.; Cai, T.; Liang, J. X-band phase-gradient metasurface for high-gain lens antenna application. *IEEE Trans. Antennas Propag.* **2015**, *63*, 5144–5149. [CrossRef]
22. Tuloti, S.H.R.; Rezaei, P.; Hamedani, F.T. High-efficient wideband transmitarray antenna. *IEEE Antennas Wirel. Propag. Lett.* **2018**, *17*, 817–820. [CrossRef]
23. Li, Z.; Su, J.; Li, Z. Design of high-gain lens antenna based on phase-gradient metasurface. In Proceedings of the 2016 11th International Symposium on Antennas, Propagation and EM Theory (ISAPE), Guilin, China, 18–21 October 2016; pp. 135–138.
24. Abdelrahman, A.H.; Yang, F.; Elsherbeni, A.Z.; Nayeri, P. Analysis and design of transmitarray antennas. *Synth. Lect. Antennas* **2017**, *6*, 1–175. [CrossRef]
25. Jiang, M.; Chen, Z.N.; Zhang, Y.; Hong, W.; Xuan, X. Metamaterial-based thin planar lens antenna for spatial beamforming and multibeam massive MIMO. *IEEE Trans. Antennas Propag.* **2016**, *65*, 464–472. [CrossRef]
26. Li, H.; Wang, G.; Liang, J.; Gao, X.; Hou, H.; Jia, X. Single-layer focusing gradient metasurface for ultrathin planar lens antenna application. *IEEE Trans. Antennas Propag.* **2016**, *65*, 1452–1457.
27. Luo, Q.; Gao, S.; Sobhy, M.; Yang, X. Wideband transmitarray with reduced profile. *IEEE Antennas Wirel. Propag. Lett.* **2018**, *17*, 450–453. [CrossRef]
28. An, W.; Xu, S.; Yang, F.; Li, M. A double-layer transmitarray antenna using malta crosses with vias. *IEEE Trans. Antennas Propag.* **2015**, *64*, 1120–1125. [CrossRef]
29. Pozar, D. Flat lens antenna concept using aperture coupled microstrip patches. *Electron. Lett.* **1996**, *32*, 2109–2111. [CrossRef]
30. Fletcher, R. *Practical Methods of Optimization*; John Wiley & Sons: Hoboken, NJ, USA, 2013.

31. Horst, R.; Pardalos, P.M.; van Thoai, N. *Introduction to Global Optimization*; Springer Science & Business Media: Berlin, Germany, 2000.
32. Haupt, R.L. Thinned arrays using genetic algorithms. *IEEE Trans. Antennas Propag.* **1994**, *42*, 993–999. [CrossRef]
33. Boeringer, D.W.; Werner, D.H. Particleswarm optimization versus genetic algorithms for phased array synthesis. *IEEE Trans. Antennas Propag.* **2004**, *52*, 771–779. [CrossRef]
34. Robinson, J.; Rahmat-Samii, Y. Particle swarm optimization in electromagnetics. *IEEE Trans. Antennas Propag.* **2004**, *52*, 397–407. [CrossRef]
35. Xu, S.; Rahmat-Samii, Y. Boundary conditions in particle swarm optimization revisited. *IEEE Trans. Antennas Propag.* **2007**, *55*, 760–765. [CrossRef]
36. Nayeri, P.; Yang, F.; Elsherbeni, A.Z. Design of single-feed reflectarray antennas with asymmetric multiple beams using the particle swarm optimization method. *IEEE Trans. Antennas Propag.* **2013**, *61*, 4598–4605. [CrossRef]
37. Balanis, C.A. *Antenna Theory: Analysis and Design*; John Wiley & Sons: Berlin, Germany, 2016.
38. Abdelrahman, A.H.; Nayeri, P.; Elsherbeni, A.Z.; Yang, F. Bandwidth improvement methods of transmitarray antennas. *IEEE Trans. Antennas Propag.* **2015**, *63*, 2946–2954. [CrossRef]
39. Cai, T.; Wang, G.M.; Fu, X.L.; Liang, J.G.; Zhuang, Y.Q. High-efficiency metasurface with polarization-dependent transmission and reflection properties for both reflectarray and transmitarray. *IEEE Trans. Antennas Propag.* **2018**, *66*, 3219–3224. [CrossRef]
40. Abbas-Azimi, M.; Mazlumi, F.; Behnia, F. Design of broadband constant-beamwidth conical corrugated-horn antennas [Antenna designer's notebook]. *IEEE Antennas Propag. Mag.* **2009**, *51*, 109–114. [CrossRef]

sensors

MDPI

Article

A Novel Dual-Band (38/60 GHz) Patch Antenna for 5G Mobile Handsets

Marwa H. Sharaf *, Amira I. Zaki, Radwa K. Hamad and Mohamed M. M. Omar

Electronics and Communications Department, College of Engineering and Technology, Arab Academy for Science, Technology & Maritime Transport, Alexandria 21937, Egypt; amzak10@aast.edu (A.I.Z.); radwa_hamad@aast.edu (R.K.H.); abuahmad.omar@aast.edu (M.M.M.O.)

* Correspondence: marwasharaff@aast.edu; Tel.: +20-0100-311-3158

Received: 1 April 2020; Accepted: 24 April 2020; Published: 29 April 2020

Abstract: A compact dual-frequency (38/60 GHz) microstrip patch antenna with novel design is proposed for 5G mobile handsets to combine complicated radiation mechanisms for dual-band operation. The proposed antenna is composed of two electromagnetically coupled patches. The first patch is directly fed by a microstrip line and is mainly responsible for radiation in the lower band (38 GHz). The second patch is fed through both capacitive and inductive coupling to the first patch and is mainly responsible for radiation in the upper frequency band (60 GHz). Numerical and experimental results show good performance regarding return loss, bandwidth, radiation patterns, radiation efficiency, and gain. The impedance matching bandwidths achieved in the 38 GHz and 60 GHz bands are about 2 GHz and 3.2 GHz, respectively. The minimum value of the return loss is −42 dB for the 38 GHz band and −47 for the 60 GHz band. Radiation patterns are omnidirectional with a balloon-like shape for both bands, which makes the proposed single antenna an excellent candidate for a multiple-input multiple-output (MIMO) system constructed from a number of properly allocated elements for 5G mobile communications with excellent diversity schemes. Numerical comparisons show that the proposed antenna is superior to other published designs.

Keywords: MIMO; 5G mobile handsets; dual-band antenna; microstrip patch antenna; millimeter-wave

1. Introduction

Fifth Generation (5G) technology has been a matter of controversy due to the enormous increase of users connected with their smartphones to the network relative to the scarcity of the available current bandwidth of the Fourth Generation (4G) technology [1]. This has led to the severe need for larger capacity and faster data rates [2–6], which are currently 100 times faster, and are expected to be 1000 times faster by 2030 [7]. Wireless channel capacity can be increased without the necessity for additional spectra or power in environments rich in scattering when introducing 5G mobile communication techniques [8–10]. This can be applied by increasing the number of antennas at the transmitter and/or the receiver of the wireless link [11–19]. As a consequence of requiring multiple antennas in a compact system like a mobile handset, the single antenna should be compact and operate at different required bands as required [20,21]. The present paper offers a solution for such low profile antennas operating at 38 and 60 GHz frequency bands.

The unused millimeter-wave electromagnetic spectrum (30–300 GHz) has attracted attention due to its multi-gigabit/s transmission rate exploiting widely available bandwidth to meet the demands of 5G applications which require high quality and low latency transmission [22–26]. The frequency bands centered at 28, 38, 60, and 73 GHz have been allocated for 5G mobile communications by the International Telecommunications Union (ITU) [27]. Bands of 59–64 GHz are allocated by the Federal Communications Commission (FCC) as an unlicensed band for short-range and wireless communications at high speeds [28]. The frequency band around 60 GHz has particular importance because of the wide 7–9 GHz frequency range of unlicensed spectra available. However, significant attenuation is caused by oxygen molecules in the atmosphere to react with radio frequency (RF) signals, reaching up to 10 dB/km. Due to this defect, the 60 GHz band is not a suitable band for wireless communication applications and long-range radar. For this reason, the dual-band (38/60 GHz) microstrip patch antenna proposed in the present work could be a good solution for forthcoming 5G antenna systems that utilize the 60 GHz frequency band, and provides the 38 GHz frequency band (a potential candidate for next generation communication) as an additional operational band due to its low oxygen absorption rates relative to the 60 GHz band [29].

Recently, a significant amount of research work has provided numerous designs for mobile handset antennas for 5G applications. In [29], a 28 GHz four-port MIMO antenna is proposed, where each antenna has an end-fire gain of about 10 dBi by employing an array of metamaterial unit cells. The work of [30] introduces a 60 GHz antenna that consists of H-shaped and E-shaped slots on the radiating patch. A two-port 5G dual-band (28/38 GHz) MIMO antenna system is proposed in [31]. This antenna is realized using two arrays, each consisting of three elements. In [32], a dual-band (38/54 GHz) microstrip patch antenna is proposed. The work of [27] proposes a 38 GHz slotted patch antenna for 5G wireless applications. A circularly polarized magneto-electric dipole antenna with high efficiency based on printed ridge gap waveguide is proposed in [33] to operate in the frequency range of 29.5–37 GHz. The work of [34] presents a compact antenna design with polarization and pattern diversity operating in the frequency band (34–38 GHz) for MIMO-based 5G mobile communication systems.

The present work proposes a novel (38/60 GHz) dual-band microstrip patch antenna for 5G mobile phones. The proposed patch can be used either as a single antenna or an element to construct compact MIMO antenna systems. In addition, due to the very weak coupling between adjacent elements of the proposed dual-band antenna, it can be used to construct high-gain compact arrays for 5G applications requiring smart antennas with capabilities of beam steering and detection of direction of arrival. The proposed antenna is composed of two patches: the first is inset-fed by a microstrip line, whereas the second is fed using both capacitive and inductive coupling to the first patch. The dimensions of the first patch are set so that the patch operates at the 38 GHz band, whereas the second patch is designed to operate at the 60 GHz band. Combining these two bands in a single compact antenna will save the power and space needed in a mobile handset. The radiation patterns produced by this antenna are shown to be suitable for 5G mobile communications. It should be noted that the electromagnetic simulations presented in the present work were performed using the commercially available CST Studio Suite® software package.

The novelty of this work lies in the duality usage of the 38 and 60 GHz bands. This is unprecedented in previous work.

The remaining part of the paper is organized as follows. Section 2 describes the evolutionary design (stages) and an explanation of the principles of operation of the dual-band microstrip patch antenna proposed for 5G mobile handsets. Section 3 displays and discusses the obtained numerical results and the experimental measurements concerning the performance assessment of the proposed patch. Some comparisons with the results of other published work are presented for comparative assessment. Finally, Section 4 is devoted to the main conclusions of this work.

2. The Proposed Dual-Band Microstrip Patch Antenna

The proposed dual-band microstrip patch antenna is shown in Figure 1. The proposed antenna is based on the design of two patches operating at different bands. The first patch is excited directly by a microstrip line through an inset feed, whereas the second patch is fed indirectly through a capacitive (edge) coupling with the first patch. The dimensions of the first patch are set so that the patch operates at 38 GHz, whereas the dimensions of the second patch are designed to operate at 60 GHz.

At 38 GHz the radiation pattern is produced by the first patch, which is omnidirectional in the azimuth plane (xy) and has a balloon-like shape in the (elevation) E- and H-planes. The second patch does not contribute to the radiation at 38 GHz due to its small dimensions.

At 60 GHz, the radiation pattern is obtained by the second patch. On the other hand, the first patch contributes to the radiation through a higher-order mode, which results in a radiation pattern with undesirable nulls and side lobes. The cuts made in the first patch at the corners on the sides of the microstrip feed line are made to diminish the surface currents and slot fields near these regions and, thereby, prevent the formation of such higher-order modes in the cavity below the radiating patch. Thus, the radiation pattern at 60 GHz has a shape that is suitable for a diversity scheme that enhances the performance of a MIMO antenna system constructed from a number of properly allocated elements of such an antenna.

The geometry of the proposed dual-band microstrip patch antenna is illustrated in Figure 1, where LP_1, WP_1 are the length and width of the first patch, respectively. Similarly, LP_2, WP_2 are the length and width of the second patch, respectively, added to the length of the inset feed (L_{ins}), the gap width of the inset feed (W_{ins}), and the depth of the gap (GP) cut in the first patch. The dimensions are stated in Table 1. The proposed microstrip patch antenna is printed on Rogers RO3003^{TM} substrate of dielectric constant $\varepsilon_r = 3.0$; dimensions of the substrate are 15×25 mm, height $h = 0.25$ mm, and loss tangent $\delta = 0.001$. The conducting surface is made of copper with conductivity $\sigma = 5.6 \times 10^7$ S/m.

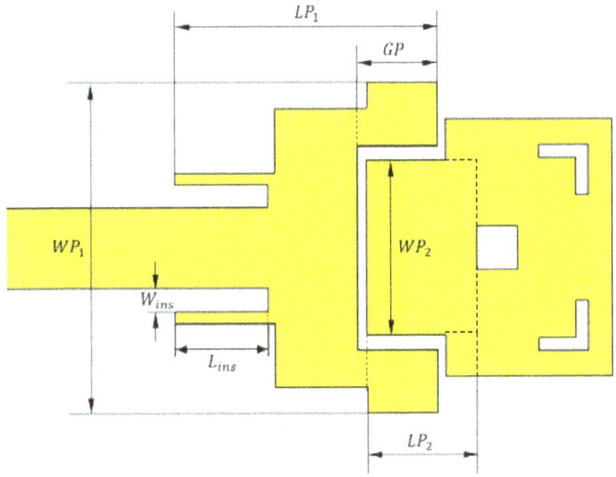

Figure 1. Geometry of the proposed dual-band (38/60 GHz) microstrip patch antenna.

Table 1. Dimensions of the dual-patch microstrip patch antenna.

Dimension	LP_1	LP_2	WP_1	WP_2	GP	L_{ins}	W_{ins}
Value (mm)	2.165	1.22	2.57	1.36	0.66	0.76	0.18

3. Simulations and Results

In this section, the performance of the dual-band patch antenna is investigated and the corresponding numerical results are presented and discussed. The presented results are concerned with investigating the return loss and radiation patterns of the single element antenna using the commercially available CST Studio Suite® software package.

3.1. Development of the Dual-Band Microstrip Patch antenna Design

It may be interesting to demonstrate the stages of the proposed antenna design to combine multiple radiation mechanisms to construct a single antenna operating in the dual-frequency band. Preliminarily, the simple idea of producing dual-band radiation can be implemented using two rectangular patches as shown in Figure 2.

The mutual coupling between the two patches results in the frequency response of the return loss $|S_{11}|$ presented in Figure 2. The resonance frequencies of the two patches are shifted (particularly the resonance frequency of the upper band) due to the capacitive load caused by the other patch resulting in resonance frequencies of about 37.76 GHz and 57.26 GHz for the first and second patches, respectively. For further explanation of the radiation mechanisms at each frequency, the surface current distributions are presented in Figure 3a,c at 37.76 GHz and 57.26 GHz, respectively. The corresponding radiation patterns in the orthogonal elevation planes ($\phi = 0°$ and $\phi = 90°$) are presented in Figure 3b,d. These radiation patterns have similar balloon-like shapes, which means that the radiation is omnidirectional in the azimuth planes.

Both resonance frequencies of the first patch (at 37.76 GHz and 57.26 GHz) are first-order, as the active region of the patch (the area over which the surface current has significant magnitude) at each of these frequencies is concentrated in the close vicinity of the junction between the feed line and the patch. Moreover, the area of the active region at each frequency is proportional to the square of the effective wavelength ($\lambda_0 / \sqrt{\varepsilon_{r\,eff}}$). As a consequence, the radiation pattern at each of these frequencies has the conventional balloon-like shape of a rectangular patch radiating in its first-order resonance.

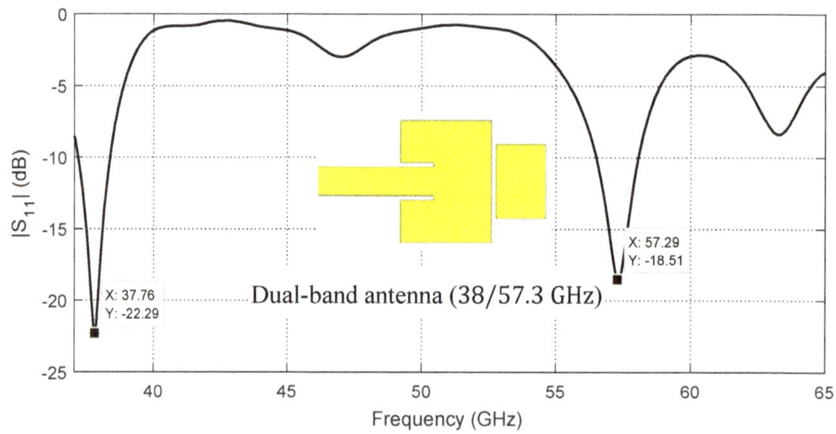

Figure 2. Simulated dependence of the return loss, $|S_{11}|$, on the frequency over a wide band of the frequency for the indicated dual-patch antenna (initial design).

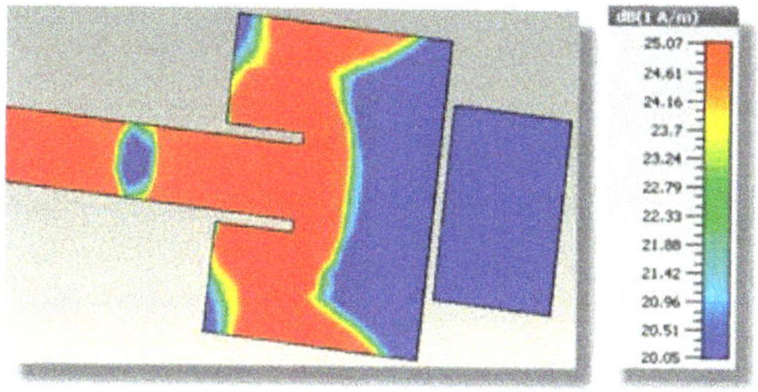

(**a**) Surface current distribution at 37.76 GHz.

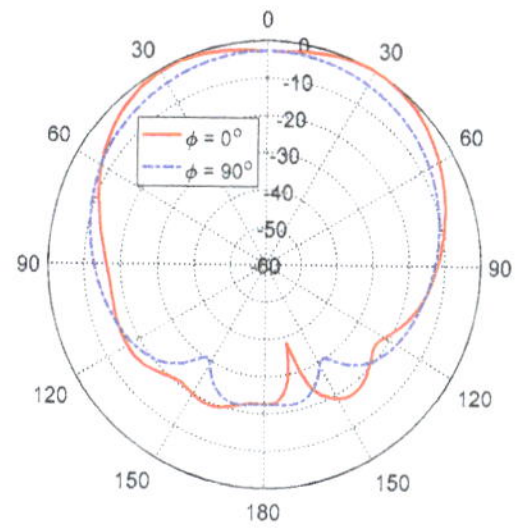

(**b**) Radiation pattern at 37.76 GHz.

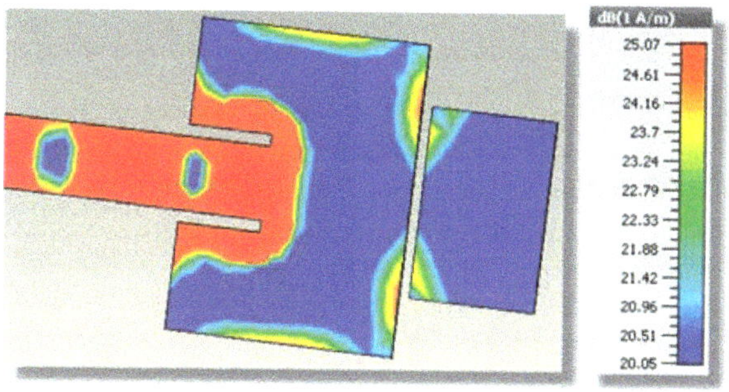

(**c**) Surface current distribution at 57.26 GHz.

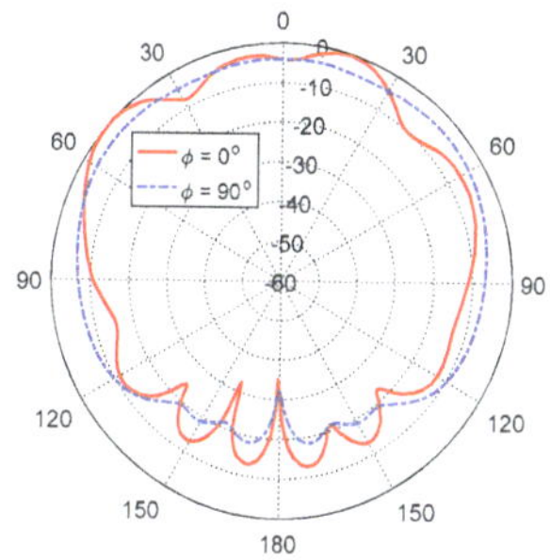

(**d**) Radiation pattern at 57.26 GHz.

Figure 3. Simulated surface current distributions and the corresponding radiation patterns of the proposed dual-band patch antenna at 37.76 GHz and 57.26 GHz.

This design is modified by introducing an inductance to reduce the capacitive load caused by edge coupling between the two patches. This is achieved by subtending the second patch between two arms which are made as side extensions of the first patch as shown in Figure 4. The resulting "C" shape of the first patch can have the effect of a semi-turn producing the required compensating inductance which, in turn, reduces the reactive impedance caused by the capacitive edge coupling. For more inductive effect, the conducting part indicated as the green colored surface shown in Figure 4 is adhered to the second patch to complete the semi-turn made by the first patch and its extended arms. This modification results in the frequency response of the return loss, $|S_{11}|$, depicted in Figure 4. It is apparent that the resonance frequency at the center of the upper band is significantly corrected to be 59.56 GHz.

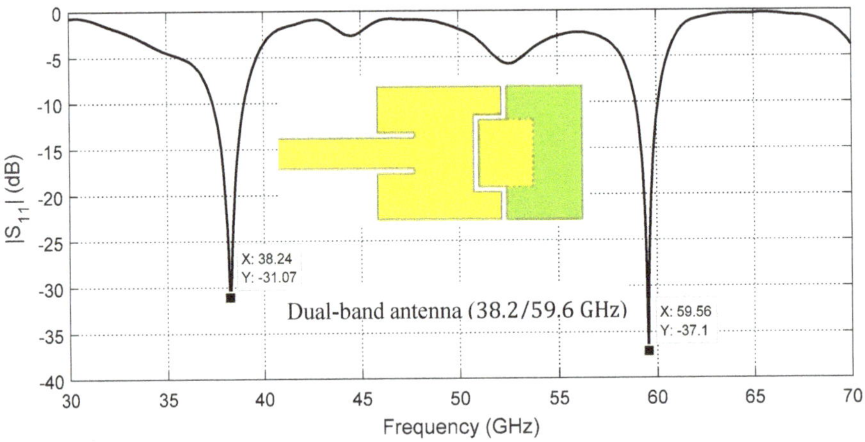

Figure 4. Simulated dependence of the return loss, $|S_{11}|$, on the frequency over a wide band of the frequency for the indicated dual-patch antenna (intermediate stage for the final design).

At 38.24 GHz, the dual-band antenna presented in Figure 4 has the surface current distribution presented in Figure 5a and the corresponding radiation patterns presented in Figure 5b. As shown in Figure 5b, the radiation pattern at 38.24 GHz has an acceptable balloon-like shape. This is attributed to the current distribution, which is still concentrated in a continuous region around the inset feed as shown in Figure 5a. On the other hand, the surface current, the tangential magnetic field distribution at 59.56 GHz, and the corresponding radiation patterns are presented in Figure 5c–e, respectively. Due to increasing the coupling between the first and second patches, the current flowing on the second patch is considerably increased in comparison to that presented in Figure 3c at the upper-frequency band. However, the radiation pattern of this patch at 59.56 GHz, shown in Figure 5e, is not acceptable. The difference between the maximum and minimum radiation in the elevation plane $\phi = 90°$ is greater than 11 dB. In addition, the level of radiation in the plane $\phi = 90°$ is globally much greater than the corresponding level in the plane $\phi = 0°$. This indicates that the radiation pattern is not omnidirectional in the azimuth planes, which can be made clear by mentioning that the surface current, which is distributed over a wide area, extends on both the first and second patches. Moreover, Figure 5d shows that the tangential magnetic field is mainly concentrated in two regions (surrounded by blue dashed oval shapes) where the tangential field has opposite horizontal directions leading to a considerable reduction of the radiated field in the direction normal to the patch surface ($\theta = 0°$). Hence, the antenna design presented in Figure 4 needs some modifications to increase the concentration of the surface current on the second patch (responsible for radiation in the upper-frequency band).

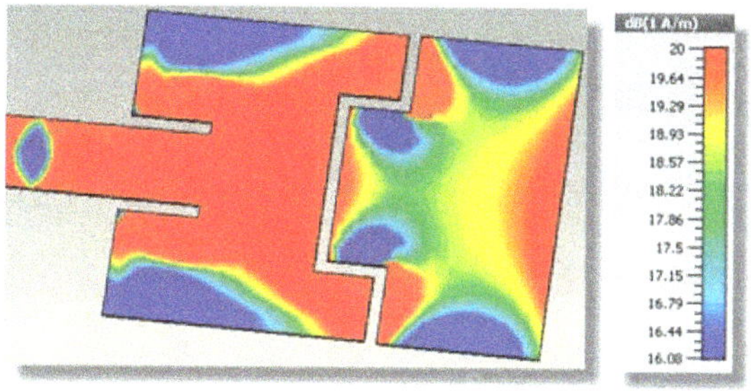

(**a**) Surface current distribution at 38.24 GHz.

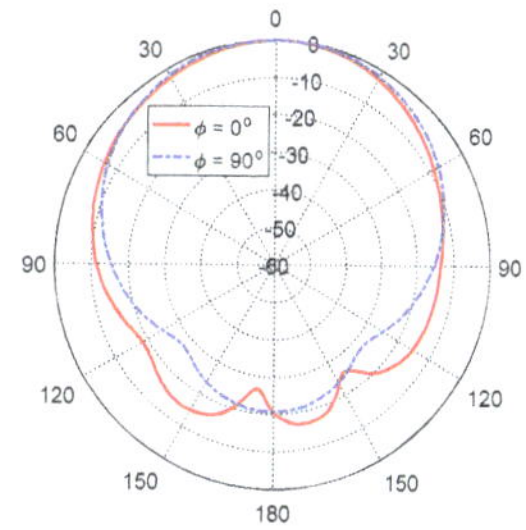

(**b**) Radiation patterns at 38.24 GHz.

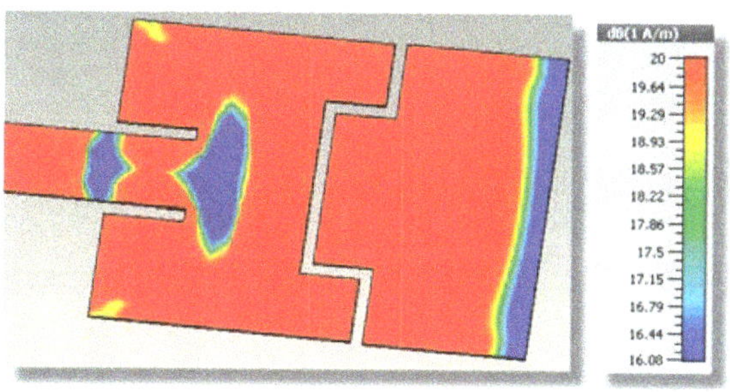

(**c**) Surface current distribution at 59.56 GHz.

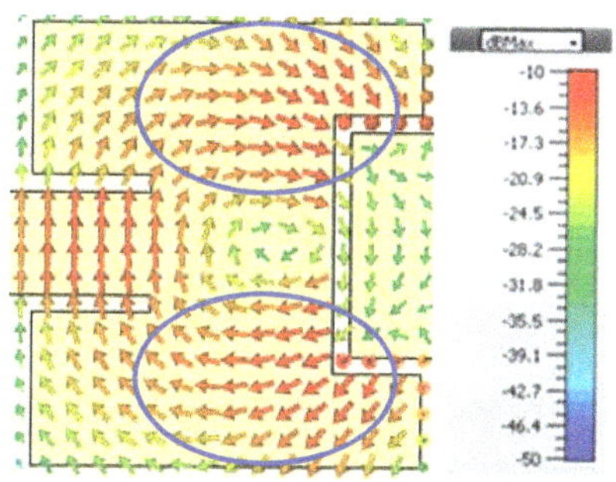

(**d**) Tangential magnetic field distribution at 59.56 GHz.

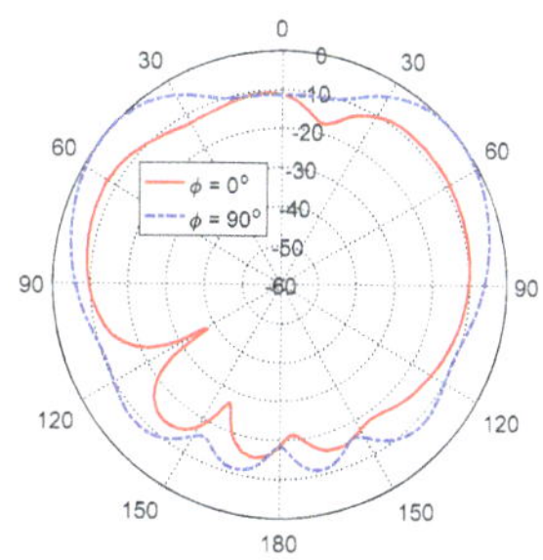

(**e**) Radiation patterns at 59.56 GHz.

Figure 5. Simulated surface current, tangential magnetic field distributions, and the corresponding radiation patterns of the proposed dual-band patch antenna at 38.24 GHz and 59.56 GHz.

The dimensions of the dual-band patch antennas introduced in the initial and intermediate design stages are shown in Figure 6.

Further improvements to the geometry of the dual-band microstrip patch antenna can be made to achieve a radiation pattern shape at 60 GHz that is suitable for MIMO applications. As shown in Figure 7, some cuts in the first patch can be made at the corners on the sides of the microstrip feed line (shown in green color) to weaken the surface currents and slot fields near these regions. This prevents the formation of higher-order modes in the cavity below the radiating patch. Additionally, undesired grating lobes or nulls in the radiation pattern at 60 GHz resulting from possibly flowing currents on the large surface area of the second patch can be canceled by making other cuts in this patch. A central square gap and two corner-shaped slots can be cut as shown in Figure 7 to reduce the area of the active region of this patch by disturbing the paths of possible higher-order currents flowing near the corners. The geometrical parameters of the dual-patch antenna according to the final design are presented in Section 3.3 (after demonstrating a procedural parametric scan for selecting their optimal values).

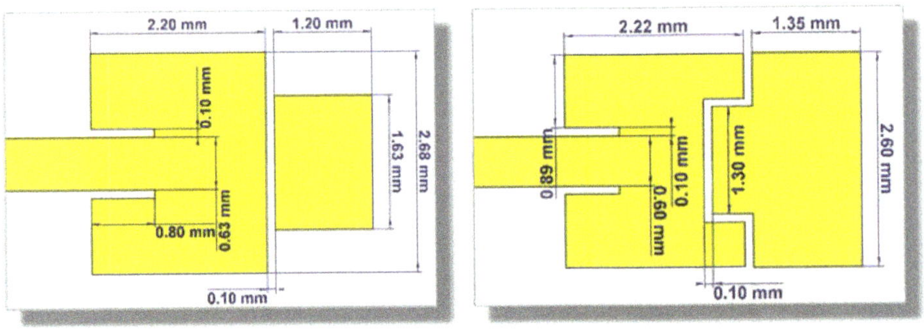

(a) Initial design stage.

(b) Intermediate design stage.

Figure 6. Dimensions of the dual-patch for the initial and intermediate design stages.

These modifications of the antenna geometry result in the frequency response of the return loss, $|S_{11}|$, presented in Figure 7. As shown in this figure, the antenna impedance is perfectly matched at the two frequencies 38 and 60 GHz with return loss −42 dB and −47 dB, respectively. At 38 GHz, the bandwidth is about 1.2 GHz (37.34–38.54 GHz), while, at 60 GHz, the bandwidth is about 2.52 GHz and can operate with matched impedance over the frequency range (58.01–60.53 GHz). The wide bandwidth at 60 GHz can be explained due to the second patch (responsible for radiation at this frequency) being excited through indirect feeding, where both capacitive and inductive coupling mechanisms are implemented. This leads to stabilizing the patch impedance over a much wider frequency band than that of the first patch, which is directly fed through a conventional microstrip line.

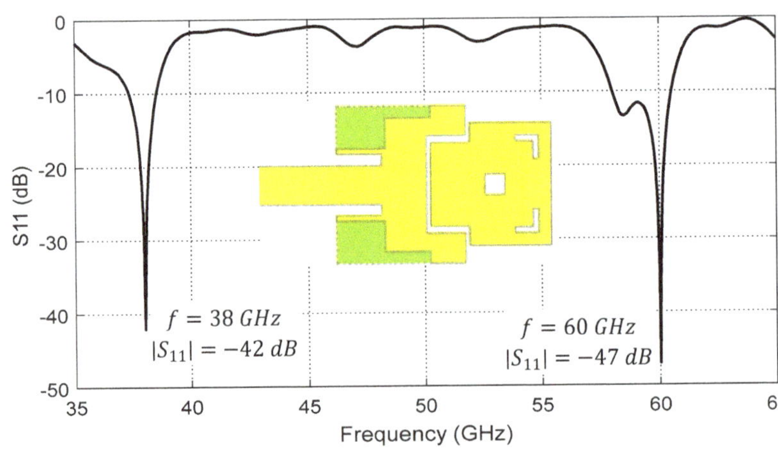

Figure 7. Simulated dependence of the return loss, $|S_{11}|$, on the frequency over a wide band of the frequency for the proposed dual-patch antenna (final design).

The surface current distributions and the corresponding radiation patterns at 38 and 60 GHz are presented in Figure 8. The radiation patterns for the proposed dual-band antenna at 38 GHz and 60 GHz are presented in Figure 8b,e, respectively, in the E-plane ($\phi = 0°$) and H-plane ($\phi = 90°$).

It is obvious that the radiation patterns are almost balloon-like in the elevation planes and exhibit omnidirectional radiation in the azimuth planes. It should be noted that the geometrical modifications of the patch design shown in Figure 4 result in the tangential magnetic field being mainly concentrated in the central region (surrounded by blue dashed oval shape), as shown in Figure 8d where the magnetic field has unified vertical directions. When compared with the tangential magnetic field directions presented in Figure 5d, it becomes clear that the geometrical modifications of the patch have led to an increase in the radiated field in the direction normal to the patch surface ($\theta = 0°$), resulting in a more appropriate shape of the radiation pattern as shown in Figure 8e. The maximum gain is about 6.5 dBi at 38 GHz and 5.5 dBi at 60 GHz. Such radiation patterns (in both the lower- and upper-frequency bands) have suitable shapes for efficient diversity schemes when utilized to construct a 5G mobile handset MIMO system of properly allocated elements of such an antenna.

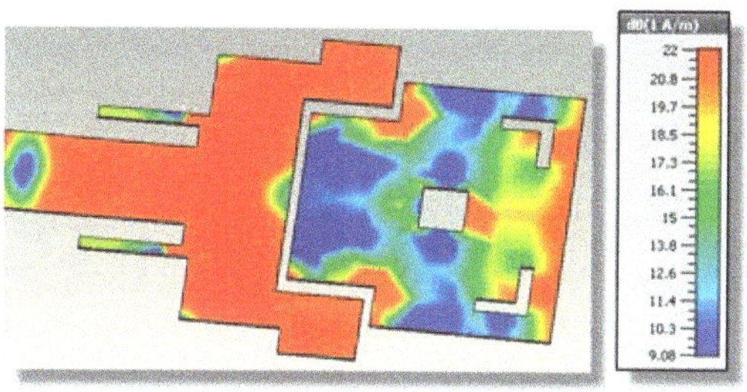

(**a**) Surface current distribution at 38 GHz.

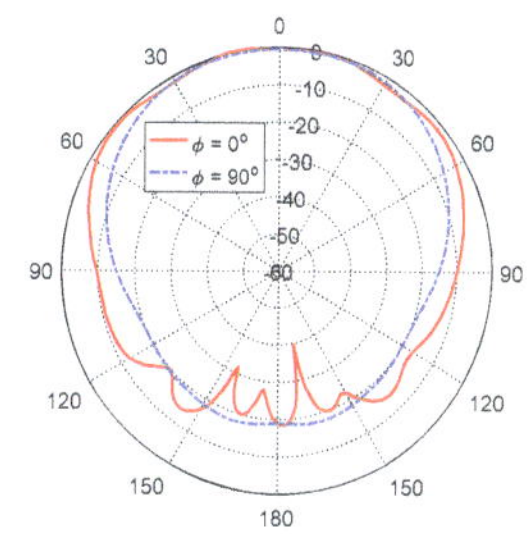

(**b**) Radiation pattern at 38 GHz.

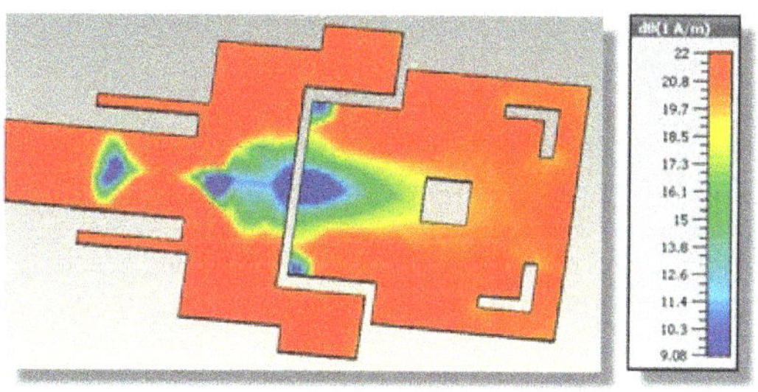

(**c**) Surface current distribution at 60 GHz.

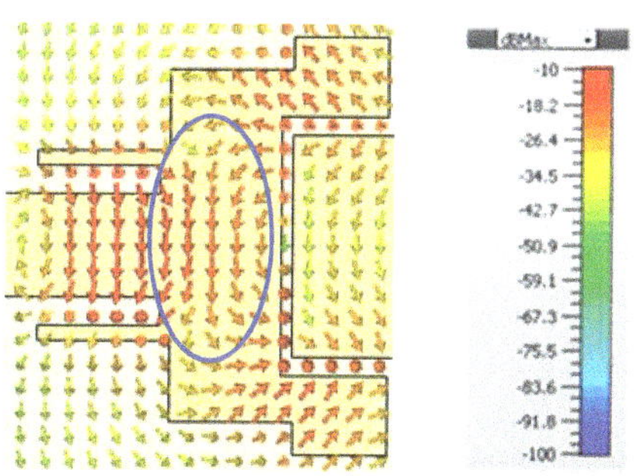

(**d**) Tangential magnetic field distribution at 60 GHz.

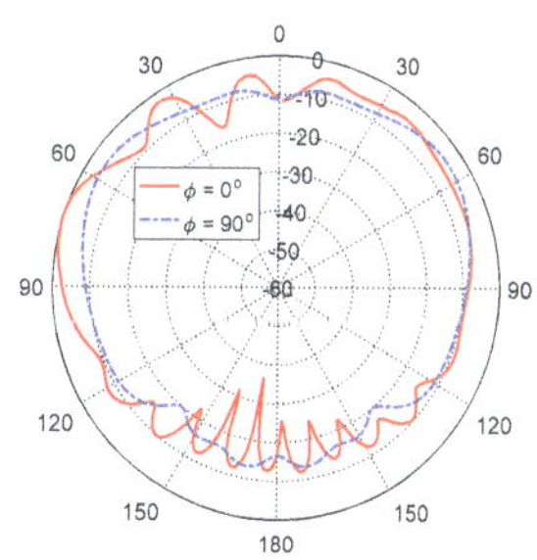

(**e**) Radiation pattern at 60 GHz.

Figure 8. Simulated surface current, tangential magnetic field distributions, and the corresponding radiation patterns of the proposed dual-band patch antenna at 38 GHz and 60 GHz.

3.2. Parametric Study for the Selection of the Optimal Parameters of the Dual-Band Patch Antenna

The proposed dual-band microstrip patch antenna should satisfy the required operational and performance measures to be suitable as a 5G mobile handset antenna. The proposed patch antenna operates at 38 and 60 GHz with satisfactory performance, including impedance matching (low return loss), wide bandwidth, and proper shape of the radiation patterns in both the lower and upper bands of operation. This may be achieved using parametric study, including of the most important dimensional parameters of the patch antenna shown in Figure 1, such as the length and width of each radiating patch (LP_1, WP_1 for the first patch, LP_2, WP_2 for the second patch), length of the inset feed (L_{ins}), gap width for the inset feed (W_{ins}), and the depth of the gap (GP) cut in the first patch.

3.2.1. Effect of the Parasitic Patch Width

The center frequency of the upper band (around 60 GHz) of the proposed patch antenna is very sensitive to any change in the second patch width, WP_2. This is clear in Figure 9, which presents the frequency response of the return loss, $|S_{11}|$, for different values of WP_2. The center frequency of the upper band changes from $f_U = 58.6$ GHz for $WP_2 = 1.0$ mm to $f_U = 61.6$ GHz for $WP_2 = 1.57$ mm. Moreover, the center frequency of the lower band f_L has a weak dependence on WP_2 such that it changes from $f_L = 37.7$ GHz for $WP_2 = 1.0$ mm to $f_L = 38.5$ GHz for $WP_2 = 1.57$ mm. Thus, changing this parameter enables the tuning of the center frequencies of the lower and upper bands. As shown in Figure 9, setting $WP_2 = 1.36$ mm results in satisfying the required frequencies $f_L = 38$ GHz and $f_U = 60$ GHz at the same time.

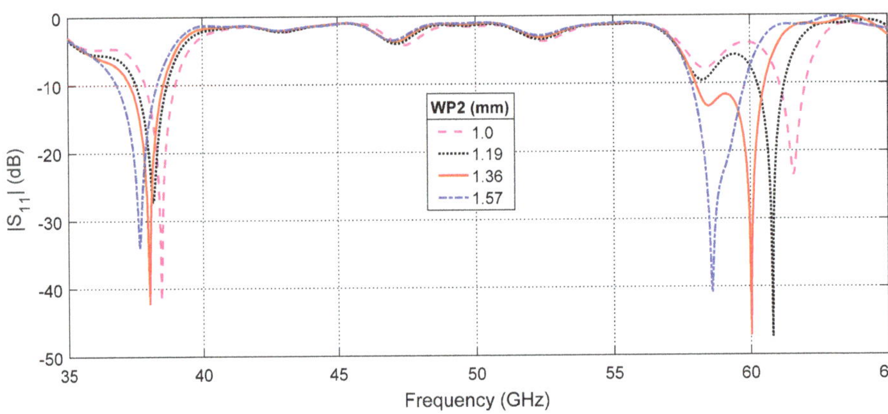

Figure 9. Changing the width of the second patch, WP_2, leads to coarse tuning of the center frequency of the upper band (60 GHz) and fine tuning of the center frequency of the lower band.

3.2.2. Effect of the Inset Length

One of the antenna design parameters that has a great effect on the input impedance of the patch antenna is the feeding inset length, L_{ins}; see Figure 1. The effect of changing L_{ins} on the frequency response of the return loss, $|S_{11}|$, at the antenna input port relative to 50 Ω characteristic impedance of the microwave source is presented in Figure 10. As shown in this figure, $|S_{11}|$ is strongly dependent on L_{ins} at 60 GHz for the indicated range of lengths. The optimum value of $|S_{11}|$ is −47 dB and is obtained for $L_{ins} = 0.757$ mm. On the other hand, the value of $|S_{11}|$ at 38 GHz seems to be independent of L_{ins} (through the indicated range of lengths).

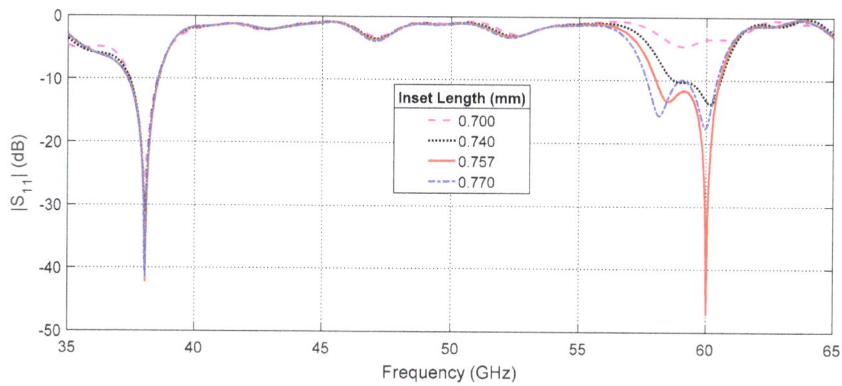

Figure 10. Effect of changing L_{ins} on the frequency response of the return loss $|S_{11}|$ at the antenna input port relative to 50 Ω characteristic impedance of the microwave source.

3.3. Optimal Design of the Dual-Band Microstrip Patch Antenna

According to the parametric study performed in Section 3.2, the dimensional design parameters to achieve the best performance of the proposed dual-band patch antenna are indicated in Figure 11. The following subsections are concerned with the theoretical and experimental investigations of the return loss and the radiation patterns of the dual-band antenna with its optimal design parameters.

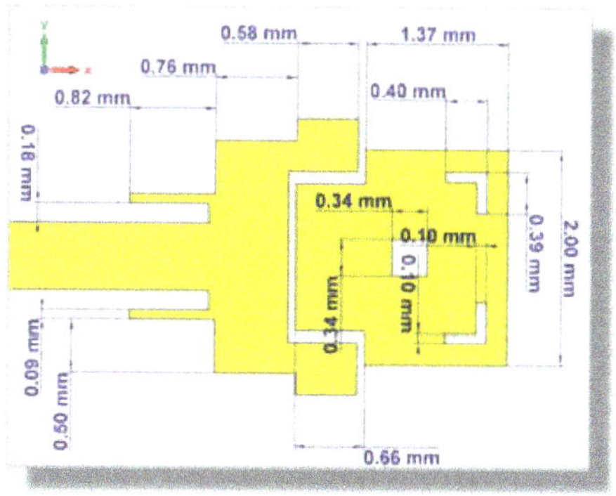

Figure 11. Optimum dimensional parameters of the dual-band microstrip patch antenna for operation.

The azimuth radiation patterns for the proposed dual-band antenna at 38 GHz and 60 GHz are presented in Figure 12a,b, respectively, in the planes $\theta = 30°$ and $\theta = 60°$. As shown in the figures, the radiation patterns are almost omnidirectional in the azimuth planes.

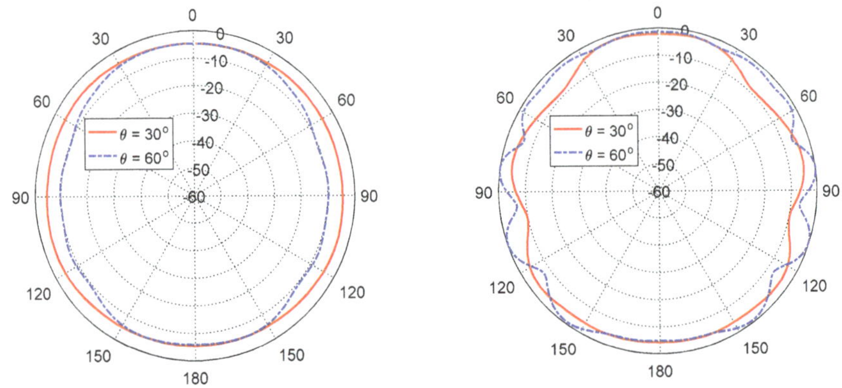

(**a**) Radiation pattern at 38 GHz. (**b**) Radiation pattern at 60 GHz.

Figure 12. Simulated radiation patterns in the azimuth planes for the dual-band microstrip patch antenna whose dimensions are shown in Figure 11.

The proposed dual-band antenna with the optimized design has the radiation and total efficiencies listed in Table 2 at the center frequencies of the lower and upper bands of operation. The antenna efficiency at the lower frequency is better than that at the higher frequency as the losses in the dielectric substrate increase with the frequency.

Table 2. Radiation and total efficiencies of the dual-band microstrip patch antenna whose dimensions are shown in Figure 11 (The listed values have been obtained through electromagnetic simulation using CST®).

	Frequency	
	38 **GHz**	60 **GHz**
Total Efficiency	89.53%	79.42%
Radiation Efficiency	89.57%	79.87%

3.4. *Dual-Band Patch Antenna Fabrication and Experimental Assessment*

This section is concerned with the presentation of the experimental measurements of the dual-band microstrip patch antenna. To confirm the accuracy of the assessed performance for the proposed antenna, the measurement results are compared to those obtained by electromagnetic simulation using the CST™ software package.

3.4.1. Fabrication of the Antenna Prototype

The prototype shown in Figure 13a was fabricated for the purpose of experimental assessment of the performance of the proposed dual-band microstrip patch antenna. The circuit was fabricated by photolithography. A photomask of the circuit layout was prepared and the top layer of the copper-coated substrate was covered with photoresist by spin coating. The photomask was placed next to the top layer of the substrate and exposed to an intense UV light to remove the photoresistant layer from the unwanted copper areas. In etching, a liquid chemical agent removed the uppermost layer of the substrate in the areas that were not protected by the photoresistant layer.

3.4.2. Measurements of the Return Loss

A vector network analyzer (VNA; Rhode and Schwartz model ZVA67) was used for measuring the frequency response of the return loss S_{11} and the corresponding voltage standing wave ratio (VSWR).

A 1.85 mm end-launch connector from Southwest Microwave Inc. was used for connecting the antenna to the VNA as shown in Figure 13b.

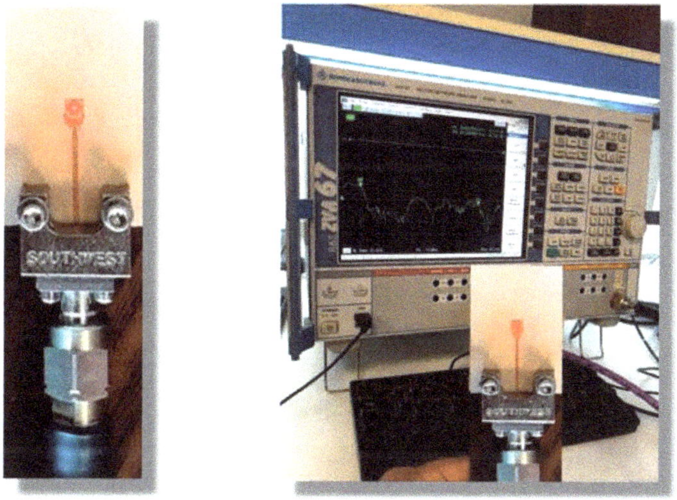

(**a**) Fabricated prototype. (**b**) Measurement of the S-parameters.

Figure 13. Measurement of the scattering parameter $|S_{11}|$ of the proposed dual-band patch antenna using the vector network analyzer (VNA; Rhode and Schwartz model ZVA67).

The frequency response of the return loss $|S_{11}|$ as measured by the VNA is presented in Figure 14. Comparisons between the simulation and measurement results for the frequency dependencies of $|S_{11}|$ and the corresponding VSWR are presented in Figure 15, and show excellent agreement. The impedance matching bandwidths (for $|S_{11}| < -10$dB) obtained through measurements are shown to be better than those obtained by simulation. At 38 GHz, the measured bandwidth is about 2.0 GHz, whereas the simulated bandwidth is about 1.2 GHz. In addition, at 60 GHz, the measured bandwidth is about 3.2 GHz, whereas the simulated is about 2.52 GHz. Moreover, the upper-frequency bandwidth obtained by measurements is shown to be centered at about 60 GHz, whereas the upper band obtained by simulation is shown to be shifted and centered at about 59.5 GHz.

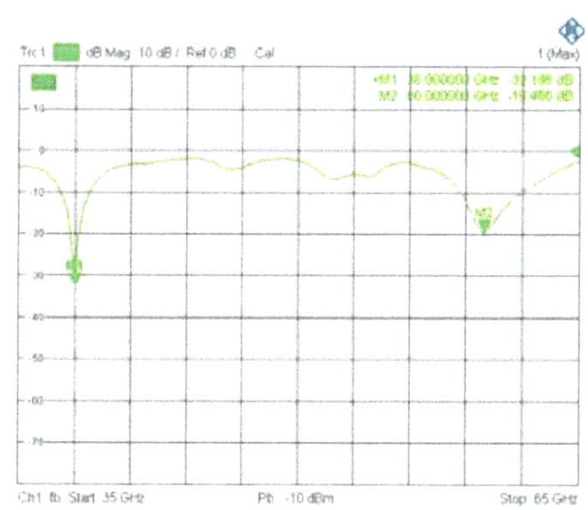

Figure 14. The measured frequency response of the scattering parameter $|S_{11}|$ of the VNA.

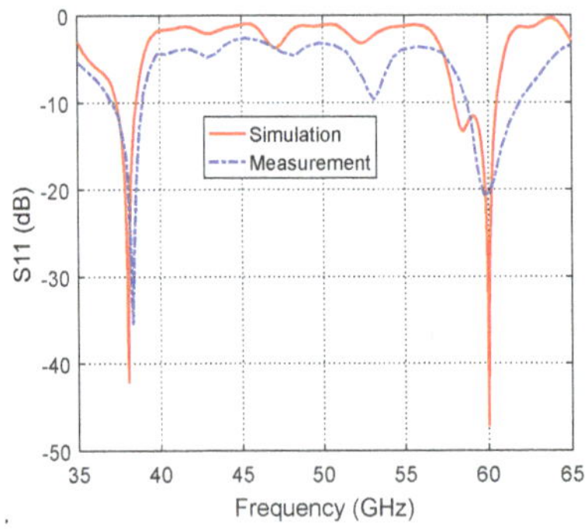

(**a**) The frequency response of $|S_{11}|$.

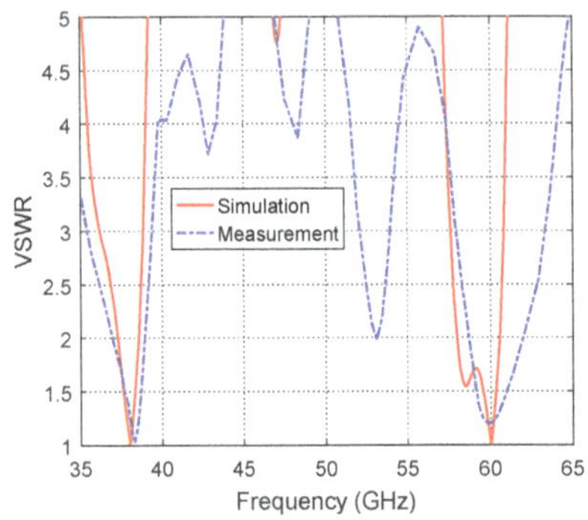

(**b**) The frequency response of the voltage standing wave ratio (VSWR).

Figure 15. Measured frequency responses of the return loss, $|S_{11}|$, and the corresponding VSWR of the proposed dual-band microstrip patch antenna (relative to 50 Ω characteristic impedance of the microwave source) compared with the simulation results.

3.4.3. Measurement of the Radiation Patterns and Maximum Gain

The experimental setup for measuring the radiation patterns and the maximum gain of the proposed antenna is presented in Figure 16. The VNA operating in two-port measurement mode was used for this purpose by measuring the transmission coefficient $|S_{21}|$ through the antenna under test and the reference-gain linearly-polarized horn antennas, models LB-018400 (for 38 GHz band) and LB-12-10-A (for 60 GHz band).

Figure 16. Experimental setup for measuring the radiation pattern and gain of the dual-band antenna.

The measured radiation patterns in the elevation planes $\phi = 0°$ and $\phi = 90°$ corresponding to the E-plane and H-plane, respectively, of the proposed patch antenna, are presented in Figures 17 and 18 at the frequencies 38 GHz and 60 GHz, respectively. It is shown that the measured and simulated radiation patterns are close to each other and show good agreement. The measured maximum gain values are 6.2 dBi and 4.9 dBi at 38 GHz and 60 GHz, respectively, which are close to the measured values (6.5 dBi and 5.5 dBi at 38 GHz and 60 GHz, respectively).

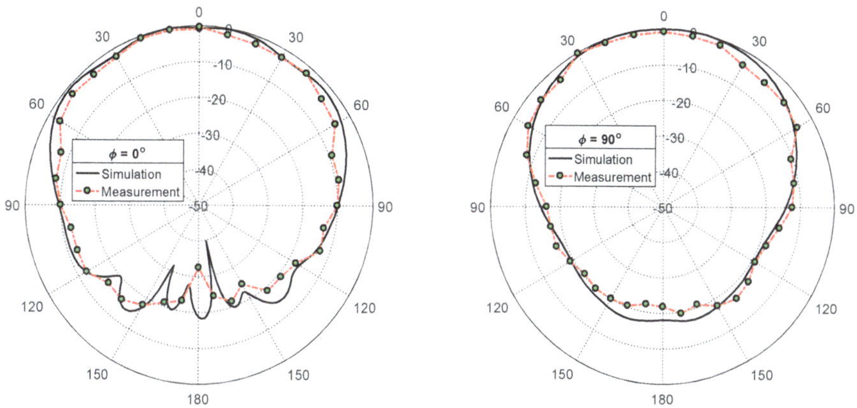

(**a**) E-plane ($\phi = 0°$). (**b**) H-plane ($\phi = 90°$).

Figure 17. Measured radiation patterns of the proposed dual-band microstrip patch antenna at 38 GHz compared with the simulation results.

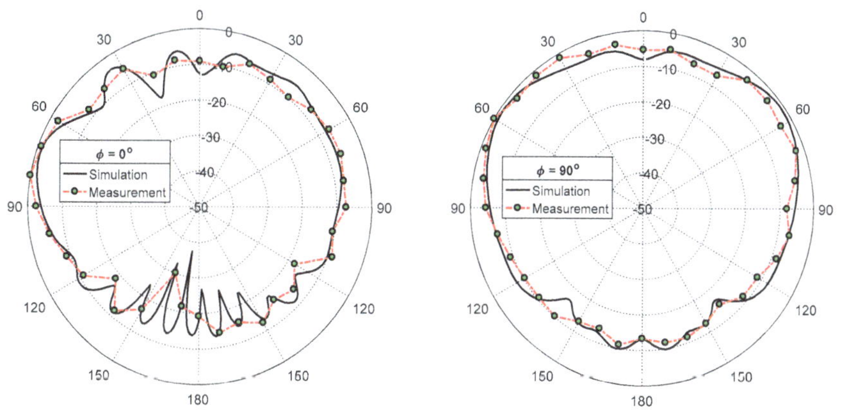

(**a**) E-plane ($\phi = 0°$). (**b**) H-plane ($\phi = 90°$).

Figure 18. Measured radiation patterns of the proposed dual-band microstrip patch antenna at 60 GHz compared with the simulation results.

3.5. Performance Comparison with Published Work

The performance of the proposed dual-band microstrip regarding the impedance matching and the corresponding bandwidth can be compared to other published results for comparative assessment. For example, the frequency response of the measured $|S_{11}|$ for the antenna proposed in the present work is compared to that obtained in [23] and [32] as shown in Figure 19a. The corresponding frequency responses of the VSWR are compared to each other as shown in Figure 19b. The minimum value of the return loss achieved in the present work is about -35 dB (VSWR $\approx$ 1) whereas that achieved in [32] is about -17 dB (VSWR $\approx$ 1.3) and that achieved in [23] is about -21 dB (VSWR $\approx$ 1.2). Moreover, for the antenna proposed in the present work, the bandwidth (BW) of operation for $|S_{11}| < -10$ dB is about 2 GHz and is centered exactly at 38 GHz, whereas that obtained in [32] is centered at about 38.3 GHz (a shift of about 300 MHz) with nearly the same bandwidth. The impedance matching bandwidth obtained in [23] is about 1 GHz and centered at 37 GHz (with about 1 GHz frequency shift from the desired frequency).

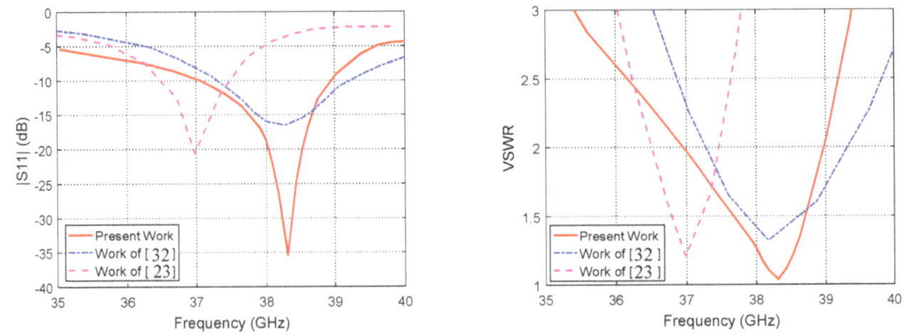

(**a**) Frequency response of $|S_{11}|$. (**b**) Frequency response of the VSWR.

Figure 19. Comparisons between the frequency responses of the $|S_{11}|$ and VSWR obtained in the present work for the proposed dual-band antenna and those obtained in [23] and [32] at 38 GHz.

In the upper band centered at 60 GHz, the frequency response of $|S_{11}|$ measured for the antenna proposed in the present work is compared to those presented in [35] and [36] as shown in Figure 20. The minimum value of the return loss achieved in the present work is about −21 dB, whereas those achieved in [35] and [36] are about −19 dB and −20 dB, respectively. The impedance matching bandwidth achieved in the present work is about 3.3 GHz (for VSWR≤ 2.0 or, equivalently, $|S_{11}| \leq -10$ dB), whereas those achieved in [35] and [36] are 0.9 GHz and 0.35 GHz, respectively.

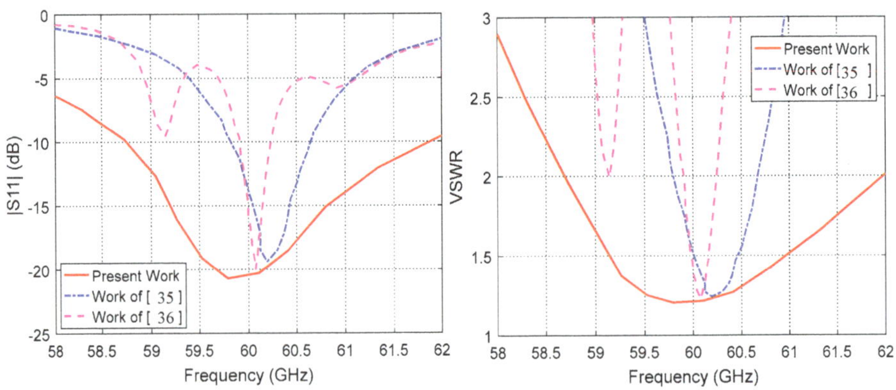

(**a**) Frequency response of $|S_{11}|$. (**b**) Frequency response of the VSWR.

Figure 20. Comparisons between the frequency response of the $|S_{11}|$ obtained in the present work for the proposed dual-band antenna and those obtained in [35] and [36] at 60 GHz.

A comparative summary with the antenna performance achieved in [23,32,35,36] is listed in Table 3. It is shown that the dual band microstrip patch antenna proposed in the present work has superior performance regarding the impedance matching bandwidth in all the cases, and particularly for stringent matching conditions (VSWR ≤ 1.5 and VSWR ≤ 1.25).

Table 3. Comparison between the bandwidth achieved in the present work to that achieved in other published work.

Performance Measure		Performance at 38 GHz			Performance at 60 GHz		
		Present Work	**Work of [32]**	**Work of [23]**	**Present Work**	**Work of [35]**	**Work of [36]**
Center Frequency		38.0	38.3	37.0	60.5	60.3	60.1
BW	VSWR ≤ 1.25	2.0	2.0	1.0	3.3	0.9	0.35
	VSWR ≤ 2.0	1.0	0.75	0.45	1.85	0.5	0.2
	VSWR ≤ 1.5	0.5	0	0.07	0.8	0	0.02
Patch area (mm^2)		2.0 × 3.0	1.0 × 1.0	3.55 × 3.55	2.0 × 3.0	1.4 × 1.4	Array only
Patch Gain (dBi)		6.5	7.4	7.7	5.5	6.2	NA

4. Conclusions

The present work introduces a microstrip patch antenna with a novel design to operate in the 38/60 GHz dual-band. The single antenna is constructed as first and second rectangular patches with some geometrical modifications to achieve perfect impedance matching and balloon-like radiation patterns over the lower- and upper-frequency bands of operation. A microstrip line is the main feeder for the first patch, which is responsible for the lower band (38 GHz) radiation, and capacitive and inductive feeding are responsible for the upper band (60GHz) radiation. The performance of the single-element antenna is assessed through numerical and experimental investigations. It is shown that the simulation results agree with the experimental measurements and both show the good performance of the proposed dual-band patch antenna. The bandwidths achieved around 38 GHz and 60 GHz are about 2.0 GHz and 3.2 GHz, respectively. The return loss has minimum values of −42 dB and −47 dB at 38 GHz and 60 GHz, respectively. The maximum gain is 6.5 dBi and 5.5 dBi at 38 GHz and 60 GHz, respectively. The experimentally assessed frequency band of the proposed antenna is shown to be better than that obtained by other published work for operation at 38 GHz.

Author Contributions: Conceptualization: M.H.S., A.I.Z., R.K.H. and M.M.M.O.; methodology, M.H.S.; software, M.H.S.; validation, M.H.S., formal analysis, M.H.S.; investigation, M.H.S.; resources, M.H.S., A.I.Z., R.K.H. and M.M.M.O.; data curation, M.H.S.; writing—original draft preparation, M.H.S.; writing—review and editing, A.I.Z., R.K.H. and M.M.M.O.; visualization, M.H.S.; supervision, A.I.Z., R.K.H., and M.M.M.O.; project administration, M.H.S. All authors have read and agreed to the published version of the manuscript.

Funding: This research has no external funding.

Conflicts of Interest: The authors declare no conflict of interest.

References

1. Jilani, S.F.; Alomainy, A. Millimetre-wave T-shaped MIMO antenna with defected ground structures for 5G cellular networks. *IET Microw. Antennas Propag.* **2018**, *12*, 672–677. [CrossRef]
2. Diawuo, H.A.; Jung, Y.B. Broadband proximity-coupled microstrip planar antenna array for 5G cellular applications. *IEEE Antennas Wirel. Propag. Lett.* **2018**, *17*, 1286–1290. [CrossRef]
3. Das, G.; Sharma, A.; Gangwar, R.K.; Sharawi, M.S. Performance improvement of multiband MIMO dielectric resonator antenna system with a partially reflecting surface. *IEEE Antennas Wirel. Propag. Lett.* **2019**, *18*, 2105–2109. [CrossRef]
4. Lu, D.; Wang, L.; Du, Y.; Wu, X.; Wang, G. Compact dual-band bow-tie MIMO antennas with fragment-type isolation structure. In Proceedings of the 12th European Conference on Antennas and Propagation, London, UK, 9–13 April 2018; pp. 761–764.
5. Abdullah, M.; Kiani, S.H.; Abdulrazak, L.F.; Iqbal, A.; Bashir, M.A.; Khan, S.; Kim, S. High-performance multiple-input multiple-output antenna system for 5G mobile terminals. *Electronics* **2019**, *8*, 1090. [CrossRef]
6. Youcheng, W.; Yanjiao, Y.; Qingxi, C.; Hucheng, P. Design of a compact ultra-wideband MIMO antenna. *J. Eng.* **2019**, *2019*, 6487–6489. [CrossRef]

7. Muhammad, S.; Yaro, A.S.; Ya'u, I.; Salawudeen, A.T. Design of 5G mobile millimeter wave antenna. *ATBU J. Sci. Technol. Educ.* **2019**, *7*, 178–184.
8. Ojaroudi Parchin, N.; Jahanbakhsh Basherlou, H.; Alibakhshikenari, M.; Ojaroudi Parchin, Y.; Al-Yasir, Y.I.; Abd-Alhameed, R.A.; Limiti, E. Mobile-phone antenna array with diamond-ring slot elements for 5G massive MIMO systems. *Electronics* **2019**, *8*, 521. [CrossRef]
9. Sa'don, S.N.; Jamaluddin, M.H.; Kamarudin, M.R.; Ahmad, F.; Yamada, Y.; Kamardin, K.; Idris, I.H. Analysis of graphene antenna properties for 5G applications. *Sensors* **2019**, *19*, 4835. [CrossRef]
10. Ikram, M.; Nguyen-Trong, N.; Abbosh, A. Multiband MIMO microwave and millimeter antenna system employing dual-function tapered slot structure. *IEEE Trans. Antennas Propag.* **2019**, *67*, 5705–5710. [CrossRef]
11. Li, Y.; Luo, Y.; Yang, G. Multiband 10-antenna array for sub-6 GHz MIMO applications in 5-G smartphones. *IEEE Access* **2018**, *6*, 28041–28053. [CrossRef]
12. Chen, C.M.; Volski, V.; Van der Perre, L.; Vandenbosch, G.A.; Pollin, S. Finite large antenna arrays for massive MIMO: Characterization and system impact. *IEEE Trans. Antennas Propag.* **2017**, *65*, 6712–6720. [CrossRef]
13. Ojaroudi Parchin, N.; Jahanbakhsh Basherlou, H.; Al-Yasir, Y.I.; Abdulkhaleq, A.M.; Patwary, M.; Abd-Alhameed, R.A. A new CPW-Fed diversity antenna for MIMO 5G smartphones. *Electronics* **2020**, *9*, 261. [CrossRef]
14. Yang, B.; Yu, Z.; Dong, Y.; Zhou, J.; Hong, W. Compact tapered slot antenna array for 5G millimeter-wave massive MIMO systems. *IEEE Trans. Antennas Propag.* **2017**, *65*, 6721–6727. [CrossRef]
15. Khalid, M.; Iffat Naqvi, S.; Hussain, N.; Rahman, M.; Mirjavadi, S.S.; Khan, M.J.; Amin, Y. 4-Port MIMO antenna with defected ground structure for 5G millimeter wave applications. *Electronics* **2020**, *9*, 71. [CrossRef]
16. Layegh, M.A.; Ghobadi, C.; Nourinia, J.; Samoodi, Y.; Mashhadi, S.N. Adaptive Neuro-Fuzzy Inference System approach in bandwidth and mutual coupling analyses of a novel UWB MIMO antenna with notch bands applicable for massive MIMOs. *Aeu-Int. J. Electron. Commun.* **2018**, *94*, 407–417. [CrossRef]
17. Zhang, S.; Zhinong, Y. MIMO Antennas for Mobile Terminals. Available online: https://scholar.google.com/scholar?cluster=16108449328272360019&hl=zh-CN&as_sdt=0,5 (accessed on 20 April 2020).
18. Marzouk, H.M.; Ahmed, M.I.; Shaalan, A.E. Novel dual-band 28/38 GHz MIMO antennas for 5G mobile applications. *Prog. Electromagn. Res.* **2019**, *93*, 103–117. [CrossRef]
19. Luo, X.S.; Weng, Z.B.; Zhang, W.J.; Yang, L. Compact planar multiband MIMO antenna based on composite right/left-handed transmission line for mobile phone applications. *Microw. Opt. Technol. Lett.* **2018**, *60*, 1505–1511. [CrossRef]
20. Niu, B.J.; Tan, J.H. Compact SIW cavity MIMO antenna with enhanced bandwidth and high isolation. *Electron. Lett.* **2019**, *55*, 631–632. [CrossRef]
21. Thummaluru, S.R.; Kumar, R. Chaudhary RK.Isolation and frequency reconfigurable compact MIMO antenna for wireless local area network applications. *IET Microw. Antennas Propag.* **2019**, *13*, 519–525. [CrossRef]
22. Abirami, M. A review of patch antenna design for 5G. In Proceedings of the 2017 IEEE International Conference on Electrical, Instrumentation and Communication Engineering (ICEICE), Karur, India, 27–28 April 2017; IEEE: Piscataway, NJ, USA, 2017; pp. 1–3.
23. Khan, J.; Sehrai, D.A.; Ali, U. Design of dual band 5G antenna array with SAR analysis for future mobile handsets. *J. Electr. Eng. Technol.* **2019**, *14*, 809–816. [CrossRef]
24. Dzagbletey, P.A.; Jung, Y.B. Stacked microstrip linear array for millimeter-wave 5G baseband communication. *IEEE Antennas Wirel. Propag. Lett.* **2018**, *17*, 780–783. [CrossRef]
25. Rachakonda, A.; Bang, P.; Mudiganti, J. A compact dual band MIMO PIFA for 5G applications. *Mater. Sci. Eng. Conf. Ser.* **2017**, *263*, 052034. [CrossRef]
26. Haraz, O.M. Broadband and 28/38-GHz dual-band printed monopole/elliptical slot ring antennas for the future 5G cellular communications. *J. InfraredMillim. Terahertz Waves* **2016**, *37*, 308–317. [CrossRef]
27. Şeker, C.; Güneşer, M.T. A single band antenna design for future millimeter wave wireless communication at 38 GHz. *Eur. J. Eng. Form. Sci.* **2018**, *2*, 35–39.
28. Hong, W.; Baek, K.H.; Ko, S. Millimeter-wave 5G antennas for smartphones: Overview and experimental demonstration. *IEEE Trans. Antennas Propag.* **2017**, *65*, 6250–6261. [CrossRef]
29. Wani, Z.; Abegaonkar, M.P.; Koul, S.K. A 28-GHz antenna for 5G MIMO applications. *Prog. Electromagn. Res.* **2018**, *78*, 73–79. [CrossRef]

30. Saini, J.; Agarwal, S.K. Design a single band microstrip patch antenna at 60 GHz millimeter wave for 5G application. In Proceedings of the 2017 international conference on Computer, Communications and Electronics (Comptelix), Jaipur, India, 1–2 July 2017; IEEE: Piscataway, NJ, USA, 2017; pp. 227–230.
31. Saad, A.A.; Mohamed, H.A. Printed millimeter-wave MIMO-based slot antenna arrays for 5G networks. *AEU-Int. J. Electron. Commun.* **2019**, *99*, 59–69. [CrossRef]
32. Imran, D.; Farooqi, M.M.; Khattak, M.I.; Ullah, Z.; Khan, M.I.; Khattak, M.A.; Dar, H. Millimeter wave microstrip patch antenna for 5G mobile communication. In Proceedings of the 2018 International Conference on Engineering and Emerging Technologies (ICEET), Lahore, Pakistan, 22–23 February 2018; IEEE: Piscataway, NJ, USA, 2018; pp. 1–6.
33. Dadgarpour, A.; Sorkherizi, M.S.; Kishk, A.A. High-efficient circularly polarized magnetoelectric dipole antenna for 5G applications using dual-polarized split-ring resonator lens. *IEEE Trans. Antennas Propag.* **2017**, *65*, 4263–4267. [CrossRef]
34. Lin, H.S.; Lin, Y.C. Millimeter-wave MIMO antennas with polarization and pattern diversity for 5G mobile communications: The corner design. In Proceedings of the 2017 IEEE International Symposium on Antennas and Propagation & USNC/URSI National Radio Science Meeting, San Diego, CA, USA, 9–14 July 2017; IEEE: Piscataway, NJ, USA, 2017; pp. 2577–2578.
35. Cabrol, P.; Pietraski, P. 60 GHz patch antenna array on low cost Liquid-Crystal Polymer (LCP) substrate. In Proceedings of the IEEE Long Island Systems, Applications and Technology (LISAT) Conference 2014, Farmingdale, NY, USA, 2 May 2014; IEEE: Piscataway, NJ, USA, 2014; pp. 1–6.
36. Zhang, G.; Pu, S.; Xu, X.; Liu, Y.; Wang, C. Design of 60-GHz microstrip antenna array composed through circular contour feeding line. In Proceedings of the 2016 Asia-Pacific International Symposium on Electromagnetic Compatibility (APEMC), Shenzhen, China, 17–21 May 2016; IEEE: Piscataway, NJ, USA, 2016; Volume 1, pp. 1010–1013.

sensors

MDPI

Article

Design of Multi-Mode Antenna Array for Use in Next-Generation Mobile Handsets

Naser Ojaroudi Parchin [1,*], Haleh Jahanbakhsh Basherlou [2] and Raed A. Abd-Alhameed [1,3]

[1] Faculty of Engineering and Informatics, University of Bradford, Bradford BD7 1DP, UK; R.A.A.Abd@bradford.ac.uk

[2] Bradford College, Bradford BD7 1AY, UK; Hale.Jahanbakhsh@gmail.com

[3] Department of Communication and Informatics Engineering, Basra University College of Science and Technology, Basra 61004, Iraq

* Correspondence: N.OjaroudiParchin@bradford.ac.uk; Tel.: +44-734-143-6156

Received: 31 March 2020; Accepted: 24 April 2020; Published: 25 April 2020

Abstract: In this study, a new design of a tri-band multiple-input–multiple-output (MIMO) antenna array is proposed for fifth-generation (5G) cellular systems. Its structure is composed of eight identical planar-inverted F antenna (PIFA) elements placed at different edge corners of the handset mainboard with overall dimensions of 150×75 mm^2. The PIFA elements and ground plane of the MIMO antenna system are arranged on the back layer of the platform, which makes the design easy to integrate with the handset circuit. For $S_{11} \leq -10$ dB, the radiation elements of the MIMO design operate at the frequency ranges of 2.5–2.7 GHz, 3.4–3.75 GHz, and 5.6–6 GHz covering the long-term evolution (LTE) 41, 42/43, and 47 operation bands, respectively. The array achieves better than 15 dB return loss results across the three operating bands. The presented antenna array not only exhibits multi-band operation but also generates the polarization diversity characteristic, which makes it suitable for multi-mode operation. The proposed antenna array was simulated and experimentally tested. Fundamental characteristics of the proposed design are investigated. It offers three band S-parameters with acceptable isolation and dual-polarized radiation with quite good efficiency and gain results. Besides this, the total active reflection coefficient (TARC) and envelope correlation coefficient (ECC) results of the PIFAs are very low over the bands. In addition, the radiation characteristics of the MIMO antenna in the presence of the user and handset components are studied. Moreover, a new and compact phased array millimeter-wave (MM-Wave) antenna with broad bandwidth and end-fire radiation is introduced which can be easily integrated into the smartphone antenna system. Due to its good performance and simple structures, the proposed smartphone antenna array design is a good candidate for future multi-mode 5G cellular applications.

Keywords: 5G; future handsets; modified PIFA; multi-antenna system; multi-band operation

1. Introduction

With the rapid evolution of wireless communications, the 5G network has received a great deal of attention from both academia and industry, with many reported efforts and research outputs [1–3]. Significant improvements will be made in different areas, including the data rate speed and resolution, mobility, latency, etc. Multiple-input–multiple-output (MIMO) technology with multiple antennas is a promising technology to obtain the requirements of 5G communications [4–6]. To date, 2×2 MIMO systems have been successfully employed for 4G mobile networks, and a larger number of antenna elements is expected to be applied for 5G communications [7,8]. The 5G system is predicted to possess an aggregate data rate 1000 times faster than 4G, and it has better link reliability. Thus, compared with the 4G MIMO antenna systems, at least six to eight antenna elements are integrated into a mobile terminal for 5G massive MIMO to provide good diversity and multiplexing gain [9]. This can enhance the channel capacity and link system reliability [10,11]. The greater number of antennas could make it more resistant to intentional jamming and interference. Through spatial diversity and spatial multiplexing, larger channel capacity and better communication reliability can be achieved. Therefore, the multi-antenna system is much more capable of resisting multipath fading and improving data throughput [12]. The 5G network also needs fundamental technologies to enable small cells, beamforming, full duplexing, MIMO, and millimeter-wave (MM-Wave).

For sub-6 GHz 5G cellular communications, LTE band-41 (2.6 GHz), band-42 (3.5 GHz), band-43 (3.7 GHz), and band-47 (5.8 GHz) are the main important candidate frequency bands [13]. Due to the available radio frequency (RF) circuit and test system, 2.6 GHz LTE can be considered as a default for future mobile communications, and it has recently attracted a great deal of interest. Besides this, 3.4–3.8 GHz (LTE band 42/43) is also recognized by many countries as a first step in demonstrating 5G systems. To further support more potential sub-6 GHz frequency bands, LTE band 47, which is also known as the wireless wide area network (WLAN) operation band, can be considered for 5G massive MIMO antenna design [14,15].

Several smartphone antenna designs with MIMO systems have been proposed recently [16–32]. However, all of these designs either cover only a single-band operation frequency or use a few antenna elements with large sizes which could occupy a huge space of the mainboard. The 5G handset antenna designs introduced in [16–24] only cover a single frequency band. In [25–29], dual-band or wideband arrays are proposed to support two 5G spectrums. Only a few handset antennas with tri-band function are reported in [30–32] for handset applications. However, these antennas have double/quad antenna elements or do not cover important bands such as 2.6 GHz. In this study, we introduced a new MIMO antenna with eight-element planar-inverted F antenna (PIFA) elements which, unlike the reported designs, can cover multi-frequency bands simultaneously. In addition, due to the large number of radiators, the proposed handset antenna can be applied for massive MIMO communication [33]. Furthermore, the proposed handset antenna generates the polarization diversity characteristic, which supports both vertical and horizontal polarizations [34–36]. The modified PIFA radiation elements of the design are employed at four corners of the printed circuit board (PCB) to operate at three different frequencies covering the LTE 2600, 42/43, and 47 operation bands. The proposed PIFA array system operates at three different bands—2.6, 3.6, and 5.8 GHz—of sub-6 GHz 5G cellular networks. It exhibits good properties in terms of the fundamental characteristics and could be used in future handsets.

Apart from the sub-6-GHz spectrum, 5G smartphones are also expected to support the MM-Wave spectrum [37]. Compact antennas arranged as an array can be employed in different portions of a smartphone PCB to form linear phased arrays with high gain and directional radiation beams [38–40]. In contrast to conventional antennas, such as patch, slot, or monopole antennas, end-fire antennas are more suitable to achieve the required radiation coverage [41,42]. Phased array antennas with high performance are highly desirable for MM-Wave 5G communications as they can increase the radiation and the connectivity of the system [43–45]. In addition to the proposed MIMO antenna, a new and compact phased array millimeter-wave (MM-Wave) antenna with broad bandwidth and end-fire radiation is introduced for 28 GHz applications. Its configuration is composed of eight loop dipole resonators with pairs of directors arranged in a linear form which can be easily integrated into the smartphone antenna system. The following sections present the design details, single-element performance, characteristics of the tri-band MIMO antenna, and the 28 GHz phased array.

2. Design and Configuration of the Proposed 5G Antenna Array

The schematic of the designed MIMO handset antenna is plotted in Figure 1. As shown, it is composed of four PIFA pairs that have been deployed at different corners of the mainboard. The presented handset antenna is designed on an FR-4 dielectric with a relative permittivity of 4.4, loss tangent of 0.026 and a thickness of h_x = 1.6 mm. Each PIFA element is fed by a 50 ohm discrete feeding technique extended from the ground plane to the antenna feedline. The values of the design details are listed in Table 1.

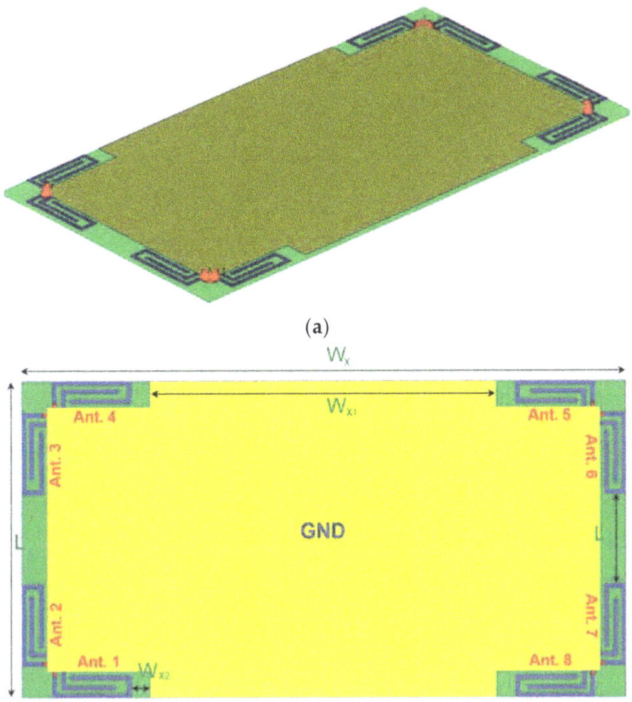

(b)

Figure 1. (**a**) Side and (**b**) back views of the multi-mode antenna array.

Table 1. The dimension values of the presented multi-mode array.

Parameter	W_X	L_X	W_{X1}	L_{X1}	W_{X2}	W	L	W_f
Value (mm)	150	75	83	18	5	20.5	6.5	1
Parameter	L_f	W_1	L_1	W_2	L_2	W_3	L_3	L_4
Value (mm)	1	1	15.3	3.5	14.3	16	5.5	4.5

3. Characteristics of the Single-Element/Multi-Band PIFA Resonator

The PIFA is a compact size antenna radiator that provides omnidirectional radiation patterns and can be used in hand-held devices [46–48]. The conventional PIFAs exhibit single-band operation. However, the modified designs can cover multi-frequency bands for multi-mode operation [49]. The configuration of the PIFA element is depicted in Figure 2a. Its structure is composed of an open-loop resonator with an L-shaped strip protruding from the ground plane. As shown, it has a low profile with the dimension of W × L. A 50 Ω discrete feeding port is employed to excite the antenna. The computer simulation technology (CST) software is used to investigate the properties of the designed mobile-handset antenna [50].

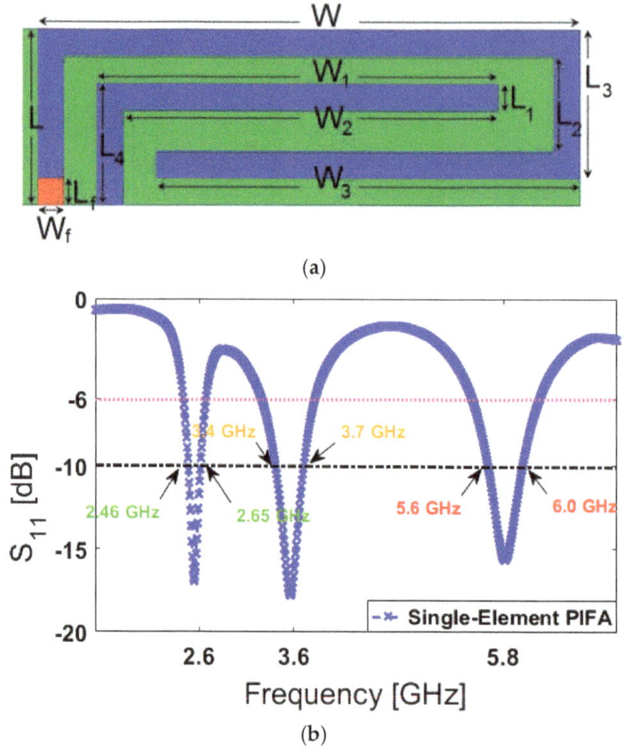

Figure 2. (**a**) Transparent view of the planar-inverted F antenna (PIFA) and (**b**) its S_{11} performance.

The main motive behind the modified PIFA is to obtain a compact antenna element which can support different frequencies and could be integrated with a mainboard circuit while occupying a small clearance. Figure 2b illustrates the simulated reflection coefficient (S_{11}) characteristic of the PIFA element. As shown, the antenna operates at 2.6, 3.6, and 5.8 GHz and provides wide impedance bandwidths at these frequencies.

In order to justify the tri-band function of the design, the simulated current densities of the modified PIFA element at different operation frequencies are illustrated in Figure 3. It is worth mentioning that the maximum scaling for all figures is the same. At 2.6 GHz (first resonance), as can be seen, the L-shaped strip has high current densities with the maximum distribution. Additionally, the current flow reverses on the interior edge of the surrounded open loop [51]. It is evident that the second resonance of the antenna S_{11} has been achieved using the open-loop resonator as it appears very active at 3.6 GHz. The third resonance can be considered as the second harmonic of the first resonance [52]. As shown in Figure 3c, the current distribution is almost equal around the L-shaped strip and the open-loop resonator. Nevertheless, some coupling and interactions between the employed parasitic strip of the modified PIFA can be discovered which could affect the frequency response of the antenna [53].

The main motive behind the PIFA design is to obtain a low-profile and multi-mode radiator with the possibility of integration in the mobile-handset mainboard. The S_{11} characteristics of the modified PIFA antenna can be adjusted by changing the values of the fundamental antenna parameters [54,55]. The first resonance (at 2.6 GHz for the low-band) is mainly determined by the L-shaped strip. The second and third resonances (at 3.6 GHz and 5.8 GHz) depends on the main resonator (open-loop).

Therefore, the circumference lengths of the resonators can satisfy the dielectric wavelength at the corresponding frequency points [56].

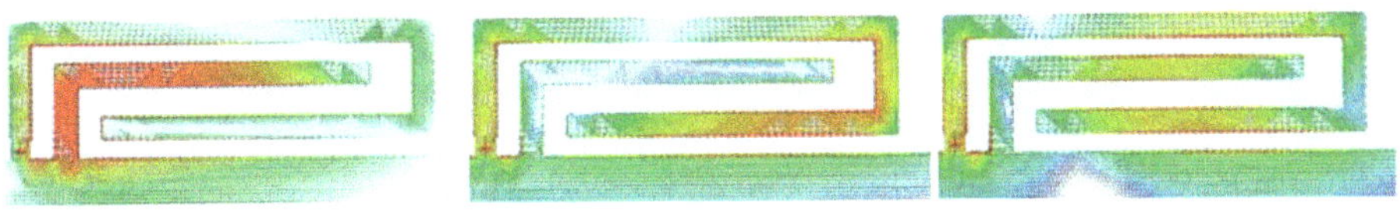

(**a**) (**b**) (**c**)

Figure 3. Surface current densities at (**a**) 2.6 GHz, (**b**) 3.6 GHz, and (**c**) 5.8 GHz.

Figure 4 illustrates the antenna S_{11} characteristic of varying design parameters including W_3, L_4, W_2, and W. In the simulation of the designed antenna, when one parameter changes, the rest of the parameters are kept the same as listed in Table 1. The antenna S_{11} results for different values of W_3 are illustrated in Figure 4a. As evident from the figure, there is very little impact on the first resonance, while the second and third resonant frequencies are influenced and tuned to higher frequencies. Figure 4b shows that L_4 (unlike W_3) has a significant impact on the first resonance frequency and little effect on the second and third resonance frequencies. Figure 4c shows the effects of W_2 (length of L-shaped strip) on the S_{11} of the antenna. It can be observed that as W_2 decreases from 12.65 to 15.65 mm, the first and third resonances at 2.6 GHz and 5.8 GHz shift up to higher frequencies, while very little variation is observed at the middle resonance frequency (3.6 GHz). The antenna S_{11} characteristic of the antenna at different frequencies is also highly dependent on the length of the open-loop resonator (W). As shown in Figure 4d, changing the value of W affects all three resonances of the antenna at different operation bands. According to the obtained results, it can be calculated that the antenna frequency response in all operation bands is very flexible to be tuned to lower or upper frequencies. In addition, its impedance matching can be also affected by changing the parameter values [57,58].

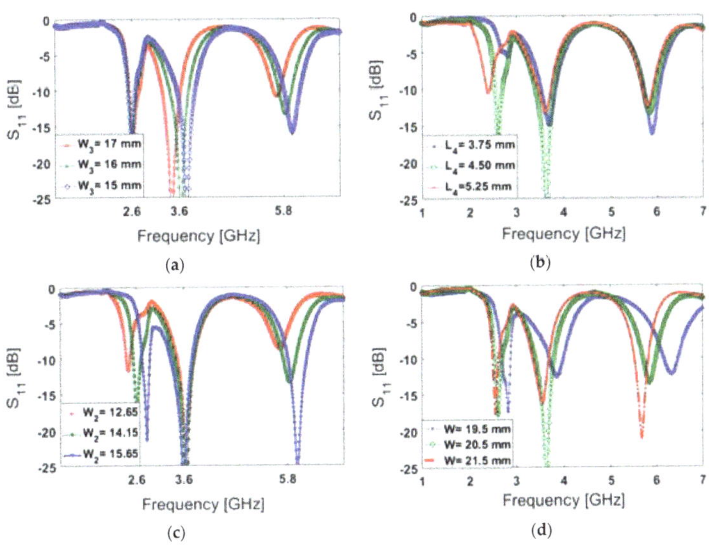

Figure 4. The S_{11} characteristics for different sizes of (**a**) W_3, (**b**) L_4, (**c**) W_2, and (**d**) W.

The fundamental radiation characteristics of the modified PIFA design including the radiation efficiency (R.E.), total efficiency (T.E.) and maximum gain (M.G.) are studied in Figure 5. In theory, the radiation and total efficiencies are related according to

$$e = e_r e_{cd} \tag{1}$$

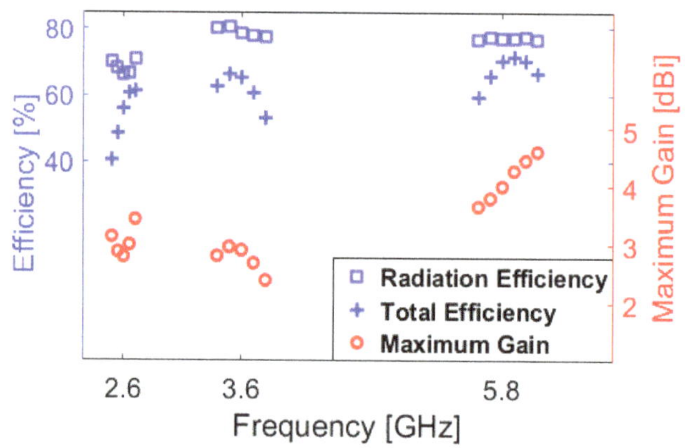

Figure 5. Radiation characteristics of the PIFA element versus its operation bands.

The antenna gain can be calculated using the radiation efficiency and the directivity as follows:

$$G_0(dB) = 10\log(e_{cd}D_0) \tag{2}$$

where e_0 is the total efficiency, e_r is the reflection (mismatch) efficiency = (1−|Γ|2), e_{cd} is the radiation efficiency and D_0 is the antenna directivity [59]. As can be observed from Figure 5, the antenna provides better than 40% and 65% radiation and total efficiencies over the three operation bands. In addition, the maximum gain of the design varies from 2.5 to 4.5 dBi.

4. Characteristics of the Handset Antenna Array

Figure 6 shows the S parameters of the designed handset antenna. As illustrated, the antenna exhibits good S parameters at three operation bands with acceptable mutual coupling at less than −10 dB [5,60]. According to the obtained results in Figure 6, slight variation can be observed in the S_{nn} results of the antenna elements, especially in the lower band (2.6 GHz). This variation is mainly due to different placements, feeding points, neighboring, and also unsymmetrical configurations of the employed elements in the main design (smartphone PCB), as shown in Figure 1. Besides this, the smartphone PCB is in a rectangular shape (with a size of 150 × 75 mm^2) which could cause some discrepancies on the frequency responses and couplings, mainly between even and odd port numbers (antennas 1 and 2, for example). However, due to the flexible frequency behavior of the antenna elements (explained in Figure 4), by modifying the design parameters of the antenna element, the frequency response can be easily adjusted to the desired frequency bands [61].

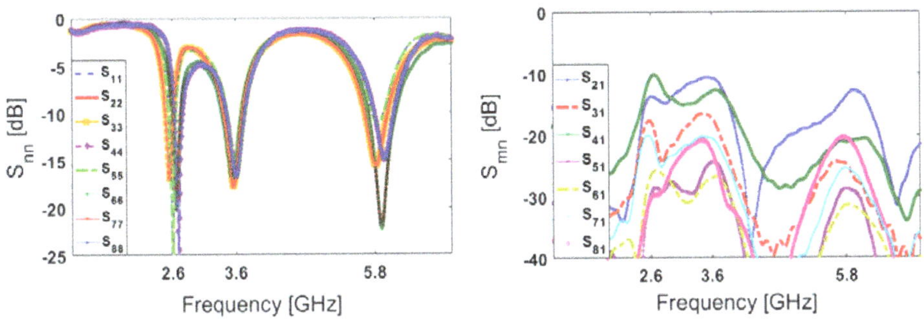

(a) (b)

Figure 6. (**a**) S_{nn} and (**b**) S_{mn} properties of the proposed 5G handset antenna.

The side view of the design radiation patterns for a single-element radiator at different operation frequencies is illustrated in Figure 7. Clearly, the antenna radiation elements exhibit high symmetric radiation patterns covering the different sides of the handset mainboard and increasing the radiation coverage [62–65]. The 3D radiation patterns for the eight PIFAs of the main design are displayed in Figure 8. As illustrated, the gain level of the design varies from 3 to more than 4 dBi. Besides, due to the placements of the PIFA element, four horizontally and vertically polarized radiation patterns are achieved to improve the MIMO performance of the design [66,67].

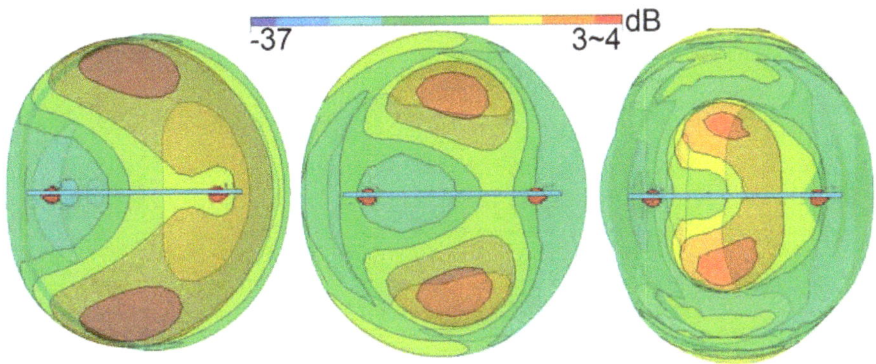

(a) (b) (c)

Figure 7. Side views of the antenna radiation patterns at (**a**) first (2.6 GHz), (**b**) second (3.6 GHz), and (**c**) third (5.8 GHz) resonances.

The envelope correlation coefficient (ECC) and total active reflection coefficient (TARC) characteristics are substantial in MIMO/diversity antenna systems [68,69]. These parameters can be extracted from the S-parameter results using the below formulas, respectively:

$$ECC = \frac{\left|S_{mm}^{*}S_{nm} + S_{mn}^{*}S_{nn}\right|^{2}}{\left(1 - |S_{mm}|^{2} - |S_{mn}|^{2}\right)\left(1 - |S_{nm}|^{2} - |S_{nn}|^{2}\right)^{*}} \quad (3)$$

$$TARC = -\sqrt{\frac{(S_{mm} + S_{mn})^{2} + (S_{nm} + S_{nn})^{2}}{2}} \quad (4)$$

The ECC and TARC results of the presented multi-mode MIMO antenna design are calculated and plotted in Figure 9. As seen from Figure 9a, the calculated ECC results of PIFA pairs are very low over the entire multi-operation bands (less than 0.01). Additionally, it can be observed from Figure 9b that the TARC value of the diverse PIFA pairs is less than −20 dB at different frequencies.

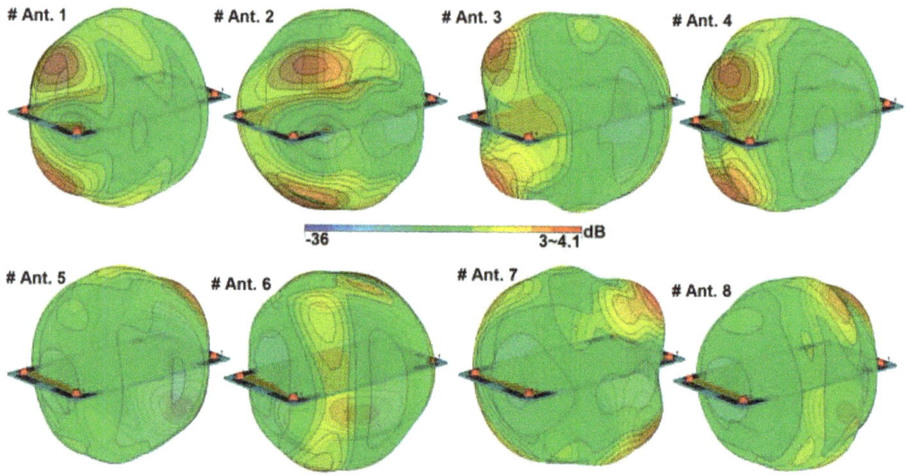

Figure 8. Radiation patterns of the PIFAs at the middle frequency (3.6 GHz).

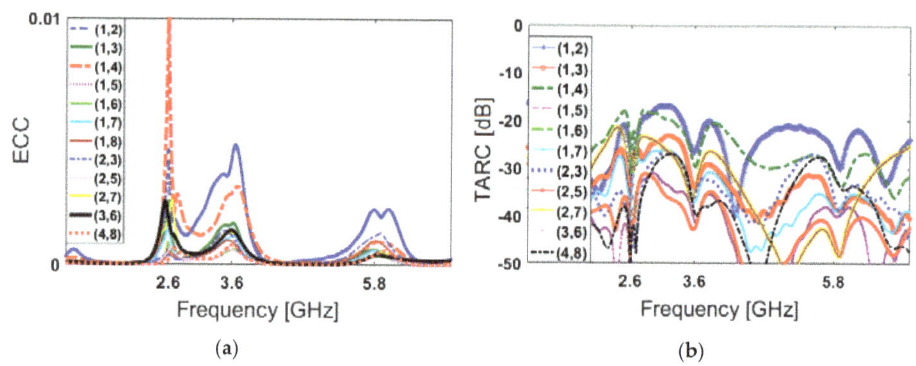

Figure 9. Calculated (**a**) ECC and (**b**) TARC results of the proposed design.

5. Fabrication and Measurements

A prototype sample of the proposed 5G handset antenna array was fabricated as illustrated in Figure 10a,b. Due to the similar placements and performances of the modified PIFA pairs, the properties of the handset antenna design for port 1 and 2 are measured and compared below. The feeding mechanism of the adjacent elements is shown in Figure 10c.

The measured and simulated results of the S-parameters are compared in Figure 11a. As seen, the S_{11}/S_{21} measurements have good agreement with the simulated results in terms of covering the required multi-operation bands: a quite good impedance bandwidth ($S_{11} < -10$) is achieved to cover the operation bands of 2.45–2.65 GHz, 3.5–3.7 GHz and 5.6–6 GHz with resonances at 2.6, 3.6, and 5.8 GHz, respectively. In addition, as shown, the mutual couplings of the adjacent PIFAs are less than −15, −10, and −13 dB at the desired frequency bands. One of the vital parameters for the MIMO performance of an antenna array is diversity gain (DG), which can be calculated using the following formula:

$$DG = 10\sqrt{1 - (ECC)^2} \tag{5}$$

(a) (b) (c)

Figure 10. (**a**) Front and (**b**) back views, and (**c**) feeding mechanism of the fabricated sample.

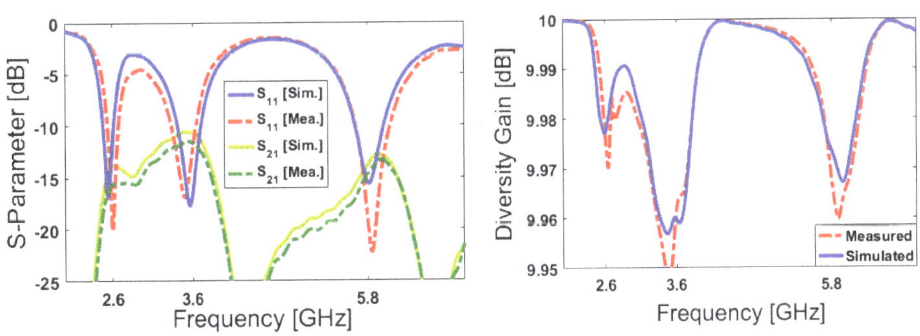

(a) (b)

Figure 11. Measured and simulated (**a**) S-parameters and (**b**) diversity gain of the two adjacent PIFAs.

The DG characteristics of the antenna are illustrated in Figure 11b. The diversity gain function of the designed antenna over its operation band is more than 9.95 dB over the operating frequency bands [70]. In Figure 12a,b, we plot and compare the calculated TARC and ECC results of the PIFA pairs from simulated and measured results. As shown, the ECC function is very low (less than 0.05) over the different frequency bands of interest. Besides, the obtained TARC results are less than −18 dB.

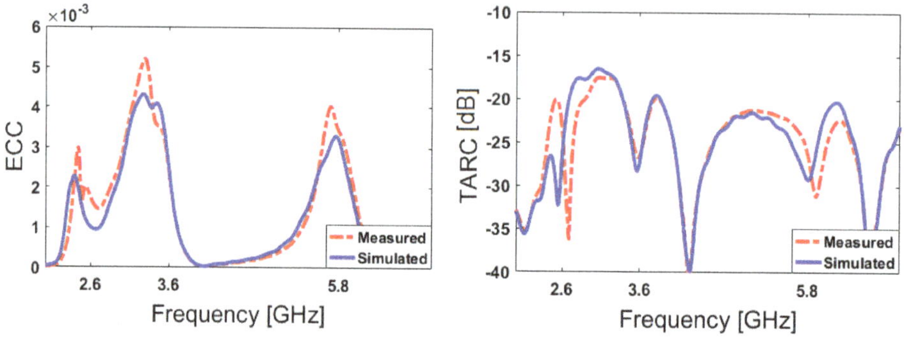

(a)

(b)

Figure 12. Calculated (**a**) ECC and (**b**) TARC results of the adjacent elements.

Measured and simulated radiation patterns (H-plane) of the modified PIFA element at different frequencies are shown in Figure 13. It is worth noting that, during the measurement of the antenna radiation patterns, one port was kept excited while the other one was loaded with a 50 Ω load. As can be clearly seen, the sample handset antenna prototype provides good quasi-omnidirectional radiation patterns at different resonance frequencies with peak gains and acceptable agreement between simulations and measurements. It is found that when the antenna frequency increases, the gain level of the antenna is increased [71–73].

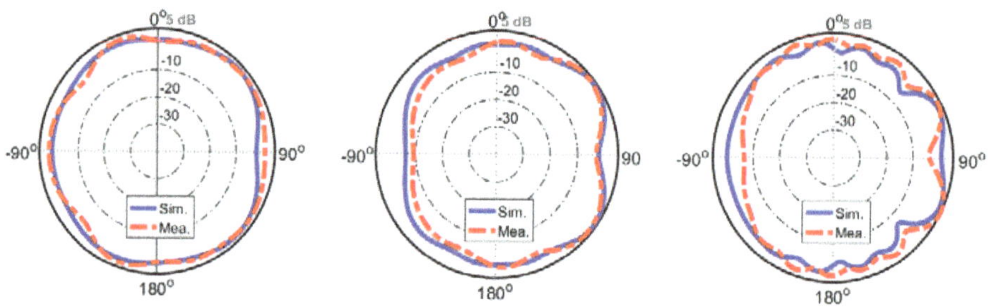

(a)

(b)

(c)

Figure 13. Two-dimensional radiation patterns at (**a**) first (2.6 GHz), (**b**) second (3.6 GHz), and (**c**) third (5.8 GHz) resonances.

6. Comparison

Table 2 compares the characteristics of the proposed handset antenna design and some reported 5G smartphone MIMO antenna designs in the literature with the recent states of the art [16–32]. To the best of our knowledge, most of the reported sub-6 GHz 5G antennas are single-band operating, and only a few works have reported on the use of the dual-band or multi-band techniques for 5G smartphone applications. The main contribution of our work is that we propose a planar 8 × 8 MIMO diversity antenna array with a triple-band function covering the LTE 41, 42/43, and 47 bands. As clearly shown, compared with the recently introduced MIMO handset antenna systems with planar and uniplanar structures, our proposed handset antenna provides better characteristics in terms of impedance matching and bandwidth and low-profile radiators.

Table 2. Comparison of the design characteristics with the referenced handset antennas. MIMO: multiple-input–multiple-output.

Reference	Antenna Type	Elements	Bandwidth (GHz)	Efficiency (%)	Overall Size (mm^2)	Isolation (dB)	ECC
			Single-Band MIMO Handset Antennas				
[16]	Coupled-Fed	8	2.55–2.68	48–63	136 × 68	12	<0.15
[17]	Petal Slot	8	2.55–2.66	-	150 × 75	10	<0.1
[18]	Loop	8	2.55–2.6	48–63	136 × 68	11	<0.15
[19]	PIFA Slot	8	3.4–3.6	62–78	140 × 70	10	<0.20
[20]	Slot-Ring	8	3.4–3.8	55–70	150 × 75	15	<0.05
[21]	Patch-Slot	8	3.55–3.65	52–76	150 × 75	11	-
[22]	Loop Element	8	3.3–3.6	40	120 × 70	15	<0.02
[23]	CPW-Fed Slot	8	3.4–4.4	65	150 × 75	16	<0.01
[24]	Monopole	8	4.55–4.75	50–70	136 × 68	10	-
			Dual-Band MIMO Handset Antennas				
[25]	Coupled-Fed Slot	10	3.4–3.8 5.15–5.92	41–84 47–79	150 × 80	12	<0.15
[26]	Dual Mode Monopole	2	3.4–3.6, 5.725–5.875	50–61, 67–80	150 × 76.6	10.5	<0.2
[27]	Cuboid Monopole	6	2.6–2.7 5.1–5.9	60–70 70–80	140 × 05	10	<0.05
[28]	Monopole	8	3.2–3.9 5–5.5	60–80 70–85	150 × 75	12	-
[29]	Folded Monopole	8	2.45–2.6 3.45–3.65	40–60 50–80	124 × 74	15	<0.2
			Tri-Band MIMO Handset Antennas				
[30]	Coupled-Fed Arm	8	3.3–3.8, 4.8–5, 5.1–5.9	55–72, 50–65, 43–73	130 × 70	10	<0.1
[31]	Yet-Decoupled Antennas	2	2.40–2.48 5.15–5.35, 5.72–5.92	44–48, 74–80, 75	150 × 75	15	<0.14
[32]	F-Shaped Monopole	4	3.3–4.2, 4.4–5, 5.15–5.85	60–80	150 × 75	12	<0.1
Proposed	**Modified Diversity PIFA**	**8**	**2.45–2.65, 3.4–3.75, 5.6–6**	**40–65, 50–70, 60–80**	**150 × 75**	**11**	**<0.01**

The proposed design achieves improvements not only around tri-band impedance bandwidth but also offers polarization diversity function at different edges of the mainboard. This is mainly due to orthogonal placements of the adjacent PIFA resonators at different corners of the PCB. In addition, unlike the reported handset antennas in the literature, the presented antenna is set on the single (back) layer of the platform which makes the design easy to integrate with the handset circuit. Furthermore, due to the small clearance of the proposed handset antenna, its characteristics in terms of data and talk modes are not changed significantly; that is, the proposed eight-element antenna array has the advantage of the comprehensive performance, meaning that it can be applied well in future 5G mobile terminals.

7. User Effects on the Characteristics of the Designed Antenna Array

The health hazards of emitted electromagnetic (EM) radiation from mobile handsets has become a point of open deliberation as the use of mobile handsets is increasing exponentially [74]. For mobile handsets, the investigation of the user-effect on the characteristics of the antenna is indispensable [75]. Below, different usage postures in the data-mode of the user-hand for right and left hands are considered and studied in Figures 14 and 15, respectively. The employed user-hand phantom in the simulation has a relative permittivity of $\varepsilon = 24$ and conductivity of $\sigma = 2$ s/m [76]. According to the obtained results, the proposed design exhibits similar radiation behavior for different hand scenarios. This is mainly due to the symmetrical schematic of the designed MIMO antenna. It is shown that the handset antenna and its modified PIFA elements exhibit good efficiencies. Besides, when the antenna frequency increases, the efficiency of the antenna is improved. The reflection coefficients (S_{nn}) of the proposed array are not affected drastically by the hand, except for some small frequency fluctuations. However, the antenna efficiencies of the proposed array are affected significantly owing to the absorption effect of the user's hand. This is because some EM energy has been absorbed by the hand [77]. Compared with the antenna performance in free space, the maximum reductions of the total efficiencies are observed for the PIFA elements that have been partially covered by the hand [78]. Additionally, as evident from the results, the antenna provides good S-parameters including S_{nn} and S_{mn} in both right and left-hand scenarios.

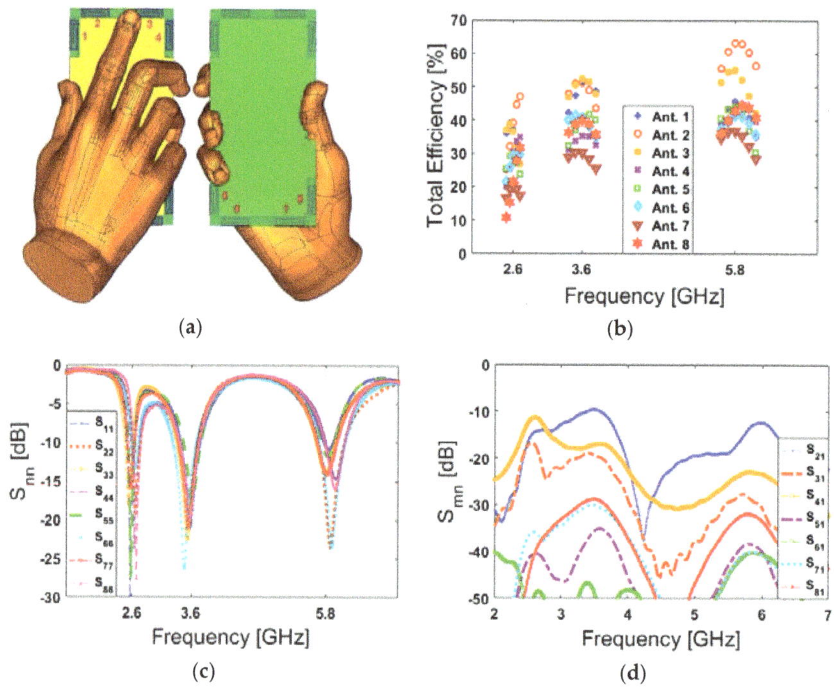

Figure 14. (**a**) Placement, (**b**) total efficiencies, (**c**) S_{nn}, and (**d**) S_{mn} results for the right-hand scenario.

Apart from the data-mode, the antenna performance in talk-mode should be also studied. Figure 16 shows the total efficiency, S_{nn} (S_{11}–S_{88}), and S_{mn} (S_{21}–S_{81}) characteristics of the designed handset antenna in the presence of the user-hand/user-head in talk-mode.

Figure 16a shows the placement of the designed antenna array in talk-mode; the array's total efficiencies are represented in Figure 16b. It is shown that the antenna elements exhibit relatively good efficiencies at different resonance frequencies. Additionally, the antenna S-parameters are depicted in Figure 16d. As can be seen, the PIFAs are operating at the target frequencies with good S_{nn} and less than −10 dB S_{mn}.

The EM energy absorbed by human body tissues can be evaluated by the specific absorption rate (SAR) [79,80]. SAR is a measure of how much power is being absorbed per unit mass. The SAR features of the MIMO design at three different operation frequencies are investigated. According to the obtained results from different investigations of the antenna elements, it is found that antenna 2 causes the maximum SAR value while the minimum SAR value is observed from antenna 6. The SAR results of antenna 2 and antenna 6 at 2.6, 3.6, and 5.8 GHz are depicted in Figure 17. According to the results, the distance between the PIFA elements and the head phantom is most important in terms of the value of the SAR function.

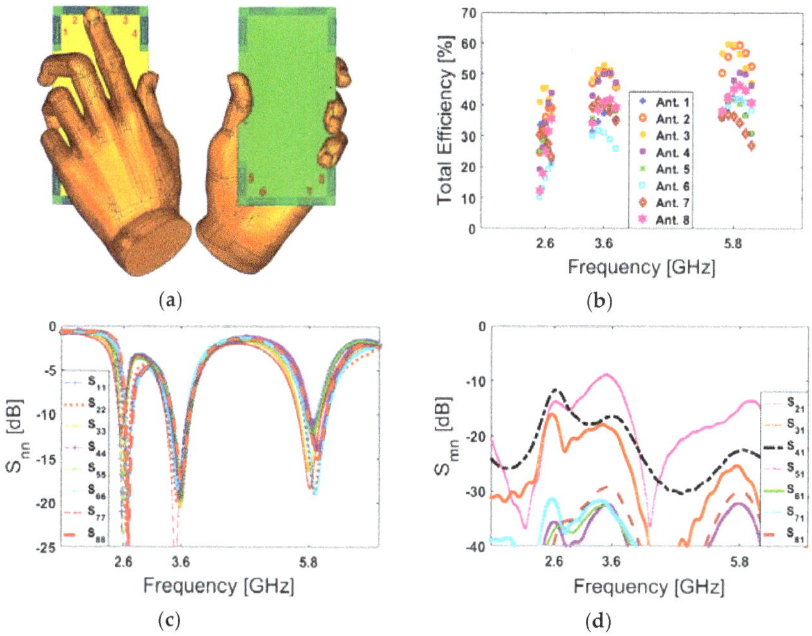

Figure 15. (**a**) Placement, (**b**) total efficiencies, (**c**) S_{nn}, and (**d**) S_{mn} results for the left-hand scenario.

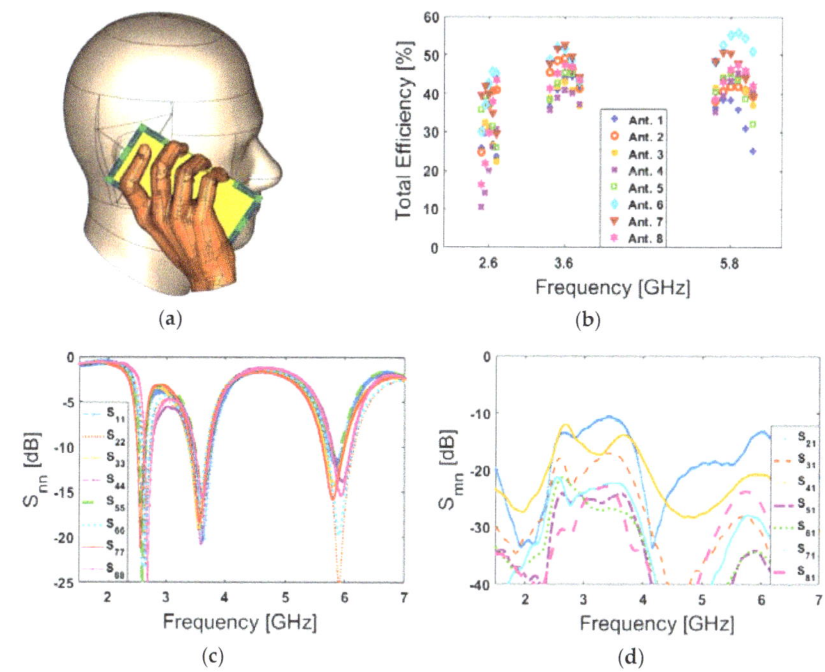

Figure 16. (**a**) Placement, (**b**) total efficiencies, (**c**) S_{nn}, and (**d**) S_{mn} results for the talk-mode scenario.

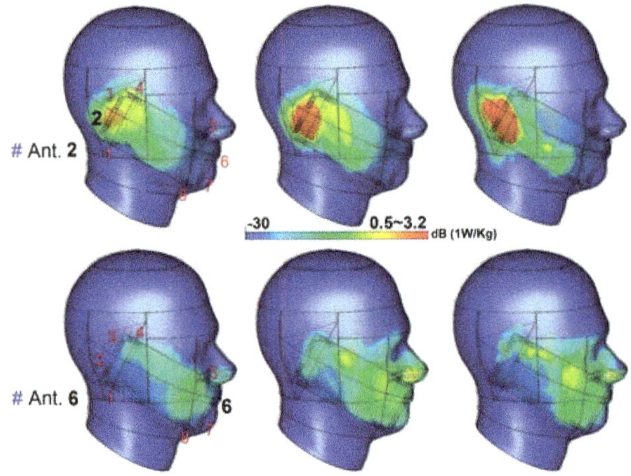

(a) (b) (c)

Figure 17. SAR for antennas 1 and 5 at (**a**) first (2.6 GHz), (**b**) second (3.6 GHz), and (**c**) third (5.8 GHz) resonances.

Below, the fundamental characteristics of the MIMO antenna including the total efficiency, S_{nn} and S_{mn} are also studied in the presence of smartphone components including the battery, speaker, camera, USB connector, and LCD screen. Table 3 lists the characteristics of the modeled components [81]. It is clearly seen from Figure 18 that the design provides consistent characteristics at the desired operating frequencies.

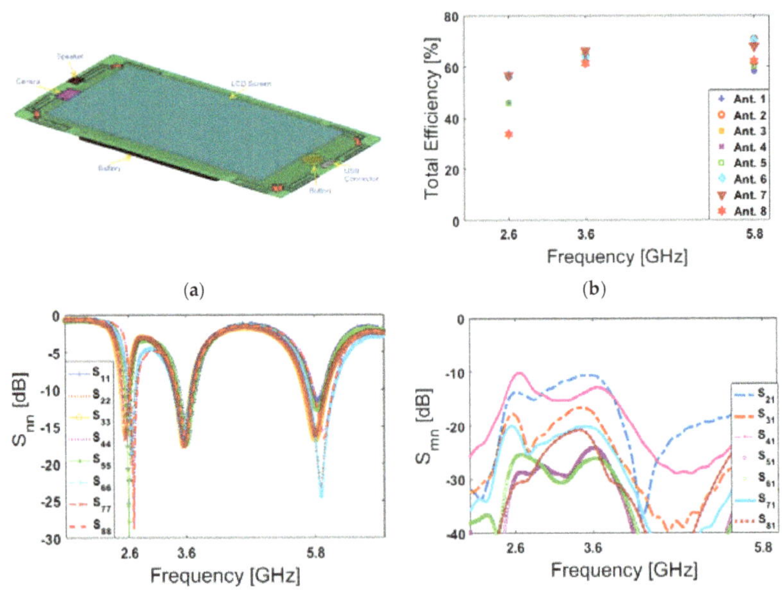

Figure 18. (**a**) Placement, (**b**) total efficiencies, (**c**) S_{nn}, and (**d**) S_{mn} results in the presence of the smartphone components.

Table 3. Characteristics of the smartphone components.

Component	Material	Permittivity
Screen	LCD film	4.8
Battery. Camera, Speaker	perfect electric conductor (PEC)	-
USB Connector	Brass (PEC)	
Button	Rubber	3.5
printed circuit board (PCB)	FR-4	4.4

8. Integration of a Compact MM-Wave Phased Array

In this section, a new and miniaturized MM-Wave phased array 5G antenna with broad bandwidth is proposed to be incorporated in a shared board. The design details of the integrated phased array are illustrated in Figure 19a. It has a very compact size, with an overall size of $W_a \times L_a = 18 \times 5$ mm^2, and it can be implemented in the same FR-4 laminate PCB with a thinner thickness of 0.8 mm. As shown, its configuration is composed of eight loop dipole resonators with pairs of directors arranged in a linear form. A discrete feeding port is applied separately for each antenna. The parameter values (in mm) are as follows: $W_{a1} = 2.8$, $W_{a2} = 0.125$, $W_{a3} = 0.15$, $W_{a4} = 1.8$, $W_{a5} = 0.4$, $W_{a6} = 0.15$, $L_{a1} = 1$, $L_{a2} = 1.5$, $L_{a3} = 0.2$, $d_a = 0.125$, $d_{a1} = 0.8$, $d_{a2} = 1.3$. The phased array is designed to work at 28 GHz—one of the promising 5G candidate bands at higher frequencies [82,83]. However, due to the broad bandwidth characteristic, the proposed phased array is also capable of covering 26 and 30 GHz 5G bands [84–86].

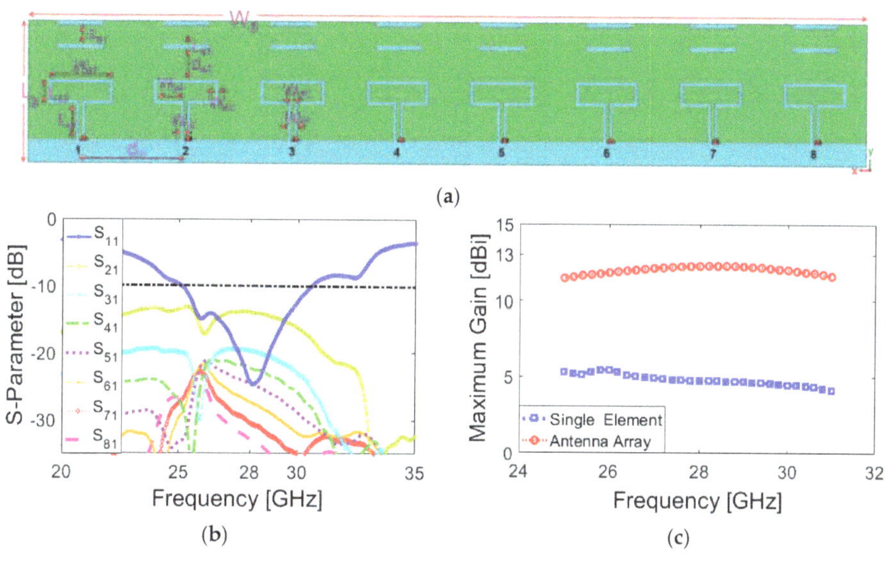

Figure 19. (**a**) Design details and configuration of the array, (**b**) its S-parameters, and (**c**) gain comparison of the array and the antenna element.

Figure 19b plots the S-parameters (S_{11}–S_{41}) of the phased array. As seen, the designed phased array provides a broad impedance bandwidth of 25–31 GHz with a central frequency of 28 GHz. Besides, less than a −15 dB mutual coupling characteristic is obtained for the antenna elements. Figure 19c compares the maximum gains of the antenna element and the phased over the operation band. It is shown that the antenna element provides 4~5 dBi gain values over the operation frequency, whereas more than 12 dBi maximum gain is achieved for the designed array. The 3D beam steering functions of the array at 28 GHz for different angles are plotted in Figure 20a. The shapes and directions of the phased array beams are determined by the relative phase amplitudes applied to each antenna element with discrete-feeding ports, as shown below [87]:

$$\psi = 2\pi\left(\frac{d}{\lambda}\right)sin\theta \tag{6}$$

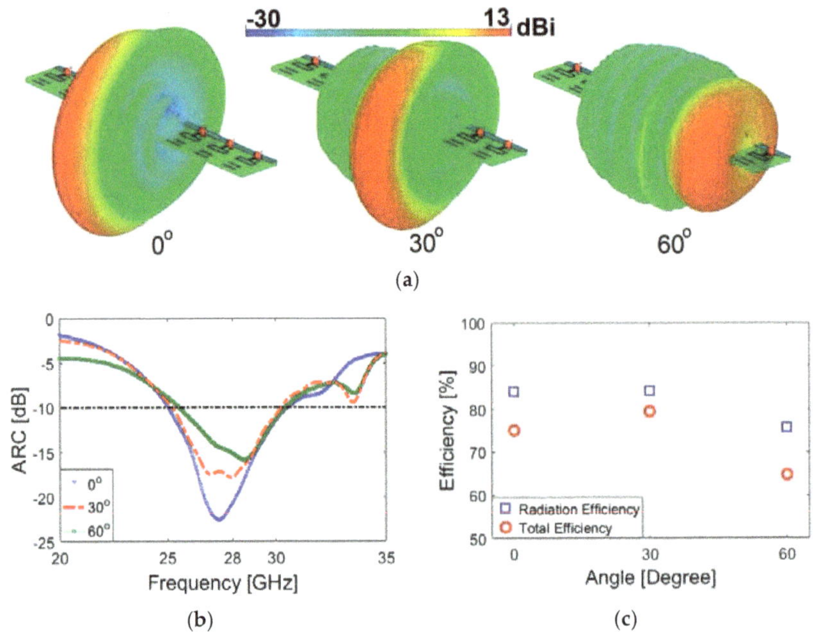

Figure 20. (**a**) Radiation beams, (**b**) active reflection coefficient, and (**c**) efficiencies at different scanning angles.

As shown, the designed phased array provides a good beam-steering function with end-fire radiation beams. As illustrated, the designed array exhibits high gain radiation beams with low sidelobes.

The active reflection coefficients (ARCs) of the design for different scanning angles are illustrated in Figure 20b. As the radiation beam of the antenna is scanned, the amount of coupling between the radiation elements changes, meaning that the active reflection coefficient curve moves slightly [88]. However, as shown, the proposed phased array antenna exhibits sufficient performance and supports the target frequency bandwidth for different scanning angles. The radiation efficiency of the proposed array antenna is more than 75% in the operating bandwidth of all switching modes. Figure 20c plots the radiation and total efficiency levels of the array at scanning angles of 0, 30, and 60. It is shown that the proposed phased array offers quite good efficiency results. Figure 21 shows the possible placements and radiation beam-steering of the proposed phased array in the configuration of the smartphone board. It can be observed that the array can be easily integrated into a small area of the PCB and could provide full radiation coverage supporting different sides of the handset [89].

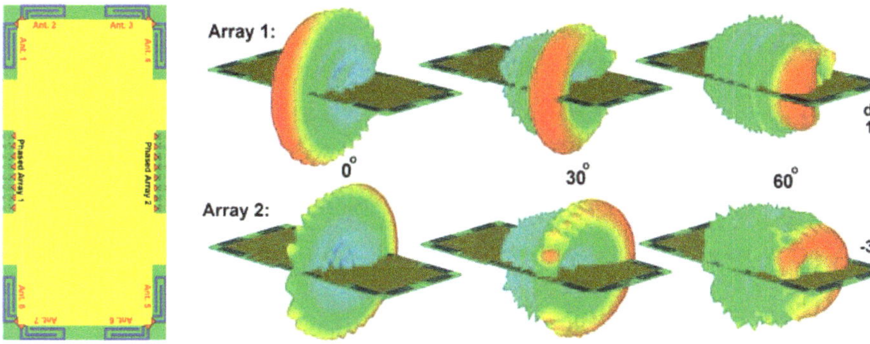

(**a**) (**b**)

Figure 21. (**a**) Different placements of the proposed phased array into the smartphone board and (**b**) its beam steering function at different angles.

9. Conclusions

The design and characteristics of a new MIMO smartphone antenna with a multi-mode operation is successfully investigated in this paper. The designed handset antenna contains eight modified PIFA elements deployed at four corners of the mainboard. The proposed design operates at 2.6, 3.6, and 5.8 GHz for sub-6 GHz 5G mobile terminals. It offers good characteristics in terms of bandwidth, isolation, and radiation patterns. In addition, quite good characteristics are observed in the presence of the user. Due to the tri-band and polarization diversity, the antenna can be considered for multi-mode future handset applications. Furthermore, due to the available space on the smartphone antenna system, a compact 28 GHz phased array is proposed to be integrated onto the 5G smartphone.

Author Contributions: Writing—original draft preparation, N.O.P., H.J.B., and R.A.A.-A.; resources, N.O.P., and R.A.A.-A.; For other cases, all authors have participated. All authors have read and agreed to the published version of the manuscript.

Funding: This project has received funding from the European Union's Horizon 2020 research and innovation program under grant agreement H2020-MSCA-ITN-2016 SECRET-722424.

Acknowledgments: The authors wish to express their thanks to the support provided by the innovation program under grant agreement H2020-MSCA-ITN-2016 SECRET-722424.

Conflicts of Interest: The authors declare no conflict of interest.

References

1. Nadeem, Q.U.A.; Kammoun, A.; Debbah, M.; Alouini, S.-M. Design of 5G full dimension massive MIMO systems. *IEEE Trans. Commun.* **2018**, *66*, 726–740. [CrossRef]
2. Ojaroudiparchin, N.; Shen, M.; Pedersen, G.F. Multi-layer 5G mobile phone antenna for multi-user MIMO communications. In Proceedings of the 23rd Telecommunications Forum Telfor (TELFOR), Belgrade, Serbia, 24–26 November 2015; pp. 559–562.
3. Osseiran, A.; Boccardi, F.; Braun, V.; Kusume, K.; Marsch, P.; Maternia, M.; Queseth, O.; Schellmann, M.; Schotten, H.; Taoka, H.; et al. Scenarios for 5G mobile and wireless communications: The vision of the METIS project. *IEEE Commun. Mag.* **2014**, *52*, 26–35. [CrossRef]
4. Yang, H.H.; Quel, Y.Q.S. Massive MIMO Meet Small Cell. *Springer Briefs Electr. Comput. Eng.* **2017**. [CrossRef]
5. Ojaroudi, N.; Ghadimi, N. Design of CPW-fed slot antenna for MIMO system applications. *Microw. Opt. Technol. Lett.* **2014**, *56*, 1278–1281. [CrossRef]
6. Parchin, N.O.; Basherlou, H.J.; Al-Yasir, Y.I.A.; Abd-Alhameed, R.A.; Abdulkhaleq, A.M.; Noras, J.M. Recent developments of reconfigurable antennas for current and future wireless communication systems. *Electronics* **2019**, *8*, 128. [CrossRef]

7. Ojaroudiparchin, N.; Shen, M.; Pedersen, G.F. Wide-scan phased array antenna fed by coax-to- microstriplines for 5G cell phones. In Proceedings of the 21st International Conference on Microwaves, Radar and Wireless Communications, Krakow, Poland, 9–11 May 2016.
8. Hussain, R.; Alreshaid, A.T.; Podilchak, S.K.; Sharawi, M.S. Compact 4G MIMO antenna integrated with a 5G array for current and future mobile handsets. *IET Microw. Antennas Propag.* **2017**, *11*, 271–279. [CrossRef]
9. Andrews, J.G.; Buzzi, S.; Choi, W.; Hanly, S.V.; Lozano, A.; Soong, A.C.K.; Zhang, J.C. What will 5G be? *IEEE J. Sel. Areas Commun.* **2014**, *32*, 1065–1082. [CrossRef]
10. Parchin, N.O.; Al-Yasir, Y.I.A.; Abd-Alhameed, R.A. *Microwave/RF Components for 5G Front-End Systems*; Avid Science: Telangana, India, 2019; pp. 1–200.
11. Chen, Q.; Lin, H.; Wang, J.; Ge, L.; Li, Y.; Pei, T.; Sim, C.-Y.-D. Single ring slot-based antennas for metal-rimmed 4G/5G smartphones. *IEEE Trans. Antennas Propag.* **2018**, *67*, 1476–1487. [CrossRef]
12. Bonfante, A.; Giordano, L.G.; López-Pérez, D.; Garcia-Rodriguez, A.; Geraci, G.; Baracca, P.; Majid Butt, M.; Marchetti, N. 5G massive MIMO architectures: Self-backhauled small cells versus direct access. *IEEE Trans. Veh. Technol.* **2019**, *68*, 10003–10017. [CrossRef]
13. Gozalvez, J. 5G worldwide developments [mobile radio]. *IEEE Veh. Technol. Mag.* **2017**, *12*, 4–11. [CrossRef]
14. 5G in the Sub-6 GHz Spectrum Bands. Available online: http://www.rcrwireless.com/20160815/fundamentals/5g-sub-6ghztag31-tag99 (accessed on 15 August 2019).
15. 5G NR (New Radio). Available online: http://3gpp.org/ (accessed on 12 December 2018).
16. Li, M.-Y.; Xu, Z.; Ban, Y.; Yang, Q.; Zhou, Q. Eight-port orthogonally dual-polarized antenna array for 5G smartphone applications. *IEEE Trans. Antennas Propag.* **2016**, *64*, 3820–3830. [CrossRef]
17. Parchin, N.O.; Basherlou, H.J.; Al-Yasir, Y.; Abdulkhaleq, A.M.; Abd-Alhameed, R.A.; Excell, P. Eight-port MIMO antenna system for 2.6 GHz LTE cellular communications. *Prog. Electromagn. Res. C* **2020**, *99*, 49–59. [CrossRef]
18. Li, M.-Y. Eight-port orthogonally dual-polarised MIMO antennas using loop structures for 5G smartphone. *IET Microw. Antennas Propag.* **2017**, *11*, 1810–1816. [CrossRef]
19. Al-Hadi, A.A.; Ilvonen, J.; Valkonen, R.; Viikan, V. Eight-element antenna array for diversity and MIMO mobile terminal in LTE 3500 MHz band. *Microw. Opt. Technol. Lett.* **2014**, *56*, 1323–1327. [CrossRef]
20. Parchin, N.O.; Al-Yasir YI, A.; Ali, A.H.; Elfergani, I.; Noras, J.M.; Rodriguez, J.; Abd-Alhameed, R.A. Eight-element dual-polarized MIMO slot antenna system for 5G smartphone applications. *IEEE Access* **2019**, *9*, 15612–15622. [CrossRef]
21. Parchin, N.O.; Al-Yasir, Y.I.A.; Noras, J.M.; Abd-Alhameed, R.A. Dual-polarized MIMO antenna array design using miniaturized self-complementary structures for 5G smartphone applications. In Proceedings of the 13th European Conference on Antennas and Propagation (EuCAP), Krakow, Poland, 31 March–5 April 2019.
22. Jiang, W.; Liu, B.; Cui, Y.; Hu, W. High-isolation Eight-Element MIMO array for 5G smartphone applications. *IEEE Access* **2019**, *7*, 34104–34112. [CrossRef]
23. Parchin, N.O.; Basherlou, H.J.; Al-Yasir, Y.; Abdulkhaleq, A.M.; Patwary, M.; Abd-Alhameed, R.A. A new CPW-Fed diversity antenna for MIMO 5G smartphones. *Electronics* **2020**, *9*, 261. [CrossRef]
24. Xu, S.; Zhang, M.; Wen, H.; Wang, J. Deep-subwavelength decoupling for MIMO antennas in mobile handsets with singular medium. *Sci. Rep.* **2017**, *7*, 12162. [CrossRef]
25. Li, Y.; Sim, C.Y.D.; Luo, Y.; Yang, G. Multiband 10-antenna array for sub-6 GHz MIMO applications in 5-G smartphones. *IEEE Access* **2018**, *6*, 28041–28053. [CrossRef]
26. Wong, K.L.; Lin, B.-W.; Li, W.-Y. Dual-band dual inverted-F/loop antennas as a compact decoupled building block for forming eight 3.5/5.8-GHz MIMO antennas in the future smartphone. *Microw. Opt. Technol. Lett.* **2017**, *59*, 2715–2721. [CrossRef]
27. Zou, H.; Li, Y.; Shen, H.; Wang, H.; Yang, G. Design of 6 × 6 dual-band MIMO antenna array for 4.5G/5G smartphone applications. In Proceedings of the 6th Asia–Pacific Conference on Antennas and Propagation (APCAP), Xi'an, China, 16–19 October 2017; pp. 1–3.
28. Ullah, A.; Parchin, N.O.; Ullah, R.; Abdullah, A.A.S.A.; Danjuma, I.M.; Kosha, J.; Abd-Alhameed, R.A.; Elkhazmi, E.; Elfoghi, E.M.I. Dual-band MIMO antenna system for next generation smartphone applications. In Proceedings of the 2020 IMDC-SDSP, Ankara, Turkey, 8–10 April 2020.
29. Jiang, W.; Cui, Y.; Liu, B.; Hu, W. A Dual-band MIMO antenna with enhanced isolation for 5G smartphone applications. *IEEE Access* **2019**, *7*, 112554–112563. [CrossRef]

30. Wang, H.; Zhang, R.; Luo, Y.; Yang, G. Compact Eight-Element Antenna Array for Triple-Band MIMO Operation in 5G Mobile Terminals. *IEEE Access* **2020**, *8*, 19433–19449. [CrossRef]
31. Wong, K.L.; Tsai, C.Y.; Li, W.Y. Integrated yet decoupled dual antennas with inherent decoupling structures for 2.4/5.2/5.8-GHz WLAN MIMO operation in the smartphone. *Microw. Opt. Technol. Lett.* **2017**, *59*, 2235–2241. [CrossRef]
32. Ren, Z.; Wu, S.; Zhao, A. Triple Band MIMO Antenna System for 5G Mobile Terminals. In Proceedings of the 2019 International Workshop on Antenna Technology (iWAT), Miami, FL, USA, 3–6 March 2019; pp. 163–165.
33. Parchin, N.O.; Basherlou, H.J.; Alibakhshikenari, M.; Parchin, Y.O.; Al-Yasir, Y.I.A.; Abd-Alhameed, R.A.; Limiti, E. Mobile-phone antenna array with diamond-ring slot elements for 5G massive MIMO systems. *Electronics* **2019**, *8*, 521. [CrossRef]
34. Parchin, N.O.; Ullah, A.; Asharaa, A.S.; Al-Yasir, Y.I.A.; Basherlou, H.J.; Nagala, M.; Abd-Alhameed, R.A.; Noras, J.M. 8×8 MIMO antenna system with coupled-fed elements for 5G handsets. In Proceedings of the The IET Conference on Antennas and Propagation (APC), Birmingham, UK, 11–12 November 2019.
35. Parchin, N.O.; Basherlou, H.J.; Al-Yasir, Y.I.A.; Ullah, A.; Abd-Alhameed, R.A.; Noras, J.M. Frequency reconfigurable antenna array with compact end-fire radiators for 4G/5G mobile handsets. In Proceedings of the IEEE 2nd 5G World Forum (5GWF), Dresden, Germany, 30 September–2 October 2019.
36. Parchin, N.O. Multi-band MIMO antenna design with user-impact investigation for 4G and 5G mobile terminals. *Sensors* **2019**, *19*, 456. [CrossRef]
37. Rappaport, T.S.; Sun, S.; Mayzus, R.; Zhao, H.; Azar, Y.; Wang, K.; Wong, G.N.; Schulz, J.K.; Samimi, M.; Gutierrez, F. Millimeter-wave mobile communications for 5G cellular: It will work! *IEEE Access* **2013**, *1*, 335–349. [CrossRef]
38. Ojaroudiparchin, N.; Shen, M.; Pedersen, G.F. 8×8 planar phased array antenna with high efficiency and insensitivity properties for 5G mobile base stations. In Proceedings of the 10th European Conference on Antennas and Propagation (EuCAP), Davos, Switzerland, 10–15 April 2016.
39. Roh, W.; Seol, J.-Y.; Park, J.; Lee, B.; Lee, J.; Kim, Y.; Cho, J.; Cheun, K.; Aryanfar, F. Millimeter-wave beamforming as an enabling technology for 5G cellular communications: Theoretical feasibility and prototype results. *IEEE Commun. Mag.* **2014**, *52*, 106–113. [CrossRef]
40. Parchin, N.O.; Abd-Alhameed, R.A.; Shen, M. A radiation-beam switchable antenna array for 5G smartphones. In Proceedings of the PhotonIcs & Electromagnetics Research Symposium (PIERS), Xiamen, China, 17–20 December 2019.
41. Parchin, N.O.; Abd-Alhameed, R.A.; Shen, M. A substrate-insensitive antenna array with broad bandwidth and high efficiency for 5G mobile terminals. In Proceedings of the PhotonIcs & Electromagnetics Research Symposium (PIERS), Xiamen, China, 17–20 December 2019.
42. Naqvi, A.; Lim, S. Review of recent phased arrays for millimeter-wave wireless communication. *Sensors* **2018**, *18*, 3194. [CrossRef]
43. Parchin, N.O.; Abd-Alhameed, R.A.; Shen, M. Frequency-switchable patch antenna with parasitic ring load for 5G mobile terminals. In Proceedings of the International Symposium on Antennas and Propagation (ISAP), Xi'an, China, 27–30 October 2019.
44. Parchin, N.O.; Alibakhshikenari, M.; Basherlou, H.J.; Abd-Alhameed, R.A.; Rodriguez, J.; Limiti, E. MM-wave phased array quasi-Yagi antenna for the upcoming 5G cellular communications. *Appl. Sci.* **2019**, *9*, 978. [CrossRef]
45. Ojaroudiparchin, N.; Shen, M.; Pedersen, G.F. Low-cost planar mm-Wave phased array antenna for use in mobile satellite (MSAT) platforms. In Proceedings of the 23rd Telecommunications Forum Telfor (TELFOR), Belgrade, Serbia, 24–26 November 2015; pp. 528–531.
46. Parchin, N.O.; Al-Yasir, Y.; Basherlou, H.J.; Abdulkhaleq, A.M.; Sajedin, M.; Abd-Alhameed, R.A.; Noras, J.M. Modified PIFA array design with improved bandwidth and isolation for 5G mobile handsets. In Proceedings of the IEEE 2nd 5G World Forum (5GWF), Dresden, Germany, 30 September–2 October 2019.
47. Ojaroudi, N.; Ojaroudi, H.; Ghadimi, N. Quadband planar inverted-f antenna (PIFA) for wireless communication systems. *Prog. Electromagn. Res. Lett.* **2014**, *45*, 51–56. [CrossRef]
48. Salonen, P.; Sydänheimo, L.; Keskilammi, M.; Kivikoski, M. A small planar inverted-F antenna for wearable applications. In Proceedings of the IEEE International Symposium on Wearable Computers, San Francisco, CA, USA, 18–19 October 1999; pp. 96–100.

49. Ojaroudi, N.; Ghadimi, N.; Ojaroudi, Y.; Ojaroudi, S. An omnidirectional PIFA for downlink and uplink satellite applications in C-band. *Microw. Opt. Technol. Lett.* **2014**, *56*, 2684–2686. [CrossRef]
50. *CST Microwave Studio*; Version 2018; CST: Framingham, MA, USA, 2018.
51. Valizade, A.; Ghobadi, C.; Nourinia, J.; Parchin, N.O.; Ojaroudi, M. Band-notch slot antenna with enhanced bandwidth by using Ω-shaped strips protruded inside rectangular slots for UWB applications. *Appl. Comput. Electromagn. Soc. J.* **2012**, *27*, 816–822.
52. Bahmani, M.; Mazloum, J.; Parchin, N.O. A compact UWB slot antenna with reconfigurable band-notched function for multimode applications. *ACES J.* **2016**, *13*, 975–980.
53. Al-Yasir, Y.; Abd-Alhameed, R.A.; Noras, J.M.; Abdulkhaleq, A.M.; Parchin, N.O. Design of very compact combline band-pass filter for 5G applications. In Proceedings of the 2018 Loughborough Antennas & Propagation Conference (LAPC), Loughborough, UK, 12–13 November 2013; pp. 1–4.
54. Ojaroudi, N.; Ojaroudi, M.; Amiri, S. Enhanced bandwidth of small square monopole antenna by using inverted U-shaped slot and conductor-backed plane. *Appl. Comput. Electromagn. Soc.* **2012**, *27*, 685–690. [CrossRef]
55. Ojaroudi Parchin, N.; Jahanbakhsh Basherlou, H.; Al-Yasir, Y.I.A.; Abdulkhaleq, A.M.; Abd-Alhameed, R.A. Ultra-wideband diversity MIMO antenna system for future mobile handsets. *Sensors* **2020**, *20*, 2371. [CrossRef]
56. Jamesn, J.R.; Hall, P.S. *Handbook of Microstrip Antennas*; Peter Peregrinus Ltd.: London, UK, 1989.
57. Mazloum, J.; Ghorashi, A.; Ojaroudi, M.; Ojaroudi, N. Compact triple-band S-shaped monopole diversity antenna for MIMO applications. *Appl. Comput. Electromagn. Soc. J.* **2015**, *30*, 975–980.
58. Valizade, A.; Ojaroudi, M.; Ojaroudi, N. CPW-fed small slot antenna with reconfigurable circular polarization and impedance bandwidth characteristics for DCS/WiMAX applications. *Prog. Electromagn. Res. C* **2015**, *56*, 65–72. [CrossRef]
59. Kumar, G.; Ray, K.P. *Broadband Microstrip Antennas*; Artech House Inc.: Norwood, MA, USA, 2003.
60. Parchin, N.O.; Shen, M.; Pedersen, G.F. End-fire phased array 5G antenna design using leaf-shaped bow-tie elements for 28/38 GHz MIMO applications ICUWB. In Proceedings of the IEEE International Conference on Ubiquitous Wireless Broadband (ICUWB), Nanjing, China, 16–19 October 2016.
61. AI-Yasir, Y.I.A.; Ojaroudi Parchin, N.; Alabdullah, A.; Mshwat, W.; Ullah, A.; Abd-Alhameed, R. New pattern reconfigurable circular disk antenna using two PIN diodes for WiMax/WiFi (IEEE 802.11a) applications. In Proceedings of the IEEE International Conference on Synthesis, Modeling, Analysis and Simulation Methods and Applications to Circuit Design (SMACD), Lausanne, Switzerland, 15–18 July 2019.
62. Zolghadr, J.; Cai, Y.; Ojaroudi, N. UWB slot antenna with band-notched property with time domain modeling based on genetic algorithm optimization. *Appl. Comput. Electromagn. Soc. J.* **2016**, *31*, 926–932.
63. Siahkal-Mahalle, B.H.; Ojaroudi, N.; Ojaroudi, M. A new design of small square monopole antenna with enhanced bandwidth by using cross-shaped slot and conductor-backed plane. *Microw. Opt. Technol. Lett.* **2012**, *54*, 2656–2659. [CrossRef]
64. Ullah, A.; Parchin, N.O.; Abd-Alhameed, R.A.; Excell, P. Coplanar waveguide antenna with defected ground structure for 5G millimeter wave communications. In Proceedings of the IEEE Middle East & North Africa COMMunications Conference, Manama, Bahrain, 19–21 November 2019.
65. Ojaroudi, N. Design of microstrip antenna for 2.4/5.8 GHz RFID applications. In Proceedings of the German Microwave Conference, GeMic 2014, RWTH Aachen University, Aachen, Germany, 10–12 March 2014.
66. Ojaroudi, N. Circular microstrip antenna with dual band-stop performance for ultra-wideband systems. *Microw. Opt. Technol. Lett.* **2014**, *56*, 2095–2098. [CrossRef]
67. Basherlou, H.J.; Parchin, N.O.; Abd-Alhameed, R.A. MIMO monopole antenna design with improved isolation for 5G WiFi applications. *Int. J. Electr. Electron. Sci.* **2019**, *7*, 1–5.
68. Sharawi, M.S. Printed multi-band MIMO antenna systems and their performance metrics [wireless corner]. *IEEE Antennas Propag. Mag.* **2013**, *55*, 218–232. [CrossRef]
69. Elfergani, I.T.E.; Hussaini, A.S.; Rodriguez, J.; Abd-Alhameed, R. *Antenna Fundamentals for Legacy Mobile Applications and Beyond*; Springer: Cham, Switzerland, 2017; pp. 1–659.
70. Parchin, N.O.; Al-Yasir, Y.; Basherlou, H.J.; Abd-Alhameed, R.A. A closely spaced dual-band MIMO patch antenna with reduced mutual coupling for 4G/5G applications. *Prog. Electromagn. Res. C* **2020**, *101*, 71–80.
71. Parchin, N.O. Low-profile air-filled antenna for next generation wireless systems. *Wirel. Pers. Commun.* **2017**, *97*, 3293–3300. [CrossRef]

72. Parchin, N.O.; Shen, M.; Pedersen, G.F. Small-size tapered slot antenna (TSA) design for use in 5G phased array applications. *Appl. Comput. Electromagn. Soc. J.* **2017**, *32*, 193–202.
73. Ojaroudi, Y.; Ojaroudi, N.; Ghadimi, N. Circularly polarized microstrip slot antenna with a pair of spur-shaped slits for WLAN applications. *Microwave Opt. Technol. Lett.* **2017**, *57*, 756–759. [CrossRef]
74. Syrytsin, I.; Zhang, S.; Pedersen, G.F. Performance investigation of a mobile terminal phased array with user effects at 3.5 GHz for LTE advanced. *IEEE Antennas Wirel. Propag. Lett.* **2017**, *16*, 1847–1850. [CrossRef]
75. Ojaroudiparchin, N.; Shen, M.; Pedersen, G.F. Design of Vivaldi antenna array with end-fire beam steering function for 5G mobile terminals. In Proceedings of the 23rd Telecommunications Forum Telfor (TELFOR), Belgrade, Serbia, 24–26 November 2015; pp. 587–590.
76. Khan, R.; Al-Hadi, A.A.; Soh, P.J.; Kamarudin, M.R.; Ali, M.T. User influence on mobile terminal antennas: A review of challenges and potential solution for 5G antennas. *IEEE Access* **2018**, *6*, 77695–77715. [CrossRef]
77. Stuchly, M. Electromagnetic fields and health. *IEEE Potentials* **1993**, *12*, 34–39. [CrossRef]
78. Hussain, R.; Sharawi, M.S.; Shamim, A. 4-element concentric pentagonal slot-line-based ultra-wide tuning frequency reconfigurable MIMO antenna system. *IEEE Trans. Antennas Propag.* **2018**, *66*, 4282–4287. [CrossRef]
79. Ojaroudiparchin, N.; Shen, M.; Zhang, S.; Pedersen, G.F. A switchable 3-D-coverage-phased array antenna package for 5G mobile terminals. *IEEE Antennas Wireless Propag. Lett.* **2016**, *15*, 1747–1750. [CrossRef]
80. Parchin, N.O.; Al-Yasir, Y.; Basherlou, H.J.; Noras, J.M.; Abd-Alhameed, R.A. Orthogonally dual-polarized MIMO antenna array with pattern diversity for sub-6 GHz 5G mobile terminals. *IET Microw. Antennas Propaga.* **2020**, *14*, 457–467. [CrossRef]
81. Halaoui, M.E.; Kaabal, A.; Asselman, H.; Ahyoud, S.; Asselman, A. Multiband planar inverted-F antenna with independent operating bands control for mobile handset applications. *Int. J. Antennas Propag.* **2017**, *2017*, 8794039. [CrossRef]
82. Parchin, N.O.; Al-Yasir, Y.; Abdulkhaleq, A.M.; Elfergani, I.T.E.; Rayit, A.; Noras, J.M.; Rodriguez, J.; Abd-Alhameed, R.A. Frequency reconfigurable antenna array for mm-Wave 5G mobile handsets. In Proceedings of the 9th International Conference on Broadband Communications, Networks, and Systems, Faro, Portugal, 19–20 September 2018.
83. Parchin, N.O.; Shen, M.; Pedersen, G.F. Wideband Fabry-Pérot resonator for 28 GHz applications. In Proceedings of the IEEE International Conference on Ubiquitous Wireless Broadband (ICUWB), Nanjing, China, 16–19 October 2016; pp. 1–4.
84. Li, R.; Zhang, Q.; Kuang, Y.; Chen, X.; Xiao, Z.; Zhang, J. Design of a miniaturized antenna based on split ring resonators for 5G wireless communications. In Proceedings of the Cross Strait Quad-Regional Radio Science and Wireless Technology Conference (CSQRWC), Taiyuan, China, 18–21 July 2019; pp. 1–4.
85. Liu, D.; Gaucher, B. Design considerations for millimeter wave antennas within a chip package. In Proceedings of the IEEE International Workshop on Anti-counterfeiting, Security, Identification, Xiamen, China, 16–18 April 2007; pp. 13–17.
86. Ojaroudiparchin, N.; Shen, M.; Pedersen, G.F. Investigation on the performance of low-profile insensitive antenna with improved radiation characteristics for the future 5G applications. *Microw. Opt. Technol. Lett.* **2016**, *58*, 2148–2158. [CrossRef]
87. Al-Yasir, Y.I.A.; Hasanain, A.H.A.; Baha, A.S.; Parchin, N.O.; Ahmed, M.A.; Abdulkareem, S.A.; Raed, A.A. New Radiation Pattern-Reconfigurable 60-GHz Antenna for 5G Communications. IntechOpen, 2019. Available online: https://www.intechopen.com/online-first/new-radiation-pattern-reconfigurable-60-ghz-antennafor-5g-communications (accessed on 26 September 2019).
88. Gomez-Tagle, J.; Christodoulou, C.G. Broadband characterization of the active reflection coefficient of finite-sized phased array microstrip antennas. In Proceedings of the 2000 IEEE International Conference on Phased Array Systems and Technology (Cat. No.00TH8510), Dana Point, CA, USA, 21–25 May 2000; pp. 255–258.
89. Bai, T.; Heath, R. Coverage and rate analysis for millimeter wave cellular networks. *IEEE Trans. Wirel. Commun.* **2015**, *14*, 1110–1114. [CrossRef]

sensors

Article

Ultra-Wideband Diversity MIMO Antenna System for Future Mobile Handsets

Naser Ojaroudi Parchin [1,*], Haleh Jahanbakhsh Basherlou [2], Yasir I. A. Al-Yasir [1], Ahmed M. Abdulkhaleq [1,3] and Raed A. Abd-Alhameed [1,4]

1 Faculty of Engineering and Informatics, University of Bradford, Bradford BD7 1DP, UK; Y.I.A.Al-Yasir@bradford.ac.uk (Y.I.A.A.-Y.); A.Abd@sarastech.co.uk (A.M.A.); R.A.A.Abd@bradford.ac.uk (R.A.A.-A.)
2 Bradford College, West Yorkshire, Bradford BD7 1AY, UK; Hale.Jahanbakhsh@gmail.com
3 SARAS Technology Limited, Leeds LS12 4NQ, UK
4 Department of Communication and Informatics Engineering, Basra University College of Science and Technology, Basra 61004, Iraq
* Correspondence: N.OjaroudiParchin@Bradford.ac.uk; Tel.: +44-734-143-6156

Received: 12 March 2020; Accepted: 14 April 2020; Published: 22 April 2020

Abstract: A new ultra-wideband (UWB) multiple-input/multiple-output (MIMO) antenna system is proposed for future smartphones. The structure of the design comprises four identical pairs of compact microstrip-fed slot antennas with polarization diversity function that are placed symmetrically at different edge corners of the smartphone mainboard. Each antenna pair consists of an open-ended circular-ring slot radiator fed by two independently semi-arc-shaped microstrip-feeding lines exhibiting the polarization diversity characteristic. Therefore, in total, the proposed smartphone antenna design contains four horizontally-polarized and four vertically-polarized elements. The characteristics of the single-element dual-polarized UWB antenna and the proposed UWB-MIMO smartphone antenna are examined while using both experimental and simulated results. An impedance bandwidth of 2.5–10.2 GHz with 121% fractional bandwidth (FBW) is achieved for each element. However, for $S_{11} \leq -6$ dB, this value is more than 130% (2.2–11 GHz). The proposed UWB-MIMO smartphone antenna system offers good isolation, dual-polarized function, full radiation coverage, and sufficient efficiency. Besides, the calculated diversity performances of the design in terms of the envelope correlation coefficient (ECC) and total active reflection coefficient (TARC) are very low over the entire operating band.

Keywords: double-fed slot antenna; MIMO system; mobile terminals; polarization diversity; UWB technology

---

1. Introduction

Ultra-Wideband (UWB) technology provides some superiority, such as high-speed data transmission, low cost, and easy manufacture [1]. However, such a popular topic of technology suffers from multipath fading in practical applications [2]. Precisely, the multiple-input multiple-output (MIMO) technique was raised to resolve this issue [3,4]. The combination of UWB and MIMO technologies, the use of space multipath, parallel transmission of multiple signals, can obtain obvious multiplexing gain and diversity gain, and achieve a stable signal transmission in distance [5,6]. UWB technology provides wide bandwidth with low power usage. Unlike narrowband technologies, like Bluetooth and Wi-Fi, this technology might be more suitable for short-distance communications [7]. However, it is natively more precise, uses less power, and can transmit data over a wider frequency (up to several GHz). A precise angle combined with a precise distance, which leads the ability for a mobile handset to pinpoint an object to a reasonably precise location in space, as well as to recognize its surroundings. While the UWB technology is not new, its first implementation into a modern smartphone has been used by the world's smartphone makers, such as Apple [8]. It should be noted that other technologies, like Bluetooth and Wi-Fi, are still useful since they have a longer range.

Through MIMO technology with multiple antennas, the UWB can measure the angle of arrivals for the signals, as well as secure access to a myriad of the system [9]. Standard MIMO systems tend to employ two or four elements in a single physical package. However, a high number of antenna radiators are employed for the massive MIMO system [10,11]. When compared with previous generations, a large number of antenna elements, operating concurrently, is expected to be applied for 5G communications. 2 × 2 MIMO systems are successfully deployed for 4G cellular networks, while, for the massive MIMO operation of 5G communications, it is expected to employ a large number of elements, since the greater number makes the system more resistant to interferences [12,13]. Therefore, the multi-antenna structure with minimum mutual coupling employed for generating polarization and pattern diversity for reliable communication in order to load the MIMO operation into a smartphone for 5G communication. The 5G network also needs fundamental technologies to enable small cells, beamforming, full duplexing, MIMO, and millimeter-wave (MM-Wave) [14,15]. Simple structured and compact antenna elements with sufficient impedance bandwidth and isolation are desirable to be integrated into 5G smartphone platforms, in accordance with the requirement of mobile networks [16–18]. Among various MIMO antennas, microstrip antenna elements are more applicable due to their promising features, such as compact-size, Omani-directional radiation, broad impedance bandwidth, ease of integration, and manufacturing [19,20].

Several MIMO 5G smartphone antennas have been recently proposed [21–38]. However, these smartphone antennas either suffer from narrow impedance-bandwidth (either single-band or dual/multi-band) or use single-polarized uniplanar antenna resonators occupying large spaces of the mainboard, which leads to increasing the complexity of the mobile phone system. In addition, in most of the reported antennas, all of the candidate bands of sub 6 GHz 5G, including LTE Band 7 (2.550–2.650 GHz), N77 (3.3–4.2 GHz), N78 (3.3–3.8 GHz), N79 (4.4–5.0 GHz), and LTE band 46 (5150–5925 MHz) are not supported. Additionally, in the designs of many reported MIMO smartphone antennas, it is common to avoid placing elements in vertical and choose instead to place them parallel, which prevents achieving the polarization diversity. However, in this paper, the antenna elements are both perpendicular and parallel to each other to exhibit the diversity function. In [21,22], uniplanar and double-layer antenna elements with only 200 MHz bandwidths are proposed for 5G smartphone applications. Tightly arranged orthogonal-mode pairs with design complexity is proposed in [23] for 5G smartphones. In [24,25], CPW-fed antenna arrays with large-clearance are introduced for sub 6 GHz MIMO handset. In [26–28], narrow-band antenna arrays with 100 MHz are represented to be integrated at the corners of the mainboard. The proposed antennas in [29,30] use slot resonators in the ground plane, providing polarization diversity function at 3.6 GHz. A multi-element multi-layer antenna with the operation band of 4.4–4.7 GHz is proposed in [31]. The proposed antennas in [32–34] have uniplanar configurations which make them difficult to fabricate and integrate with the circuit system. In addition, the proposed smartphone antenna array in [35] with wide impedance bandwidth employing only two radiators, which is not sufficient for 5G MIMO operation. Moreover, broadband MIMO arrays with 2 GHz bandwidths are reported in [36,37]. The employed radiators are not planar and careful consideration is required in the fabrication process. In [38], an eight-element PIFA array arranged on an artificial magnetic plane with less possible screen-space is proposed.

In this manuscript, we present a new design of broadband MIMO antenna systems providing polarization diversity and ultra-wide impedance bandwidth. To the best of our knowledge, this is the first integration of multi UWB antennas with polarization and pattern diversity onto a MIMO smartphone system. The smartphone antenna is designed to work at 2.5–10.2 GHz, covering different wireless systems, including LTE-4G, sub 6 GHz 5G, WLAN, WIMAX, X-band, etc. The structure of the smartphone MIMO array design comprises four pairs of orthogonally dual-polarized slot radiators located at four edge corners of the smartphone PCB. By exciting the radiator from the different feeding ports, two orthogonally polarized waves are generated, which leads the radiation pattern and polarization diversity function. A low-cost FR-4 dielectric is used as the PCB substrate. A prototype sample of the proposed design was fabricated and then tested. Good impedance bandwidth and isolation are achieved. The characteristics of the single radiator and its MIMO package are elaborated below.

2. Dual-Polarized UWB Slot Antenna

Figure 1 provides the geometry of the designed UWB diversity antenna design. As shown, the design is composed of an open-ended circular-ring slot radiator with outer and inner radiuses of r_3 and r_2 differently fed by a pair of symmetrical semi-arc-shaped microstrip-line in orthogonal positions. As shown in Figure 1a, 50-Ohm SMA (SubMiniature version A) connectors are employed for both antenna feeding ports. Table 1 lists the dimensions of the proposed UWB diversity antenna element. The antenna is designed on an FR4 dielectric whose permittivity and loss tangent are 4.4 and 0.025, respectively. The antenna has an overall dimension of $W_S \times L_S \times h_S = 34 \times 34 \times 1.6$ mm^3. It is operating at the frequency range of 2.5 to 10.2 GHz. The impedance matching band of the proposed slot antenna is governed by several parameters, such as the dimensions of the open-ended circular-ring slot and the semi-arc-shaped radiation stub, as well as the small strip pairs protruded to increase the matching band and mutual coupling, especially at high frequencies. All of these parameters were optimized to enable an UWB operating frequency band with sufficient mutual coupling inferior to −10 dB.

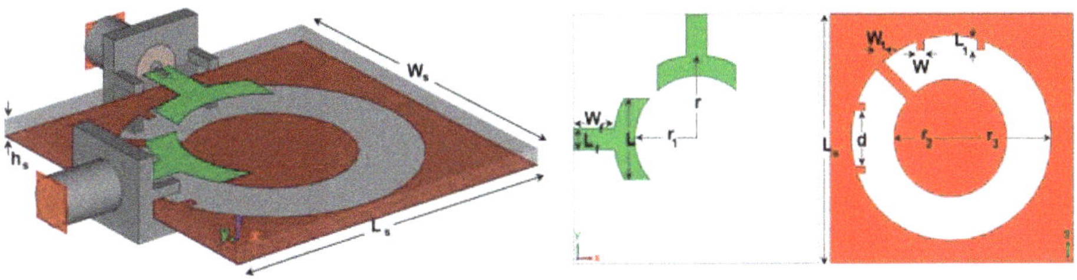

(a) (b) (c)

Figure 1. Overall view of the single-element Ultra-Wideband (UWB) antenna design, (**a**) three-dimensional (3D) view, (**b**) top, and (**c**) bottom layers.

Table 1. Parameter values of the proposed UWB diversity designs.

Parameter	W_S	L_S	h_S	W_f	L_f	d
Value (mm)	34	34	1.6	5.5	3	7.5
Parameter	h_S	W	L	W_1	L_1	r
Value (mm)	1.6	1	11	1.5	1.5	11.5
Parameter	r_1	r_2	r_3	W_{sub}	L_{sub}	H_{sub}
Value (mm)	8.5	8	14	75	150	1.6

Various configurations of the antenna are shown in Figure 2. Simulated S parameter results of the antenna design with a conventional circular-ring slot radiator (Figure 2a), with an open-ended circular-ring slot radiator (Figure 2b), and the presented design (Figure 2c) are shown in Figure 3a–c, respectively. As illustrated, by converting the circular-ring slot radiator into the open-ended circular-ring slot, not only the frequency bandwidth of the antenna but also the isolation characteristic of the design have been significantly improved. Finally, by adding pairs of small strips under the antenna feed-lines, the antenna provides good matching with improved-bandwidth and well-isolated characteristics at the desired frequency band (2.5–10 GHz).

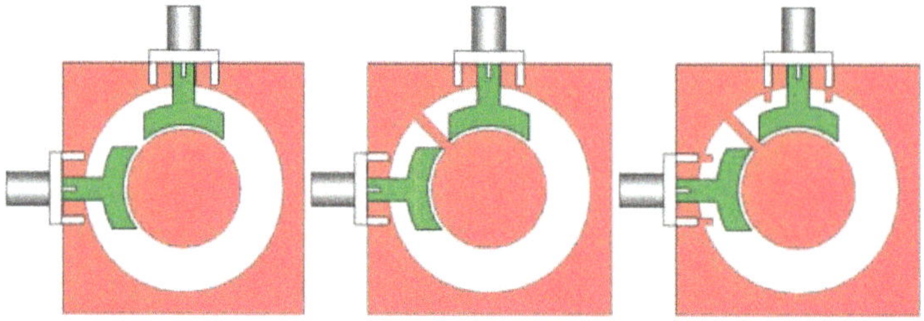

(a) (b) (c)

Figure 2. Various structures of the UWB slot antenna, (**a**) antenna with a circular-ring slot radiator, (**b**) the antenna with open-ended circular-ring slot, and (**c**) the proposed UWB antenna design.

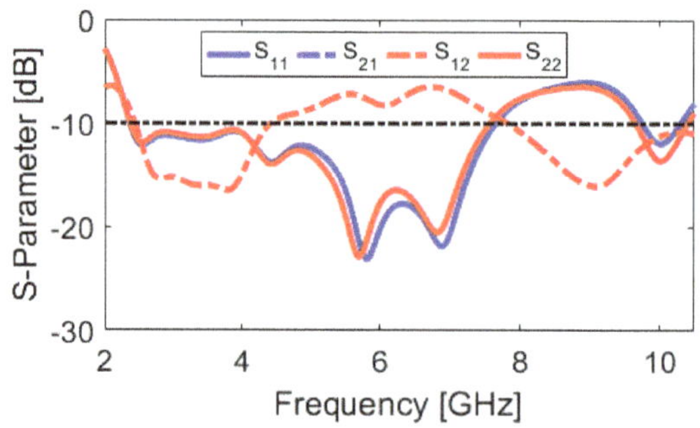

(**a**)

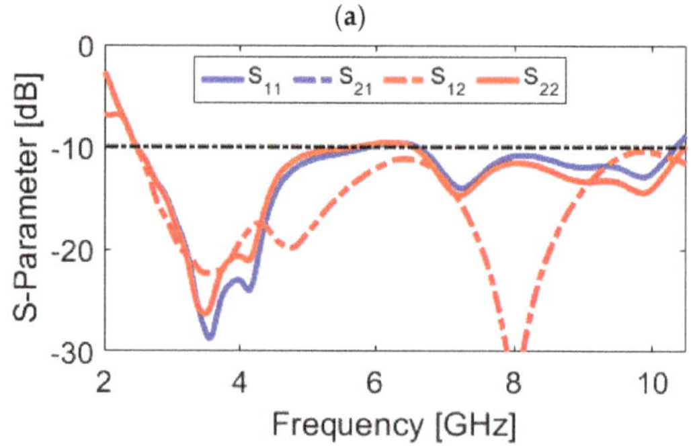

(**b**)

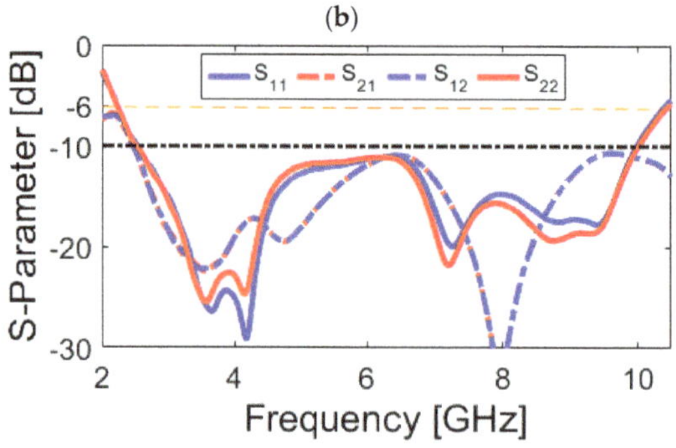

(c)

Figure 3. S parameters of the (**a**) antenna with a circular-ring slot radiator, (**b**) the antenna with open-ended circular-ring slot, and (**c**) the proposed UWB antenna design.

The current densities of the antenna (port 1) at different resonance frequencies (including 3.8, 4.2, 7.1, and 9.1 GHz) are presented in Figure 4 and discussed in the following. It is obvious that the maximum distributions of the surface current have been mainly concentrated around the ring-slot cut of the ground plane, since it is the main resonator of the design, which provides different resonances of the antenna. Additionally, the semi-arc-shaped microstrip feeding lines current is very active at different frequencies. However, the current is highly concentrated around the edge of the open-ended circular-ring at lower frequencies, as can be observed in Figure 4a,b. This can be explained from Figure 3. It is observed that by converting the circular-ring slot radiator to the open-ended circular-ring slot, the isolation characteristic of the antenna S_{11}/S_{22} has been improved significantly (from −12 to less than −25 dB), according to the obtained results from Figure 3a,b. Moreover, the semi-arc-shaped radiation stub and the protruded small rectangular strips in the ground plane appear very active with high current densities at the upper frequencies (7.1 and 9.1 GHz), as shown in Figure 4c,d. This is mainly because of their positive impact on further enhancement of the antenna performance, especially at higher frequencies, as shown in Figure 3c.

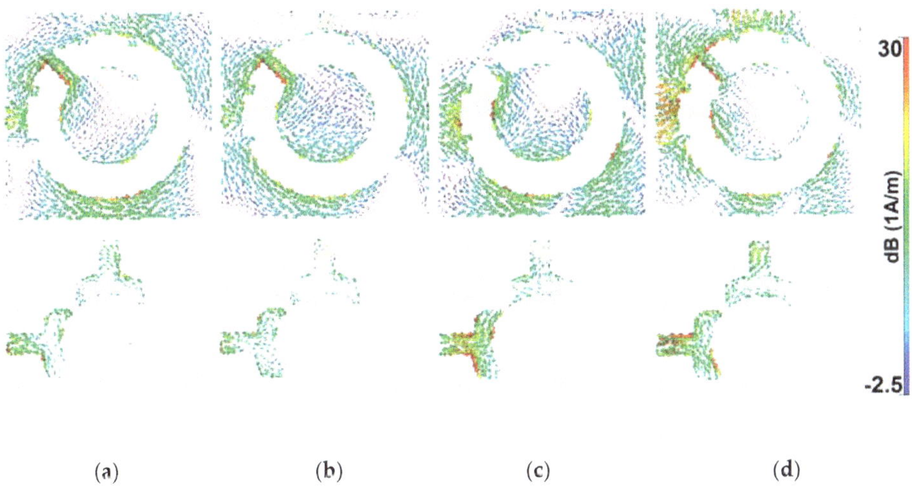

Figure 4. Current densities from Port 1 at resonances: (**a**) 3.8, (**b**) 4.2, (**c**) 7.1, and (**d**) 9.1 GHz.

The reflection and transmission coefficients (S_{11} and S_{21}) of the multi-resonance/UWB-MIMO antenna with various parameters are investigated. Figures 5–8 show the results of varying fundamental parameters r, L, r_1, and W_1. In the simulation of the designed antenna, when one parameter changes, the rest of the parameters are kept the same as the parameters that are listed in Table 1. Since the structure is symmetric, it is sufficient to only show the S_{11} and S_{21}. Figure 5 shows the effects of *r* (outer radius of the main resonator) on the S-parameters. It can be observed that as L increases, the lower frequency bandwidth shifts down slowly to a lower band and also the upper operation frequency moves fast toward a lower band, which means that the length of the slot radiator changes its capacitance characteristic [39,40]. As seen, by changing the size of r from 12 to 16 mm, the lower and upper frequencies of the antennas move from 3.5 to 1.5 GHz and 11 to 8 GHz, respectively. In addition, by changing the values of r, the mutual coupling (S_{21}) of the design is changed, especially at middle frequencies, as illustrated in Figure 5b.

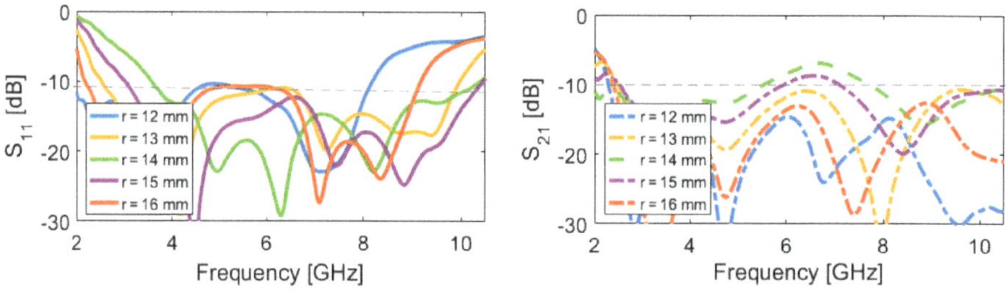

(a) (b)

Figure 5. (**a**) S_{11} and (**b**) S_{21} characteristics for different values of r.

Figures 6 and 7 represent the impact of *L* (length of arc-shaped radiation patch) and r_1 (inner radius of the slot resonator) on the S-parameters of the designed UWB dual-polarized antenna. It is found that, unlike the outer radius of the main resonator (r), L and r_1 mainly affect the lower operation band of the antenna. As shown in Figure 6a, as L decreases from 12 to 6 mm, the lower frequency bandwidth changes from 2 to 4 GHz. It should be noted that changing the values of L does not significantly impact the transmission coefficients (or mutual coupling) that are characteristic of the antenna (shown in Figure 6b). Figure 7 illustrates the simulated S-parameter results of the antenna for various values (9 to 7 mm) of the inner radius (r_1) of the slot resonator. It is clear that, when the inner radius of the slot varies from 9 to 7 mm, the lower operation bandwidth of the design tunes from 2.1 to 3.1 GHz. In addition, this parameter value changing also affects the isolation of the mutual coupling characteristic of the design, as illustrated in Figure 7b. It is evident that, by the reduction of the inner radius of the slot, the isolation of S_{21} has been significantly improved.

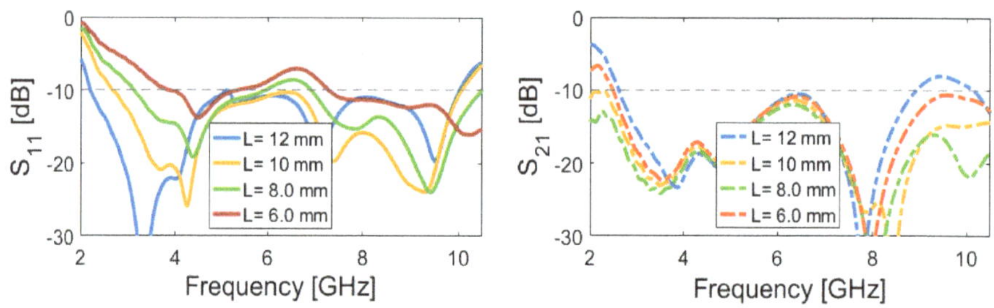

(a) (b)

Figure 6. (**a**) S_{11} and (**b**) S_{21} characteristics for different values of (**a**) L.

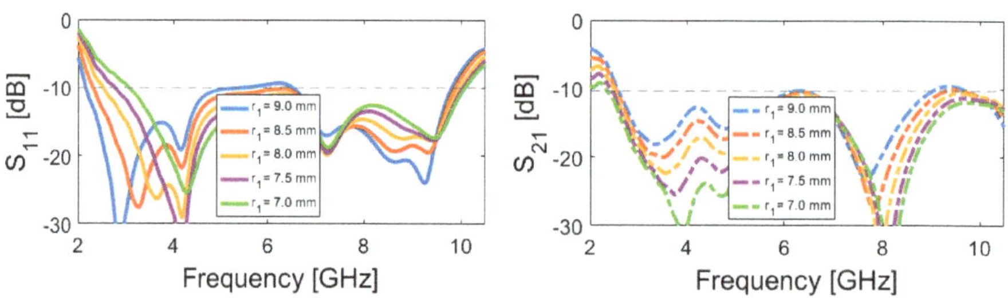

(a) (b)

Figure 7. (**a**) S_{11} and (**b**) S_{21} characteristics for different values of r_1.

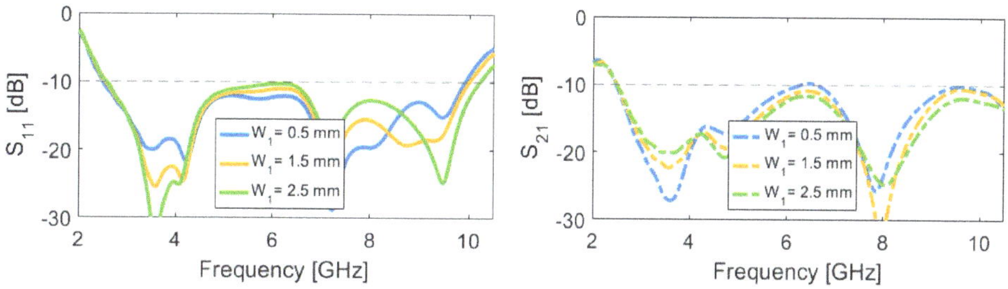

(a) (b)

Figure 8. (**a**) S_{11} and (**b**) S_{21} characteristics for different values of W_1.

The effects of another important design parameter, the employed rotated strip on the slot resonator (W_1), on antenna performance are investigated through simulation, as presented in Figure 8. It can be found from Figure 8a that the isolation of the resonant bands at lower and upper frequencies changes for different values of W_1. Meanwhile, the change of the mutual coupling with the variation of is W_1 is not significant. Figure 9a provides the simulated efficiencies (radiation and total) of the diversity antenna design. As shown from the figure that more than 75% radiation efficiency is obtained over the ultra-wide impedance bandwidth of the antenna with the maximum value of 90% at the lower frequencies. This is caused by the enhanced isolation characteristic of the antenna. Moreover, it is evident that the UWB antenna exhibits good total efficiency results of 50%~75%. Figure 9b depicts the maximum gain and directivity results versus the investigated frequency range. It is shown that, although the gain varies with the operation frequency, its level is still more than 3.1 dBi within the frequency band of interest. It is seen that the difference between directivity and maximum gain results is small, which is mainly because of the high-efficiency characteristic of the antenna. The overall variation of the gain and directivity is within 3 dBi in the entire frequency band, which is very good for wideband and multiband systems.

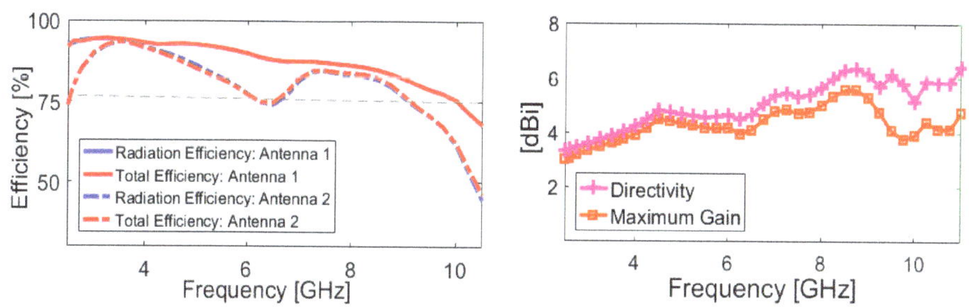

(a) (b)

Figure 9. (**a**) Radiation/total efficiencies, maximum gain and (**b**) diversity results of the diversity antenna.

In the UWB antenna system, the time-domain characteristics are equally as important as the frequency domain. In the time domain method, an important factor indicating the characteristic of the antenna is the fidelity factor. The values of the system fidelity factor vary between 0 and 100%. A system fidelity factor value of 100% shows that the received signal perfectly fits the input signal [41]. The fidelity is employed as a factor of similarity between the input and received signal and is obtained, as follows:

$$F = Max_{\tau}\left|\frac{\int_{\infty}^{+\infty} s(t)r(t-\tau)}{\sqrt{\int_{-\infty}^{+\infty} s(t)^2 dt . \int_{-\infty}^{+\infty} r(t)^2 dt}}\right| \quad (1)$$

where $s(t)$ and $r(t)$ are the input and received signals, respectively. Two identical configurations of the dual-polarized UWB antenna, including side-by-side and face-to-face orientations with a 100 mm shift of their center points, are studied. The antenna is excited by using a modulated Gaussian pulse.

A relatively good similarity has been achieved for the R_X and T_X pulses, as shown in Figure 10. In addition, using (1) the fidelity factor of the reference antenna pair are calculated and good results have been obtained (equal to 0.85 and 0.75, respectively). A prototype sample of the proposed UWB dual-polarized antenna is fabricated and tested. Figure 11 shows the placements of the fabricated prototype in measurement setups of S-parameters and radiation patterns, respectively. The measured S-parameters (S_{11}&S_{21}) of the fabricated sample is examined while using the vector network analyzer and illustrated in Figure 12a. It is found that the fabricated prototype works properly and it provides acceptable agreement with the simulations. As shown, a good frequency bandwidth ($S_{11} \leq -10$ dB) of 2.5–10 GHz is achieved for the fabricated dual-polarized UWB antenna. However, for $S_{11} \leq -6$ dB, this value could be from 2.2–11 GHz. In addition, it is evident that the mutual-coupling characteristic of the design is less than −10 dB over the entire UWB operation frequency band.

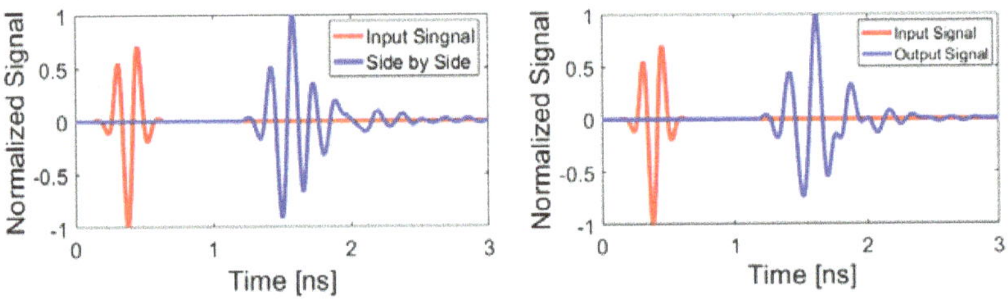

(**a**) (**b**)

Figure 10. Input/output signal pulses for (**a**) side by side and (**b**) face to face scenarios.

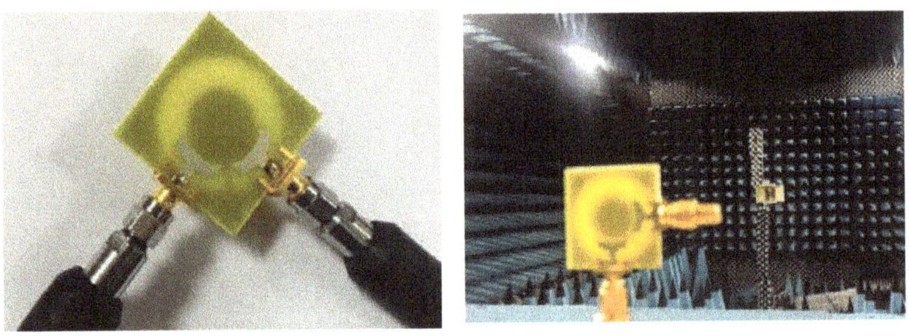

(a) (b)

Figure 11. Photographs of the antenna in measurements setup of (**a**) S-parameters and (**b**) radiation patterns.

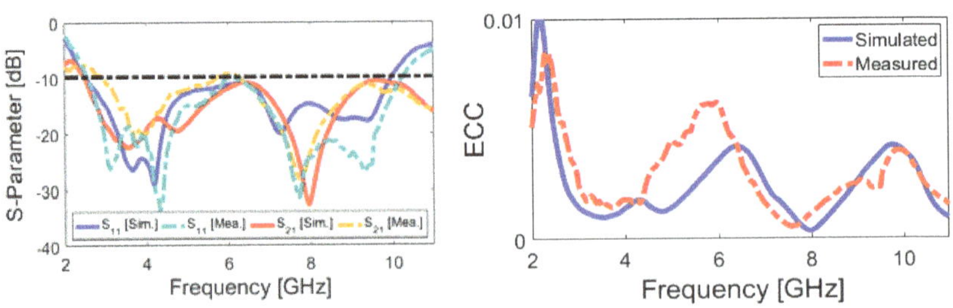

(**a**) (**b**)

Figure 12. Measured and simulated (**a**) S-parameters and (**b**) ECC results of the proposed antenna.

In order to validate the capability of a dual-polarized MIMO antenna design, envelope correlation coefficient (ECC), total active reflection coefficient (TARC), and its diversity gain (DG) properties are three important parameters to be investigated [42,43]. The acceptable limits of standards are ECC < 0.5, TARC < −10, and DG near 10 dB. These characteristics of MIMO antenna can be calculated from the S-parameter results while using the below formula:

$$ECC = \frac{\left|S^*_{mm}S_{mn} + S^*_{nm}S_{nn}\right|^2}{\left(1 - |S_{mm}|^2 - |S_{mn}|^2\right)\left(1 - |S_{nm}|^2 - |S_{nn}|^2\right)^*} \quad (2)$$

$$TARC = -\sqrt{\frac{(S_{mm} + S_{mn})^2 + (S_{nm} + S_{nn})^2}{2}} \quad (3)$$

$$DG = 10\sqrt{1 - (ECC)^2} \quad (4)$$

Figure 12b illustrates the calculated ECC characteristic of the antenna. It can be observed that the calculated ECC results are very low entire the UWB operation band (less than 0.01). Figure 13a,b illustrate the TARC and DG characteristics of the antenna, respectively. It is also found in Figure 13a that the TARC value of the dual-polarized UWB-MIMO design is less than −20 dB in the frequency range of 2.5–10 GHz. The diversity gain function of the designed antenna over its operation band is more than 9.95 dB over the entire operating frequency band, as can be seen in Figure 13b.

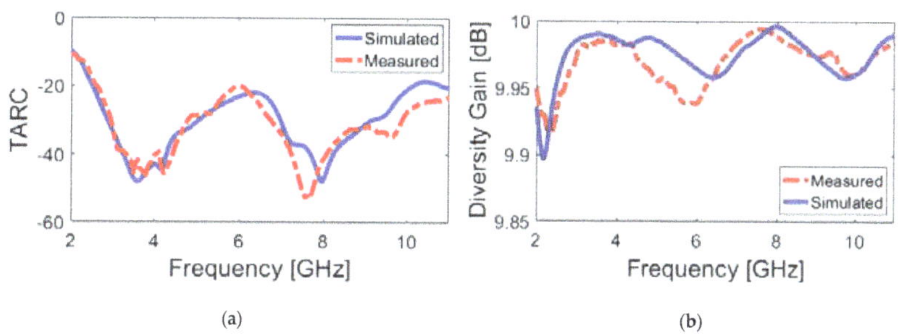

Figure 13. Calculated (**a**) TARC and (**b**) DG results from the measured and simulated results.

In addition, two-dimensional (2D) radiation patterns of the fabricated antenna have been measured. Figure 14 shows the measured/simulated radiation patterns of the antenna at selected frequencies, including 3, 6, and 9 GHz. These three frequencies are chosen form the lower, middle, and upper frequencies, respectively. In this design, the xz plane is H plane (φ = 0°) and yz-plane is E-plane (φ = 90°) for the proposed antenna. From Figure 14, we can see that the antenna can gives dumbbell-like radiation characteristics in E-plane and nearly Omni-directional patterns in H-plane [44,45]. It was found that the radiation patterns of the UWB antenna deteriorate more or less with increasing frequency. However, the radiation properties are almost stable.

The peak gains of the UWB dual-polarized antenna over its operation band are measured and compared with simulations, as illustrated in Figure 15a. Almost stable constant gain values are achieved over the antenna operation band, as can be observed. However, the antenna gain is increased from 3.2 to nearly 5.8 dB, which is caused by the deteriorated radiation at the higher band. As a well-known fact, a UWB antenna should cover a wide band. Therefore, the analysis of group delay is very important [46]. When a signal goes through a filter, the signal will distort in both amplitude and phase. This distortion depends on the characteristics of the designed filter, which can determine the communication quality. By representing the transmitting and receiving antennas as a filter, the group delay at the operation band is very important for designing UWB antennas. The group delay is defined as the measurement of the signal transition time through a device. For the UWB antennas, the altered inductance properties result in a group delay of UWB antennas. The group delay characteristic can be defined through the derivative of phase, which is expressed as:

$$\tau(\omega) = -\frac{\partial\varphi(\omega)}{\partial(\omega)} \tag{5}$$

where $\varphi(\omega)$ and ω are the phase and angular frequencies, respectively. Figure 15b illustrates the measured/simulated group delay of the designed UWB antenna. The result shows that the group delay variation is less than 1 ns in the entire operation band. This indicates a linear phase response and good pulse handling capability for the proposed antenna. Hence, the antenna is useful in the UWB impulse radio and microwave imaging.

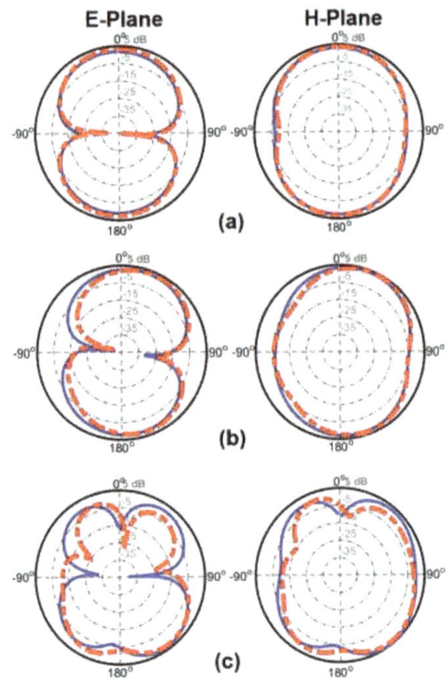

Figure 14. Measured (dash line) and simulated (solid line) results of the antenna patterns at (**a**) 3 GHz, (**b**) 6 GHz, and (**c**) 9 GHz.

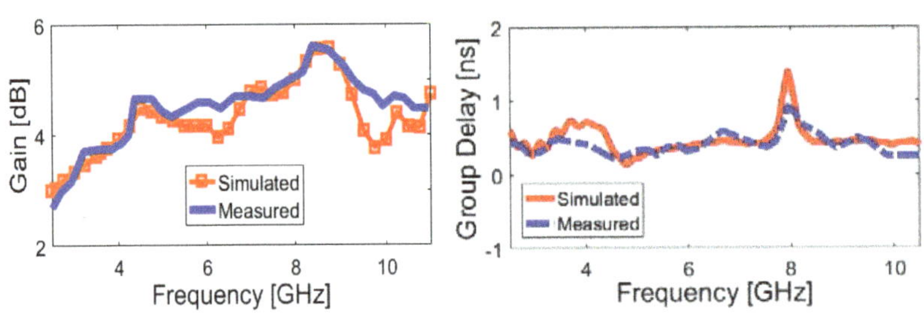

(a) (b)

Figure 15. Measured and simulated (**a**) antenna gain and (**b**) group delay over its ultra wide bandwidth.

Table 2 provides a comparative summary of the fundamental properties of the presented design with the reported dual-polarized antennas available in the literature [47–50]. When compared with reported designs, the presented antenna has a simple structure with a compact, as summarized in Table 2. In addition, a wider impedance bandwidth providing more than 120% FBW is achieved for the proposed antenna. As shown, in contrast with the reported designs in the literature, the proposed dual-port diversity antenna exhibits lower ECC results. Stable and almost constant gain values are achieved for the antenna over the antenna UWB operation band, which is very good for wideband and multiband systems. Furthermore, as explained earlier, high radiation efficiencies are achieved for the proposed design. Therefore, the antenna is suitable for different wireless applications, such as radar, microwave imaging, and cellular communications, due to these attractive features.

Table 2. Comparison between the Design and the Recently Reported Dual-Polarized UWB Antennas.

Reference	FBW (%)	Size (mm^2)	Gain (dB)	ECC
[47]	115% (2.9–11 GHz)	105 × 105	2–7	-
[48]	112% (3–11.5 GHz)	90 × 90	4–9	-
[49]	120% (3–12 GHz)	76.25 × 52.25	4–6	<0.05
[50]	107% (3–10 GHz)	72 × 72	3–6	-
[51]	120% (3–12 GHz)	66.25 × 66.25	5	<0.1
[52]	112% (3.1–11.8 GHz)	61 × 68	1–7	<0.02
[53]	114% (3–11 GHz)	57 × 57	5	<0.02
[54]	114% (3–11 GHz)	56 × 56	3–5	<0.02
[55]	120% (3–12 GHz)	40.5 × 40.5	5	<0.1
[56]	114% (3–11 GHz)	35 × 35	4.6	<0.3
Proposed Antenna	121% (2.5–10.2 GHz)	34 × 34	4–6	<0.01

3. The Proposed UWB Diversity Antenna System

Figure 16 shows the schematic of the MIMO mobile-phone antenna system. As shown, it comprises four identical pairs of compact microstrip-fed slot antennas with polarization diversity function that are placed symmetrically at different edge corners of the mainboard with a standard dimension of $75 \times 150 \times 1.6$ mm^3. However, it is also possible to arrange the design in different sizes of the handset mainboard, due to the compact sizes of the employed diversity slot elements.

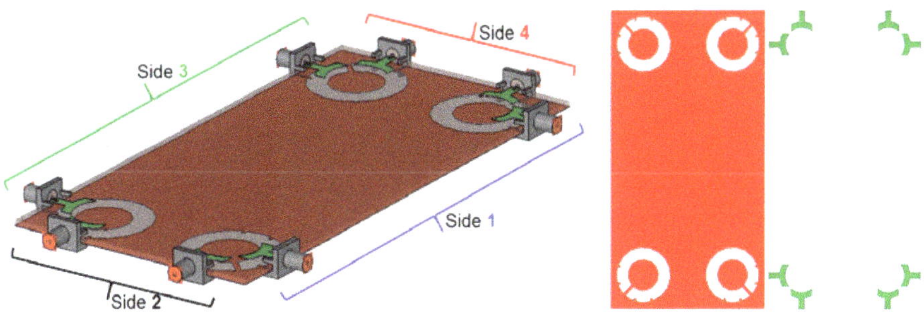

(a) (b) (c)

Figure 16. (**a**) Side, (**b**) top, and (**c**) bottom views of the UWB-multiple-input/multiple-output (UWB-MIMO) antenna system design for smartphone applications.

Figure 17 depicts the S parameters (including S_{nn} and S_{mn}). As shown, the antenna exhibits good S parameters, similar to the single-element diversity antenna, covering the frequency spectrum of 2.5–10 GHz. In addition, the mutual coupling of the design is below −10 over the operation frequency of the smartphone antenna, which meets the basic requirements for MIMO operation. As expected, the maximum mutual coupling of the MIMO design is between the closely spaced diversity elements (S_{21} (port 1 and port 2, for example)) with a fixed size of the antenna element ($W_S \times L_S$). In addition, the mutual coupling characteristics of other port pairs is less than −16 dB. This function makes the design flexible to be arranged in different sizes of PCB. Therefore, it is possible to get satisfactory results for smaller motherboard dimensions.

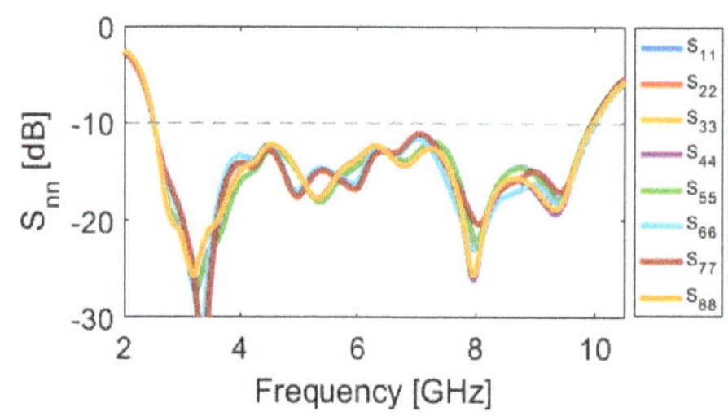

(a)

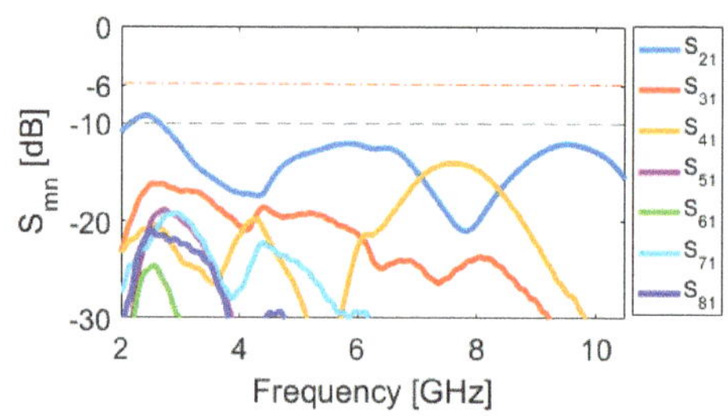

(b)

Figure 17. (**a**) S_{nn} and (**b**) S_{mn} characteristics of the simulated MIMO design.

Using (1), the fidelity factor of the MIMO antenna design is also investigated. In the proposed MIMO smartphone antenna array, the fidelity factor is the highest between the closely-spaced diversity antenna pairs with the side-by-side orientation, averaging 0.90. It should be noted that the fidelity factor can be decreased as the distance increases and vice versa [57]. Accordingly, for other possible arrangements of the antenna pairs with longer distances, the fidelity factor drops to 0.70 in most cases. However, the value is still reasonably good and acceptably low.

Figure 18 illustrates the radiation efficiency (R.E.) and total efficiency (T.E.) properties of the proposed design. According to the results, it can be concluded that the MIMO design provides sufficient efficiencies over its entire operation band [58,59]. Three-dimensional (3D) radiation patterns of each element at 6 GHz (middle frequency) have been plotted in Figure 19. It should be noted that the shapes of the radiation patterns for the antenna elements are a bit different from single-element, which makes it more suitable for the smartphone application. The radiation patterns of the smartphone antenna not only can cover different sides of the smartphone board, but also generate different polarizations, which is a unique function for MIMO design, as mentioned above [60]. This is mainly due to the corner placement of the antenna pairs as well as the big ground plane of the MIMO design. As can be observed from Figure 19, different sides (including sides 1, 2, 3, and 4, as shown in Figure 16a) of the PCB board are covered by different elements with different polarization. Therefore, in total, the designed UWB smartphone antenna contains four horizontally-polarized and four vertically-polarized radiation elements.

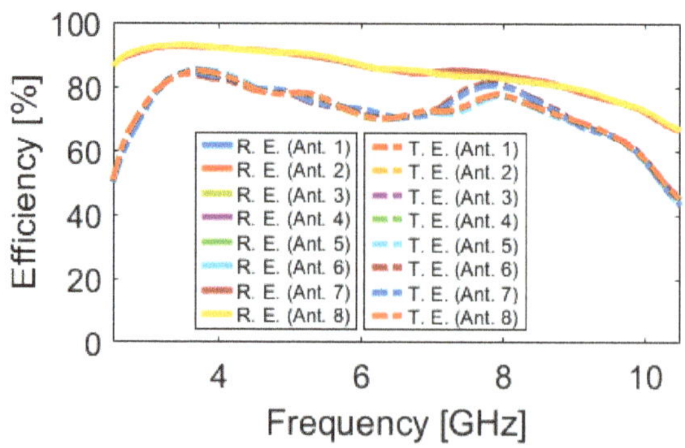

Figure 18. Efficiencies of the MIMO design.

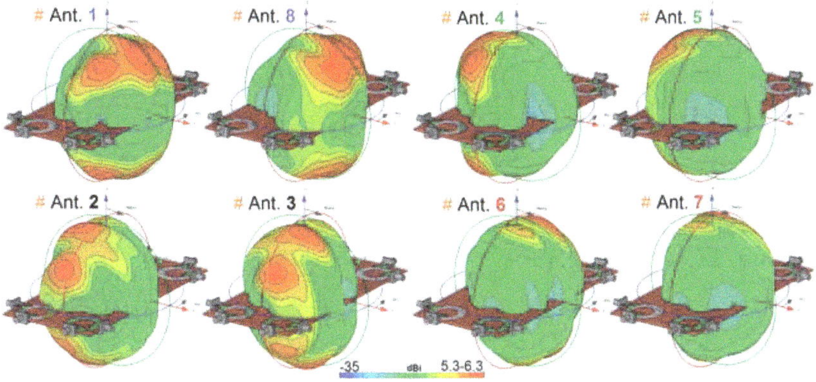

Figure 19. Radiation patterns of the diversity antenna element with directivity value at 5.5 GHz.

The proposed UWB dual-polarized antenna system was fabricated and its S-parameters, radiation patterns, and gain characteristics are measured. Figure 20a,b show the top and bottom layers of the prototype. The MIMO smartphone antenna is implemented on a cheap FR4 substrate with an overall size of 75 × 150 × 1.6 mm^3. Figure 20c illustrates the measurement setup of the fabricated antenna. As can be observed, 50-Ohm loads are employed for the elements, not under measurement in order to measure the characteristics of the MIMO system and also to eliminate the mutual effects from the other elements. Figure 21a,b depict the simulated and measured S_{nn} and S_{mn} results of the MIMO design, respectively. It should be noted that, due to the similar behavior of the elements, it is not necessary to show all of the achieved S-parameters. Similar S_{nn} results with ultra-wide bandwidth are obtained for the antenna elements, as shown in Figure 21a. In addition, all S_{mn} results of the design are less than −10 dB. Good agreement is observed for the presented MIMO smartphone antenna when compared with the simulations.

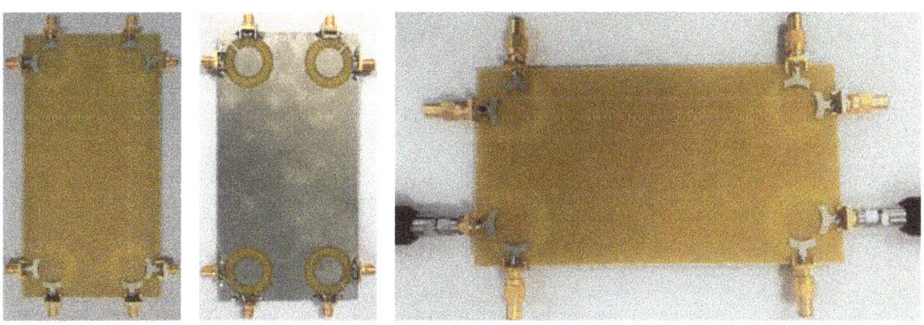

(a) (b) (c)

Figure 20. Fabricated prototype, (**a**) top view, (**b**) bottom view, and (**c**) feeding mechanism.

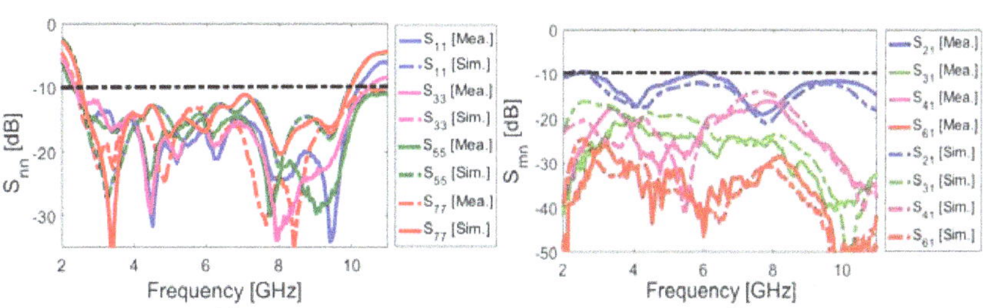

(a) (b)

Figure 21. Measured and simulated (**a**) S_{nn} and (**b**) S_{mn} characteristics.

The 2D-polar radiation patterns of two adjacent elements (including Ant. 1 and Ant. 2) are measured and compared with the simulations due to similar performances of the radiation elements of the smartphone antenna. Figure 22 plots the measured and simulated radiation patterns of the antennas at 3, 6, and 9 GHz. It is found that the fabricated smartphone antenna can provide quasi-omnidirectional radiation patterns at different frequencies of its operation bandwidth [61]. The experimental and simulation results agree well with each other, as can be observed. Next, we evaluate the TARC and ECC characteristics of the proposed UWB MIMO smartphone antenna. Since these two parameters consider the mutual effects, they can provide more meaningful measures of the designed MIMO antenna performance than the reflection coefficient.

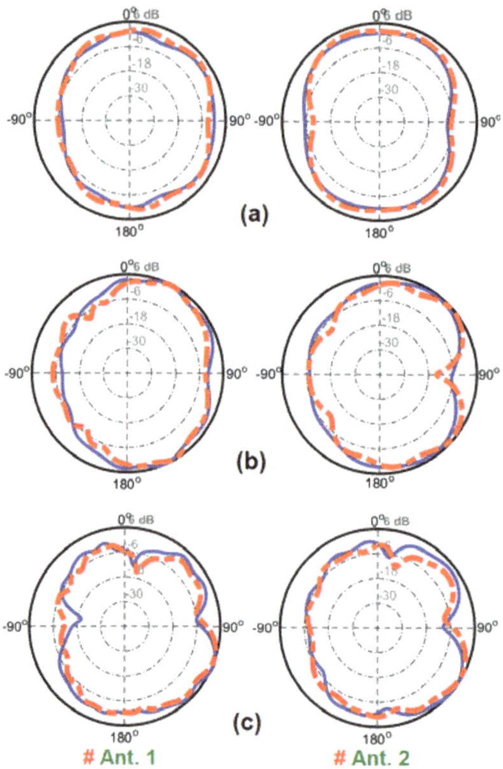

Figure 22. Measured (dash line) and simulated (solid line) radiation pattern results for Ant. 1 and Ant. 2 at (**a**) 3 GHz, (**b**) 6 GHz, and (**c**) 9 GHz.

Figure 23a,b depict the ECC and TARC parameters, as calculated from the measured/simulated results. It can be observed from Figure 23a that the designed MIMO antenna provides very low ECC (less than 0.01) through its entire operation band. Furthermore, it is clear from Figure 23b that the UWB MIMO antenna design exhibits low TARC characteristics over the ultra-wide bandwidth (2.5–10 GHz). As seen, the design has less than −20 dB TARC values at different frequencies. The fundamental properties of the proposed MIMO diversity antenna system are compared with the recently reported MIMO antenna designs and are listed in Table 3 [21–38]. As clearly seen, in contrast to the recently proposed designs, the designed MIMO antenna offers ultra-wide frequency bandwidth of 2.6–10 GHz (more than 120% FBW), a unique characteristic that none of the other MIMO antenna designs cited have. In addition, it can be seen that the presented MIMO antenna exhibits higher performances in terms of efficiency, gain level, ECC, and TARC characteristics. Furthermore, the proposed MIMO smartphone antenna exhibits radiation pattern diversity and also supports different polarizations (including vertical/horizontal) at different four sides of the mainboard, unlike most of the reported MIMO designs.

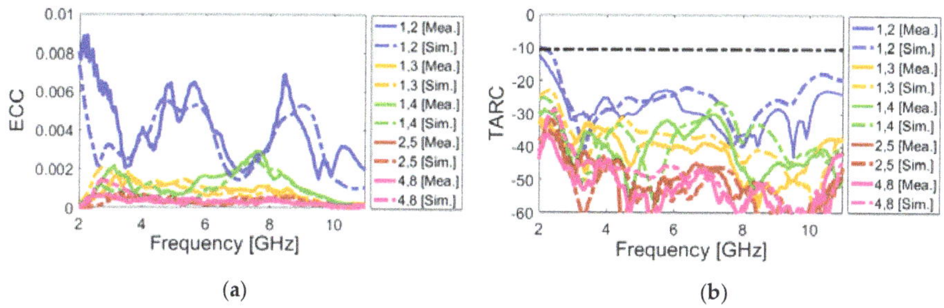

Figure 23. Calculated (**a**) ECC and (**b**) TARC results of the UWB smartphone antenna from measured/simulated results.

Table 3. Comparison between the Proposed Smartphone Antenna and the Reported MIMO Antennas.

Ref.	Design Type	B.W. (GHz)	Efficiency (%)	Size (mm^2)	ECC
[21]	Inverted-F	3.4–3.6	-	120 × 70	-
[22]	Monopole	3.4–3.6	35–50	150 × 75	<0.40
[23]	Tightly Arranged Pairs	3.4–3.6	50–70	150 × 73	<0.07
[24]	Integrated Waveguide	3.4–3.6	50–80	150 × 75	<0.2
[25]	Modified PIFA	3.25–3.85	40–75	150 × 75	<0.01
[26]	Loop	2.55–2.6	48–63	136 × 68	<0.15
[27]	SCS patch-slot	3.55–3.65	52–76	150 × 75	-
[28]	Monopole-Slot	2.55–2.68	48–63	136 × 68	<0.15
[29]	Slot	3.4–3.8	50–75	150 × 75	<0.01
[30]	circular-slot loop	3.3–3.9	60–80	150 × 75	<0.01
[31]	Monopole	4.55–4.75	50–70	136 × 68	-
[32]	Inverted-F	3.4–3.6	55–60	100 × 50	-
[33]	Self-Isolated Monopole	3.4–3.6	60–70	150 × 75	<0.015
[34]	Inverted-L Monopole	3.3–5	40–60	136 × 68	<0.2
[35]	Identical Monopole	2–6	30–60	124 × 64	-
[36]	U-Slot	3.3–6	40–75	150 × 75	<0.12
[37]	Loop	3.3–5	40–70	150 × 75	<0.1
[38]	Inverted-F	3.3–4.5	20–75	150 × 75	<0.6
This Work	Diversity Ring-Slot	2.6–10.2 (122%)	60–80	150 × 75	<0.007

Another important function is the channel capacity loss (CCL) of the MIMO system [43], which mainly depends on the S-parameters of the MIMO system with the accepted limit of ≤0.4 bps/Hz and can be calculated while using the below formulas:

$$CCL = -log_2 \det\left(\psi^R\right) \tag{6}$$

$$\psi^R = \begin{bmatrix} \rho_{11} & \cdots & \rho_{18} \\ \vdots & \ddots & \vdots \\ \rho_{81} & \cdots & \rho_{88} \end{bmatrix} \tag{7}$$

where,

$$\rho_{ii} = 1 - \left(|S_{ii}|^2 + \left|S_{ij}\right|^2\right)$$
$$\rho_{ij} = \left(S_{ii}^* S_{ij} + S_{ji}^* S_{ij}\right) \quad \text{for} \quad i,j - 1, \ldots, 8$$

Figure 24a illustrates the calculated channel capacity loss from the measured and simulated results of the proposed UWB-MIMO smartphone antenna. It is shown that the MIMO antenna design offers a very low CCL band: less than 0.3 bps/Hz over the operation band of 2.7~10 GHz is obtained. Furthermore, the calculated channel capacity (CC) is investigated in Figure 24b in order to further study the MIMO operation of the presented UWB antenna system. The CC is defined as the maximum possible transmission rate, such that the probability of error is arbitrarily small. It is defined as:

$$CC = E\left\{log_2\left[det\left(I + \frac{SNR}{n_T}\right)H_{scale}H_{scale}^T\right]\right\} \quad (8)$$

where the channel matrix H_{scale} can be calculated using:

$$H_{scale} = \sqrt{\rho_{scale,RX}}\,H_{i.i.d}\,\sqrt{\rho_{scale,TX}} \quad (9)$$

As shown, the calculated CC within the entire operating frequency is around 40 bps/Hz, whereas, for the ideal case, it is about 46 bps/Hz [62]. According to the obtained ECC, TARC, and CCL results, it can be calculated that the proposed UWB smartphone antenna has provided worthy MIMO performance.

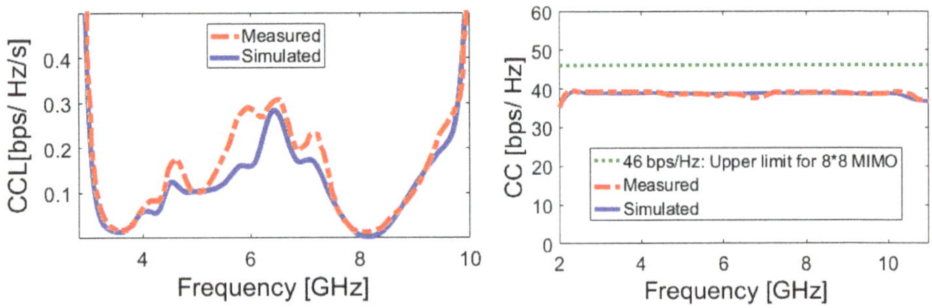

(a) (b)

Figure 24. Calculated (**a**) channel capacity loss (CCL) and (**b**) channel capacity (CC) of the smartphone antenna.

4. Conclusions

A new design structure of eight-port UWB-MIMO mobile-phone antenna is proposed for future mobile handsets. Its configuration consists of four pairs of diversity slot antenna elements located at four edge corners of the PCB mainboard with an FR-4 dielectric. The circular-ring slot radiator is converted to an open-ended slot by adding a rectangular strip in order to enhance the port isolation and increase the impedance-matching of the closely-spaced ports. The antenna elements exhibit ultra-wide impedance bandwidth, covering 2.6–10 GHz ($S_{11} \leq -10$ dB), and provide radiation pattern and polarization diversity function. Acceptable fundamental properties in terms of input-impedance, antenna gain, efficiency, radiation patterns, and user-effect are achieved for the proposed design. Low ECC, TARC, DG, and CCL characteristics have been obtained. Moreover, the performance of the designed UWB-MIMO smartphone antenna for different user-hand/user-head scenarios has been studied and sufficient efficiency performances are obtained. The antenna has a simple and planar structure under the premise of covering different frequencies, which makes it suitable for MIMO operation in future smartphone applications.

Author Contributions: Writing—original draft preparation, N.O.P., H.J.B., Y.I.A.A.-Y., A.M.A. and R.A.A.-A.; writing—review and editing, N.O.P. and R.A.A.-A.; investigation, N.O.P., H.J.B., Y.I.A.A.-Y. and A.M.A.; resources, N.O.P. and R.A.A.-A.; for other cases, all authors have participated. All authors have read and agreed to the published version of the manuscript.

Funding: This project has received funding from the European Union's Horizon 2020 research and innovation programme under grant agreement H2020-MSCA-ITN-2016 SECRET-722424.

Acknowledgments: Authors wish to express their thanks to the support provided by the innovation programme under grant agreement H2020-MSCA-ITN-2016 SECRET-722424.

Conflicts of Interest: The authors declare no conflict of interest.

References

1. Song, Y.; Guo, N.; Qiu, R.C. Towards a real-time UWB MIMO test bed for sensing and communications. *Proc. IEEE Southeastcon.* **2011**, 59–63.
2. Federal Communication Commission. *First Report and order Revision of part 15 of the Commission's Rules Regarding Ultrawideband Transmission System*; Federal Communication Commission: Washington, DC, USA, 2002.
3. Chandel, R.; Gautam, A.K.; Rambabu, K. Design and packaging of an eye-shaped multiple-input–multiple-output antenna with high isolation for wireless UWB applications. *IEEE Trans. Compon. Packag. Manuf. Technol.* **2018**, *8*, 635–642. [CrossRef]
4. Mazloum, J.; Ghorashi, S.A.; Ojaroudi, M.; Ojaroudi, N. Compact triple-band S-shaped monopole diversity antenna for MIMO applications. *Appl. Comput. Electromagn. Soc. J.* **2015**, *30*, 975–980.
5. Rajagopalan, A.; Gupta, G.; Konanur, A.S.; Hughes, B.; Lazzi, G. Increasing channel capacity of an ultrawideband MIMO system using vector antennas. *IEEE Trans. Antennas Propag.* **2007**, *55*, 2880–2887. [CrossRef]
6. Ojaroudi, N.; Ghadimi, N. Dual-band CPW-fed slot antenna for LTE and WiBro applications. *Microw. Opt. Technol. Lett.* **2014**, *56*, 1013–1015. [CrossRef]
7. Zhang, R.; Song, L.; Jaiprakash, A.; Talty, T.; Alanazi, A.; Alghafis, A.; Tonguz, O. Using ultra-wideband technology in vehicles for infrastructure-free localization. *arXiv* **2019**, arXiv:1902.07764.
8. Mearian, L. Ultra Wideband (UWB) Explained (and Why It's in the iPhone 11). Available online: https://www.computerworld.com/article/3490037 (accessed on 31 December 2019).
9. Swedberg, C. Mobile phone companies to demonstrate UWB payments. *RFID J.* **2020**.
10. Rao, Q.; Wilson, K. Design, modeling, and evaluation of a multiband MIMO/diversity antenna system for small wireless mobile terminals. *IEEE Trans. Compon. Packag. Manuf. Technol.* **2011**, *59*, 410–419. [CrossRef]
11. Jensen, M.; Wallace, J. A review of antennas and propagation for MIMO wireless communications. *IEEE Trans. Antennas Propag.* **2004**, *52*, 2810–2824. [CrossRef]
12. Larsson, E. Massive MIMO for next generation wireless Systems. *IEEE Commun. Mag.* **2014**, *52*, 186–195. [CrossRef]
13. Parchin, N.O.; Basherlou, H.J.; Al-Yasir, Y.I.; Abdulkhaleq, A.M.; Abd-Alhameed, R.A.; Excell, P.S. Eight-port MIMO antenna system for 2.6 GHz LTE cellular communications. *Progress Electromagn. Res. C* **2020**, *99*, 49–59. [CrossRef]
14. Al-Yasir, Y.I.A.; Abdullah, A.S.; Ojaroudi Parchin, N.; Abd-Alhameed, R.A.; Noras, J.M. A new polarization-reconfigurable antenna for 5G applications. *Electronics* **2018**, *7*, 293. [CrossRef]
15. Ojaroudiparchin, N.; Shen, M.; Pedersen, G.F. Small-size tapered slot antenna (TSA) design for use in 5G phased array applications. *Appl. Comput. Electromagn. Soc. J.* **2018**, *32*, 193–202.
16. Zhang, Z. *Antenna Design for Mobile Devices*; Wiley-IEEE Press: Hoboken, NJ, USA, 2011.
17. Ojaroudi, N. Circular microstrip antenna with dual band-stop performance for ultra-wideband systems. *Microw. Opt. Technol. Lett.* **2014**, *56*, 2095–2098. [CrossRef]
18. Ojaroudi, N. Design of microstrip antenna for 2.4/5.8 GHz RFID applications. In Proceedings of the GeMic 2014, German Microwave Conference, RWTH Aachen University, Aachen, Germany, 10–12 March 2014.
19. Valizade, A.; Ghobadi, C.; Nourinia, J.; Ojaroudi, N.; Ojaroudi, M. Band-notch slot antenna with enhanced bandwidth by using Ω-shaped strips protruded inside rectangular slots for UWB applications. *Appl. Comput. Electromagn. Soc. (ACES) J.* **2012**, *27*, 816–822.
20. Ojaroudi, Y.; Ojaroudi, N.; Ghadimi, N. Circularly polarized microstrip slot antenna with a pair of spur-shaped slits for WLAN applications. *Microw. Opt. Technol. Lett.* **2015**, *57*, 756–759. [CrossRef]
21. Zhao, X.; Yeo, S.P.; Ong, L.C. Decoupling of inverted-F antennas with high-order modes of ground plane for 5G mobile MIMO platform. *IEEE Trans. Antennas Propag.* **2018**, *66*, 4485–4495. [CrossRef]

22. Wong, K.-L.; Lu, J.-Y.; Chen, L.-Y.; Li, W.Y.; Ban, Y.L. 8-antenna and 16-antenna arrays using the quad-antenna linear array as a building block for the 3.5-GHz LTE MIMO operation in the smartphone. *Microw. Opt. Technol. Lett.* **2016**, *58*, 174–181. [CrossRef]
23. Sun, L.; Feng, H.; Li, Y.; Zhang, Z. Compact 5G MIMO mobile phone antennas with tightly arranged orthogonal-mode pairs. *IEEE Trans. Antennas Propag.* **2018**, *66*, 6364–6369. [CrossRef]
24. Abdullah, M.; Ban, Y.L.; Kang, K.; Li, M.Y.; Amin, M. Eight-element antenna array at 3.5GHz for MIMO wireless application. *Progress Electromagn. Res. C* **2017**, *78*, 209–217. [CrossRef]
25. Parchin, N.O.; Al-Yasir, Y.I.; Basherlou, H.J.; Abd-Alhameed, R.A.; Noras, J.M. Orthogonally dual-polarised MIMO antenna array with pattern diversity for use in 5G smartphones. *IET Microw. Antennas Propag.* **2020**.
26. Li, M.-Y. Eight-port orthogonally dual-polarised MIMO antennas using loop structures for 5G smartphone. *IET Microw. Antennas Propag.* **2017**, *11*, 1810–1816. [CrossRef]
27. Parchin, N.O.; Al-Yasir, Y.I.; Noras, J.M.; Abd-Alhameed, R.A. Dual-polarized MIMO antenna array design using miniaturized self-complementary structures for 5G smartphone applications. In Proceedings of the 13th European Conference on Antennas and Propagation (EuCAP), Krakow, Poland, 31 March–5 April 2019.
28. Li, M.-Y.; Ban, Y.L.; Xu, Z.Q.; Wu, G.; Kang, K.; Yu, Z.F. Eight-port orthogonally dual-polarized antenna array for 5G smartphone applications. *IEEE Trans. Antennas Propag.* **2016**, *64*, 3820–3830. [CrossRef]
29. Parchin, N.O.; Al-Yasir, Y.I.A.; Ali, A.H.; Elfergani, I.; Noras, J.M.; Rodriguez, J.; Abd-Alhameed, R.A. Eight-element dual-polarized MIMO slot antenna system for 5G smartphone applications. *IEEE Access* **2019**, *9*, 15612–15622. [CrossRef]
30. Parchin, N.O.; Al-Yasir, Y.I.; Basherlou, H.J.; Abdulkhaleq, A.M.; Sajedin, M.; Abd-Alhameed, R.A.; Noras, J.M. Modified PIFA array design with improved bandwidth and isolation for 5G mobile handsets. In Proceedings of the 2019 IEEE 2nd 5G World Forum (5GWF), Dresedn, Germany, 30 September–2 October 2019.
31. Xu, S.; Zhang, M.; Wen, H.; Wang, J. Deep-subwavelength decoupling for MIMO antennas in mobile handsets with singular medium. *Sci. Rep.* **2017**, *7*, 12162. [CrossRef]
32. Al-Hadi, A.A.; Ilvonen, J.; Valkonen, R.; Viikan, V. Eight-element antenna array for diversity and MIMO mobile terminal in LTE 3500MHz band. *Microwave Opt. Technol. Lett.* **2014**, *56*, 1323–1327. [CrossRef]
33. Zhao, A.; Zhouyou, R. Size reduction of self-isolated MIMO antenna system for 5G mobile phone applications. *IEEE Antennas Wirel. Propag. Lett.* **2019**, *18*, 152–156. [CrossRef]
34. Jiang, W.; Liu, B.; Cui, Y.; Hu, W. High-isolation Eight-Element MIMO array for 5G smartphone applications. *IEEE Access* **2019**, *7*, 34104–34112. [CrossRef]
35. Zhekov, S.S.; Tatomirescu, A.; Franek, O.; Pedersen, G.F. Study of the interaction user head-ultrawideband MIMO antenna array for mobile terminals. *Proc. Int. Conf. Electromagn. Adv. Appl.* **2016**, 930–933.
36. Zhang, X.; Li, Y.; Wang, W.; Shen, W. Ultra-wideband 8-port MIMO antenna array for 5G metal-frame smartphones. *IEEE Access* **2019**, *7*, 72273–72282. [CrossRef]
37. Zhao, A.; Ren, Z. Wideband MIMO antenna systems based on coupled-loop antenna for 5G N77/N78/N79 applications in mobile terminals. *IEEE Access* **2019**, *7*, 93761–93771. [CrossRef]
38. Liu, D.Q.; Luo, H.J.; Zhang, M.; Wen, H.L.; Wang, B.; Wang, J. An extremely low-profile wideband MIMO antenna for 5G smart-phones. *IEEE Trans. Antennas Propag.* **2019**, *67*, 5772–5780. [CrossRef]
39. Kamalvand, A.; Ghobadi, C.; Nourinia, J.; Ojaroudi, M.; Ojaroudi, N. Omni-directional/multi-resonance CPW-fed small slot antenna for UWB applications. *ACES J.* **2013**, *28*, 829–835.
40. Ojaroudi, N.; Ojaroudi, M.; Amiri, S. Enhanced bandwidth of small square monopole antenna by using inverted U-shaped slot and conductor-backed plane. *Appl. Comput. Electromagn. Soc. (ACES) J.* **2012**, *27*, 685–690. [CrossRef]
41. Ojaroudi, N.; Ojaroudi, M.; Ghadimi, N. UWB omnidirectional square monopole antenna for use in circular cylindrical microwave imaging systems. *IEEE Antennas Wireless Propag. Lett.* **2012**, *11*, 1350–1353. [CrossRef]
42. Dastranj, A. Optimization of a printed UWB antenna: Application of the invasive weed optimization algorithm in antenna design. *IEEE Antennas Propag. Mag.* **2017**, *59*, 48–57. [CrossRef]
43. Sharawi, M.S. *Printed MIMO Antenna Engineering*; Artech House: Norwood, MA, USA, 2014.
44. Parchin, N.O.; Jahanbakhsh Basherlou, H.; Al-Yasir, Y.I.; Abdulkhaleq, A.M.; Patwary, M.; Abd-Alhameed, R.A. A new CPW-fed diversity antenna for MIMO 5G smartphones. *Electronics* **2020**, *9*, 261. [CrossRef]
45. Parchin, N.O.; Abd-Alhameed, R.A.; Shen, M. Gain improvement of a UWB antenna using a single-layer FSS. In Proceedings of the PhotonIcs & Electromagnetics Research Symposium (PIERS), Xiamen, China, 17–20 December 2019.

46. Siahkal-Mahalle, B.H.; Ojaroudi, M.; Ojaroudi, N. Enhanced bandwidth small square monopole antenna with band-notched functions for UWB wireless communications. *Appl. Comput. Electromagn. Soc. (ACES) J.* **2012**, *27*, 759–765.
47. Zhu, F.; Gao, S. Development and implementation of a dual-polarized slot-loaded patch antenna for ultra-wideband imaging systems. In Proceedings of the URSI General Assembly and Scientific Symposium (URSI GASS), Beijing, China, 16–23 August 2014.
48. Zhu, F.; Gao, S.; Ho, A.T.; Abd-Alhameed, R.A.; See, C.H.; Brown, T.W.; Xu, J. Ultra-wideband dual-polarized patch antenna with four capacitively coupled feeds. *IEEE Trans. Antennas Propag.* **2014**, *62*, 2440–2449.
49. Ram Krishnaa, R.V.S.; Kumar, R. Design and investigations of a microstrip fed open V-shape slot antenna for wideband dual slant polarization. *Eng. Sci. Technol. Int. J. (JESTECH)* **2015**, *18*, 513–523. [CrossRef]
50. Yahya, R.; Nakamura, A.; Itami, M.; Denidni, T.A. A novel UWB FSS-based polarization diversity antenna. *IEEE Antennas Wirel. Propag. Lett.* **2017**, *16*, 2525–2528. [CrossRef]
51. Augustin, G.; Chacko, B.P.; Denidni, T.A. *Dual Port Ultra Wideband Antennas for Cognitive Radio and Diversity Applications in Advancement in Microstrip Antennas with Recent Applications*; InTech: Rijeka, Croatia, 2013.
52. Kishna, R.V.R.; Kumar, R.; Kushwaha, N. An UWB dual polarized microstrip fed L-shape slot antenna. *Int. J. Microw. Wireless Technol.* **2016**, *8*, 363–368. [CrossRef]
53. Ram Krishnaa, R.V.S.; Kumar, R. A dual-polarized square-ring slot antenna for UWB, imaging, and radar applications. *IEEE Antennas Wireless Propag. Lett.* **2016**, *15*, 195–198. [CrossRef]
54. Xiong, L.; Gao, P. Compact dual-polarized slot UWB antenna with CPW-fed structure. *Proc. IEEE Antennas Propag. Soc. Int. Symp. (APSURSI'12)* **2012**, 1–2.
55. Irene, G.; Rajesh, A. A dual-polarized UWB–MIMO antenna with IEEE 802.11ac band-notched characteristics using split-ring resonator. *J. Comput. Electron.* **2018**, *17*, 1090–1098. [CrossRef]
56. Zhu, J.; Li, S.; Feng, B.; Deng, L.; Yin, S. Compact dual polarized UWB quasi-self-complementary MIMO/diversity antenna with band-rejection capability. *IEEE Antennas Wireless Propag. Lett.* **2016**, *15*, 905–908. [CrossRef]
57. Tiang, S.S.; Hathal, M.S.; Anwar, N.S.; Ain, M.F.; Abdullah, M.Z. Development of a compact wide-slot antenna for early stage breast cancer detection featuring circular array full-view geometry. *Int. J. Antennas Propag.* **2014**, 1–11. [CrossRef]
58. Ojaroudiparchin, N.; Shen, M.; Fr, G. Design of Vivaldi antenna array with end-fire beam steering function for 5G mobile terminals. In Proceedings of the 23rd Telecommunications Forum Telfor (TELFOR), Belgrade, Serbia, 24–26 November 2015; pp. 587–590.
59. Parchin, N.O.; Jahanbakhsh Basherlou, H.; Al-Yasir, Y.I.; Ullah, A.; Abd-Alhameed, R.A.; Noras, J.M. 'Multi-band MIMO antenna design with user-impact investigation for 4G and 5G mobile terminals. *Sensors* **2019**, *19*, 456. [CrossRef]
60. Ojaroudiparchin, N.; Shen, M.; Zhang, S.; Pedersen, G.F. A switchable 3D-coverage phased array antenna package for 5G mobile terminals. *IEEE Antennas Wirel. Propag. Lett.* **2016**, *15*, 1747–1750. [CrossRef]
61. Ojaroudi, N.; Ojaroudi, Y.; Ojaroudi, S. Compact ultra-wideband monopole antenna with enhanced bandwidth and dual band-stop properties. *Int. J. RF Microw. Comput. Aided Eng.* **2014**, *25*, 346–357. [CrossRef]
62. Chang, L.; Yu, Y.; Wei, K.; Wang, H. Polarization-orthogonal co-frequency dual antenna pair suitable for 5G MIMO smartphone with metallic bezels. *IEEE Trans. Antennas Propag.* **2019**, *67*, 5212–5220. [CrossRef]

sensors

MDPI

Letter

Multimode HMSIW-Based Bandpass Filter with Improved Selectivity for Fifth-Generation (5G) RF Front-Ends †

Amjad Iqbal [1], Jun Jiat Tiang [1], Sew Kin Wong [1], Mohammad Alibakhshikenari [2,*], Francisco Falcone [3,4] and Ernesto Limiti [2]

1 Centre For Wireless Technology, Faculty of Engineering, Multimedia University, Cyberjaya 63100, Malaysia; amjad730@gmail.com (A.I.); jjtiang@mmu.edu.my (J.J.T.); skwong@mmu.edu.my (S.K.W.)
2 Electronic Engineering Department, University of Rome "Tor Vergata", Via del Politecnico 1, 00133 Rome, Italy; limiti@ing.uniroma2.it
3 Electrical Engineering Department, Public University of Navara, 31006 Pamplona, Spain; francisco.falcone@unavarra.es
4 Institute of Smart Cities, Public University of Navarre, 31006 Pamplona, Spain
* Correspondence: Alibakhshikenari@ing.uniroma2.it

† This paper is an extended version of our paper published in Iqbal, A.; Tiang, J.J.; Kwang Lee, C.; Muhammad, S. Wideband Half Mode Substrate Integrated Waveguide (HMSIW) Bandpass Filter. In Proceedings of the 2020 17th International Conference on Electrical Engineering/Electronics, Computer, Telecommunications and Information Technology (ECTI-CON), Phuket, Thailand, 24–27 June 2020; pp. 710–712.

Received: 15 November 2020; Accepted: 16 December 2020; Published: 19 December 2020

Abstract: This article presents the detailed theoretical, simulation, and experimental analysis of a half-mode substrate integrated waveguide (HMSIW)-based multimode wideband filter. A third-order, semicircular HMSIW filter is developed in this paper. A semicircular HMSIW cavity resonator is adopted to achieve wide band characteristics. A U-shaped slot (acts as a $\lambda/4$ stub) in the center of a semicircular HMSIW cavity resonator and L-shaped open-circuited stubs are used to improve the out-of-band response by generating multiple transmission zeros (TZs) in the stop-band region of the filter. The TZs on either side of the passband can be controlled by adjusting dimensions of a U-shaped slot and L-shaped open-circuited stubs. The proposed filter covers a wide fractional bandwidth, has a lower insertion loss value, and has multiple TZs (which improves the selectivity). The simulated response of filter agrees well with the measured data. The proposed HMSIW bandpass filter can be integrated with any planar wideband communication system circuit, thanks to its planar structure.

Keywords: substrate integrated waveguide (SIW); transmission zeros (TZs); metallic via; coupling topology

1. Introduction

A wideband filter with high selectivity, low insertion loss, light weight, high quality factor, and high power handling capability is the key component for future wideband communication systems [1,2]. The substrate integrated waveguide (SIW) possesses most of the above requirements [3,4]. The operational principle of SIW is identical to a bulky waveguide, but SIW is a planar circuit, unlike a conventional waveguide. A planar structure of SIW provides an extra advantage when it is integrated with other planar circuits and system-in-package applications. Much more consideration is given to SIW technology for developing RF front-end components because of its several advantages over other technologies. A multimode filter is a type of filter that has more than one excited mode at once. So far, several multimode filters have been designed using planar SIW technology [5,6], multilayer resonators [7,8], and microstrip technology [9].

The wideband response is achieved by coupling together different resonators of different resonant frequencies. Several papers demonstrate the wideband performance by coupling nearby modes. In [10], a wideband filter using SIW technology is designed by coupling the modes securely with the help of U-shaped slots. The filter in this design has only 42% of the fractional bandwidth and is larger in size, using as many as five modes. In [11], a multi-mode wideband filter is designed using three modes of triangular SIW cavity. The wideband response is achieved by controlling the first resonant mode by means of an extra via hole of radius 0.5 mm in the corner of a triangular resonator. In [12], an SIW filter is designed for millimeter waves, where the length and width of the slots control the electric coupling and bandwidth of the filter. In [13], a wideband filter is designed using three modes of the circular SIW cavity. The coupling between the modes is controlled by putting an extra via hole in the center of the resonator. Degenerated modes in the bandpass filter are used in [14] to achieve wideband. Bandpass filters, having this topology, achieve relatively small sizes with low fractional bandwidth. In [15], a UWB bandpass filter, based on a square-shaped defected ground plane and microstrip resonators, is designed. It covers the whole UWB band (3.1–10.6 GHz). The operating bandwidth of the filter is enhanced by adding two short-circuited stubs. Then, an open-circuited stub is introduced to generate a notch (stopband) at 5.8 GHz. In [16], a UWB bandpass filter with a notch band is reported. The reported filter has three layers: the lower and upper layers have T-shaped microstrip resonators, and the middle layer has three circular slots for coupling the lower and upper layers. Multiple modes of the microstrip resonators are excited to obtain a large bandwidth. Then, the microstrip resonators and ground plane are modified to get a notch band at 5.8 GHz. In [17], a wideband filter is reported. The wideband response is achieved by tight coupling between the resonators. In addition, the source/load coupling with the resonators is also strong, which leads to a wideband bandpass filter. In [18], different filters for dual- and triple-band 5G applications are designed. The reported filters use open-loop SIRs for dual-band operation and uniform folded resonators for triple-band applications. The reported filters have wideband characteristics in each passband with good out-of-band rejections. The authors of [19] reported a wideband bandpass filter. The reported filter is designed by using square ring resonators, which are loaded by stubs. In [20], wideband second- and third-order bandpass filters are designed using open- and short-circuited stubs. The reported filter has a wideband (51.9% in the lower-frequency passband and 23.3% in the high-frequency passband) and low insertion loss of 0.3 dB.

In the present paper, a circular half-mode SIW (HMSIW) resonator is used to design a wideband filter. The wideband response is achieved by coupling the first three modes. In addition, the selectivity is enhanced by generating the source-load coupling topology using a U-shaped slot and L-shaped open-circuited stub resonators.

2. Full-Mode SIW and Half-Mode SIW Cavity Resonator Analysis

A circular SIW cavity resonator with a diameter of 40 mm is suggested, as depicted in Figure 1. The initial dimensions of the circular SIW cavity are obtained with the help of Equation (1) [21] and then optimized using a full-wave electromagnetic (EM) simulator (high-frequency structure simulator; HFSS). The circular cavity is developed by placing metallic vias on the edges. The radius (r) and isolation among the plated holes (d) are selected according to the guidelines ($2r/d \geq 0.5$, and $2r/\lambda \leq 0.1$) suggested in [22], so that minimal power leaks from the cavity. RT/duroid 5880 (ϵ_r = 2.2) is used as a substrate. The width of the transmission line (W_{tl}) is calculated using standard transmission line equation for 50 Ω impedance [21]. The transmission coefficient ($|S_{21}|$) of the full-mode SIW (FMSIW) is plotted in Figure 1. The first four resonant modes (TM_{11}, TM_{21}, TM_{22}, and TM_{33}) are labeled f_1 to f_4, with f_1 as the lowest-frequency mode followed by the higher-frequency modes as far as f_4. The resonant frequencies of the first four resonant modes of the circular SIW cavity resonator are located at 4.08 GHz (TM_{11} mode), 6.44 GHz (TM_{21} mode), 8.71 GHz (TM_{22} mode), and 9.56 GHz

(TM_{33} mode). The FMSIW cavity resonator is converted to a HMSIW cavity resonator by cutting along A and A′ (quasi-magnetic wall), as shown in Figure 1. The A-A′ line on the open-ended side of the cavity is a quasi perfect magnetic boundary. The $|S_{21}|$ of the HMSIW cavity resonator is also illustrated in Figure 1. The first four resonant modes of the HMSIW cavity resonators lie at 3.54 GHz (TM_{11} mode), 5.81 GHz (TM_{21} mode), 7.58 GHz (TM_{22} mode), and 8.76 GHz (TM_{33} mode). A shift towards the lower-frequency side is observed in the resonant modes of the HMSIW cavity resonator as compared to the FMSIW cavity resonator. The lower-frequency shift in the HMSIW cavity resonator is due to the fringing effects caused by the open-ended magnetic wall. The *H*-field plots of the first four modes in the FMSIW and HMSIW cavity resonators are shown in Figure 2a,b, respectively.

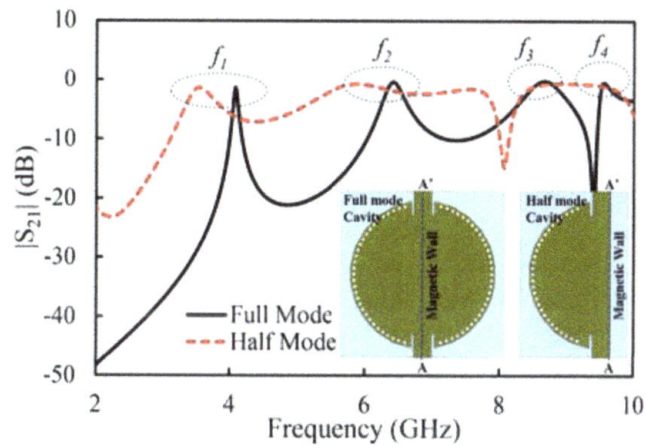

Figure 1. Transmission coefficient (S_{21}) (magnitude) of the filter in the full-mode and half-mode substrate integrated waveguide (SIW) cavity resonator.

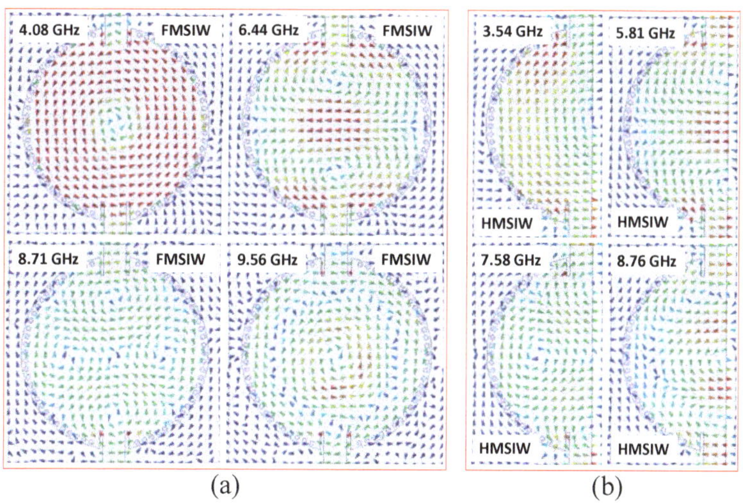

Figure 2. *H*-fields plot of the first four modes in the (**a**) full-mode circular SIW resonator, (**b**) half-mode circular SIW resonator.

$$f_r = \frac{k_{mnp}c}{2\pi r\sqrt{\epsilon_{reff}}} \quad (1)$$

where c is the speed of light in the free space, and ϵ_{reff} is the effective permittivity of the substrate. The radius of the cavity resonator is denoted by r, the resonant frequency is denoted by f_r, and the value of k_{mnp} can be computed by substituting Bessel's coefficients.

3. HMSIW Wideband Bandpass Filter Design

Figure 3a shows the design of the proposed HMSIW cavity resonator-based multimode wideband filter. The optimized dimensions can be found in the caption of Figure 3a. A rectangular inductive matching slot is designed with inset feeding to further improve the impedance matching of the filter. Therefore, the proposed filter has three excited resonant modes: TM_{11} at 3.3 GHz, TM_{21} at 4.2 GHz, and TM_{22} at 5.7 GHz, as shown in Figure 3b. The design procedure of the proposed filter, which consists of three iterations, is illustrated in Figure 3c. Initially, a semi-circular HMSIW cavity resonator is designed. Then, the filter is modified by loading a U-shaped slot; as a result, the TZ on the high-frequency side of the passband is generated, and selectivity is enhanced. Finally, in addition, two L-shaped stubs are added to further improve the selectivity. As a result, two TZs are generated on the lower-frequency side of the passband. Simulation of the proposed filter is carried out using full-wave EM simulator HFSS (solution type = driven model, frequency range = 1–12 GHz, step size = 1 MHz, type of sweep = discrete, maximum number of passes = 16, maximum delta energy = 0.05, type of ports = wave-ports, and boundary condition = radiating only in all direction). The proposed structure is designed, and wave-ports, having 50 Ω impedance, at both ends of the transmission line are assigned. A detailed design and analysis of each iteration is discussed in the following sections.

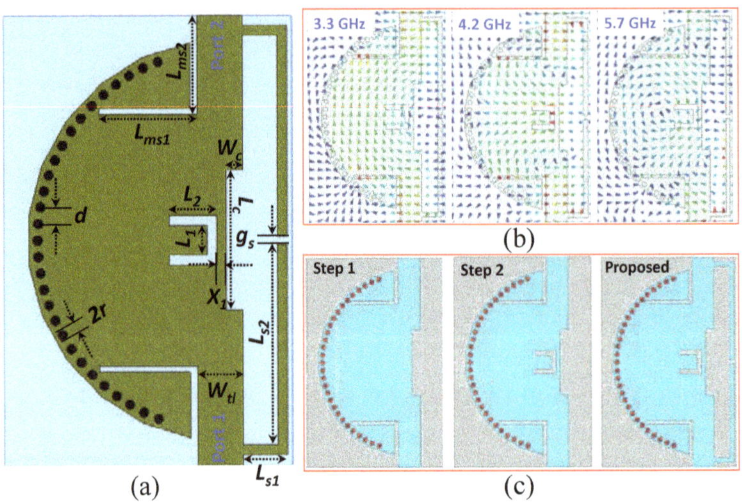

Figure 3. (**a**) Geometry of the proposed wideband filter (d = 1.65, $2r$ = 1, W_{tl} = 5, L_{s1} = 4.5, L_{s2} = 21.1, g_s = 0.8, L_c = 14, W_c = 2, L_1 = 2, L_2 = 6, L_{ms1} = 10.3, L_{ms2} = 9.3, X_1 = 2 (Unit = mm)); (**b**) H-fields plot at 3.3, 4.2, and 5.7 GHz; (**c**) design evolution steps [1].

3.1. Step 1: Wideband Filter Design with No Transmission Zero (TZ)

The maximum fields of the first four modes lie near to the arbitrary magnetic wall of the HMSIW cavity resonator (see Figure 2b). Therefore, a rectangular slot at the maximum fields side of the HMSIW cavity resonator was introduced to adjust the positions of the resonant modes. The reflection coefficient ($|S_{11}|$) and transmission coefficient ($|S_{21}|$) of the filter in the first iteration are shown in Figure 4. It can be observed that the passband is generated by the four coupled resonant modes.

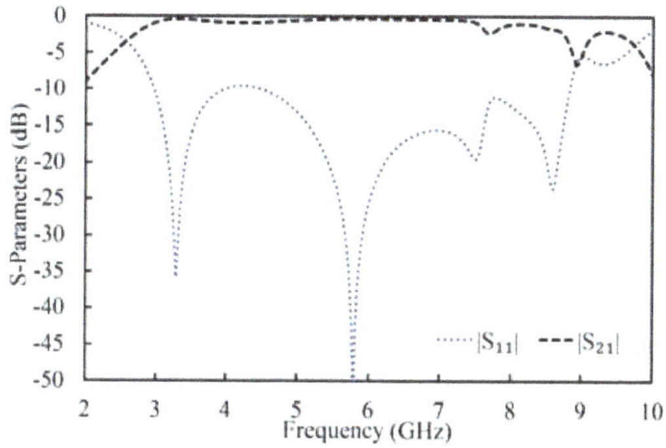

Figure 4. Simulated S-parameters (magnitude) of the filter in the first iteration [1].

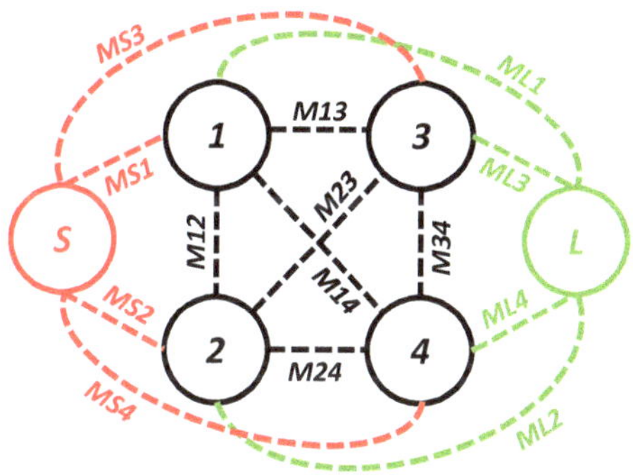

Figure 5. Coupling topology of the filter in the first iteration.

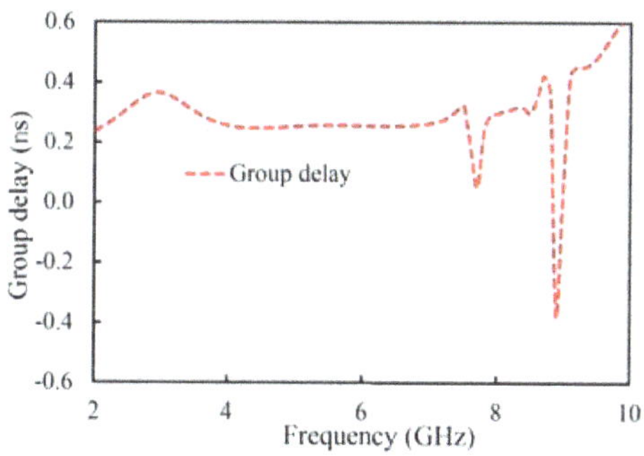

Figure 6. Simulated group delay of the filter in the first iteration and coupling topology.

During this iteration, the first four resonant modes created a passband. The coupling topology of the filter in the first iteration is shown in Figure 5. The coupling between the source and the first mode is *MS1*; between the source and the second mode it is *MS2*; between the source and the third resonator it is *MS3*; and between the source and the fourth mode it is *MS4*. From the load to the first mode it is *ML1*; from the load to the second mode it is *ML2*; from the load to the third resonator it is *ML3*; and from the load to the fourth mode it is *ML4*. From the first and second mode it is *M12*; from the first and third mode it is *M13*; from the second and third mode it is *M23*; from the second and fourth mode it is *M24*; and from the third and fourth mode it is *M34*. The passband filter, designed in the first iteration, has four poles located at 3.29, 5.77, 7.5, and 8.6 GHz. An unwanted dip in the $|S_{21}|$ graph of the filter at the passband can be seen at 7.7 GHz. In addition, a 3 dB fractional bandwidth of

117% at the center frequency of 6.12 GHz is noted. Moreover, the designed filter (first iteration) has a large bandwidth but is not suitable for applications where high selectivity is required. The selectivity of the filter can be enhanced by introducing the source-load coupling topology. The simulated group delay varies from 0.04 ns to 0.58 ns in the passband (2.51–9.73 GHz), as shown in Figure 6.

The positions of the resonant frequencies of the first four modes of the HMSIW filter were analyzed by changing W_c and L_c parameters of the rectangular slot. The impact of the changing W_c and L_c on the position of the resonant modes is illustrated in Figures 7 and 8. f_1, f_2, f_3, and f_4 show the resonant frequencies of the first mode (TM_{11} mode), second mode (TM_{21} mode), third mode (TM_{22} mode), and fourth resonant mode (TM_{33} mode), respectively. By changing the W_c value from 1.8 to 4.2 mm, the first resonant mode shifted from 3.25 to 4.5 GHz; the second resonant mode shifted from 5.8 to 7 GHz; and that of the fourth resonant mode shifted from 8.5 to 8.73 GHz. The third resonant mode had an irregular relationship with the W_c, as illustrated in Figure 7. The dependency of the first four resonant modes on the slot length (L_c) is presented in Figure 8. The positions of f_1, f_3, and f_4 were directly related to the L_c, and that of f_2 was inversely related to the L_c, as illustrated in Figure 8. The resonant frequencies of the first four modes were located at 3.2, 6.5, 7.2, and 8.1 GHz, when L_c was 16 mm. At L_c = 24 mm, the resonant frequencies of the first four modes shifted to 3.5, 5.5, 7.2, and 9 GHz. It can be observed that these parameters played a key role in positioning the first four resonant modes. All four resonant modes can be coupled together, by properly adjusting these parameters, to get a wide passband.

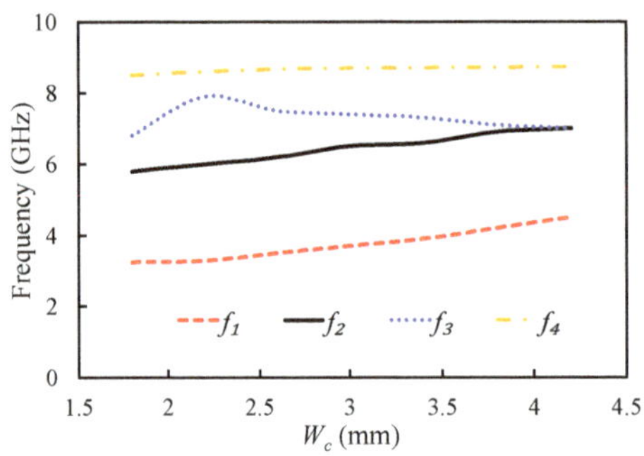

Figure 7. Mode chart for the first four resonant modes against the varying W_c.

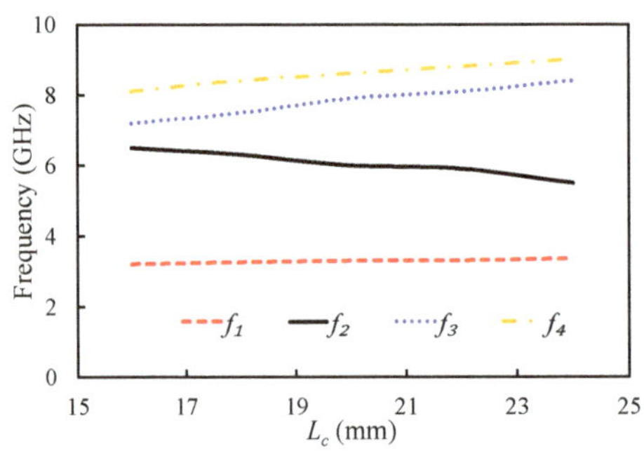

Figure 8. Mode chart for the first four resonant modes against the varying L_c.

3.2. Step 2: Wideband Filter Design with One TZ

As can be seen from Figure 4, the filter had poor selectivity in the first iteration. In order to enhance the selectivity of the filter, a U-shaped slot was etched on the upper metal layer of the HMSIW cavity resonator. The structure of the filter in the second iteration is displayed in Figure 3c. The dimensions of the U-shaped slot (L_1 and L_2) are derived using Equation (2). The slot in the resonator plays the role of a $\lambda/4$ stub, which generates transmission zero on the higher-frequency side of the passband. Therefore, the selectivity of the filter was enhanced on the higher-frequency side of the passband. As a result, a TZ on the higher-frequency side of the passband was generated.

$$f_z = \frac{c}{2L\sqrt{\epsilon_{reff}}} \quad (2)$$

In the above equation, f_z is the frequency of the TZ, L ($L = L_1 + L_2$) is the total length of the slot, c is the speed of light in a vacuum, and ϵ_{reff} is the effective dielectric constant of the substrate.

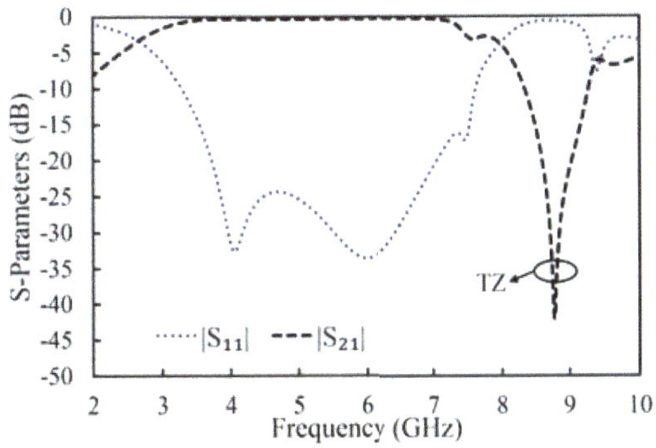

Figure 9. Simulated S-parameters (magnitude) of the filter in the second iteration [1].

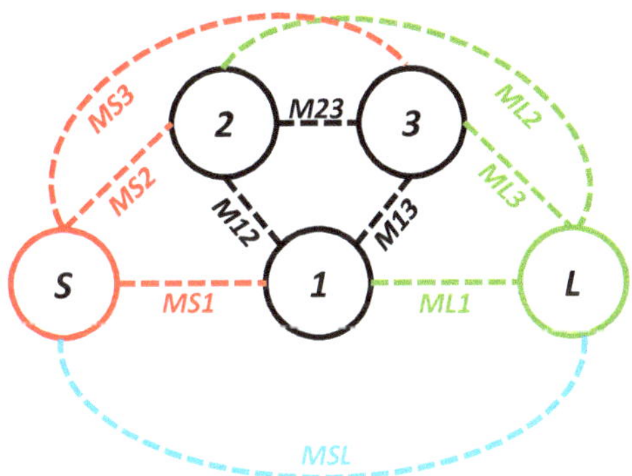

Figure 10. Coupling topology of the filter in the second iteration.

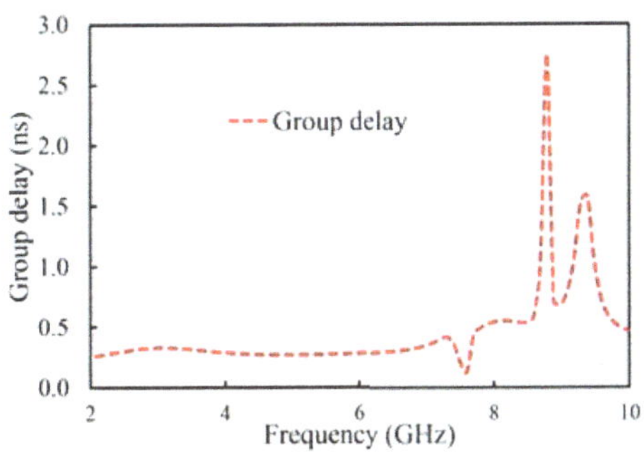

Figure 11. Simulated group delay of the filter in the second iteration.

The $|S_{11}|$ and $|S_{21}|$ of the filter in the second iteration are illustrated in Figure 9. Only the first three resonant modes reached the passband of the filter, as shown in Figure 9. The designed filter in the second iteration had its first pole at 4.08 GHz, the second pole at 6.06 GHz, and the third pole at 7.48 GHz. One TZ, generated due to the introduction of the U-shaped slot, was noted at 8.78 GHz. As a result, a 3 dB fractional bandwidth of 94% at the center frequency of 5.11 GHz was observed. The filter, designed in the second iteration, had a lower 3 dB bandwidth than the filter in the first iteration. However, the filter in the second iteration had improved selectivity at the higher-frequency side of the passband than the filter in the first iteration. The coupling topology of the filter is shown in Figure 10. The source-load coupling, which is responsible for the sharper selectivity on the higher frequency side of the passband, is obvious in the coupling topology. The simulated group delay varied from 0.25 ns to 0.41 ns in the passband (2.71–7.52 GHz), as illustrated in Figure 11.

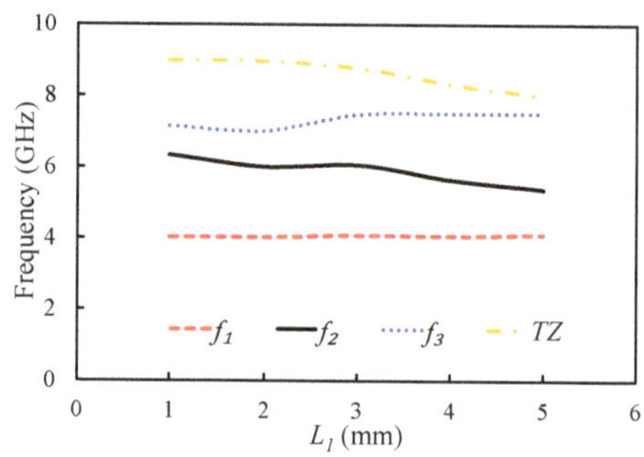

Figure 12. Mode chart for the first three resonant modes and a transmission zero (TZ) against the varying L_1.

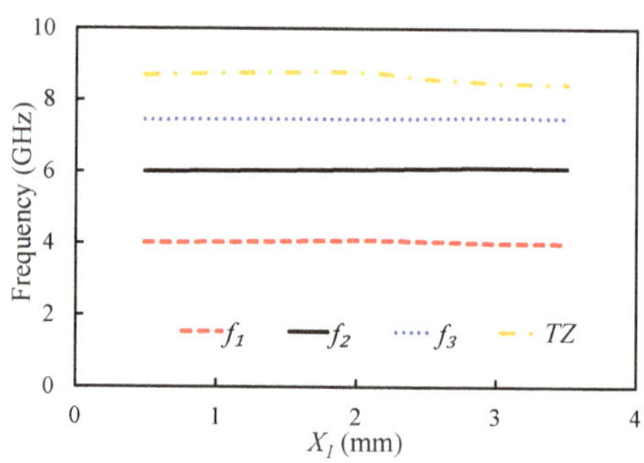

Figure 13. Mode chart for the first three resonant modes and a TZ against the varying X_1.

The impact of parameters L_1 and X_1 on the position of the first three resonant modes and TZ was analyzed as shown in Figures 12 and 13. As Equation (2) clearly shows, the slot length is related to the position of TZ. The effective length of the slot has a inverse relationship with the TZ: the TZ shifts to the lower-frequency side with an increase in the dimensions of the U-shaped slot. In fact, the position of TZ can be controlled by adjusting the length of the U-shaped slot. Figure 12 shows the impact of L_1 on the first three resonant modes and the TZ. The TZ of the filter shifted from 8.99 to 8 GHz by changing the value of L_1 from 1 to 5 mm. The selection of a suitable length for L_1 in terms of selectivity and bandwidth is crucial. If we increase L_1 to a certain extent, then the TZ can appear before the third mode. As a result, the operating bandwidth will reduce. Moreover, no significant change on the position of the first resonant mode was observed with the L_1. However, the position of the second and third resonant modes changed with the L_1. Changing the value of L_1 from 1 to 5 mm made the first resonant mode shift from 4.03 to 4.1 GHz; the second mode shifted from 6.33 to 5.38 GHz; and the third mode shifted from 7.15 to 7.51 GHz. The position of the U-shaped slot had less impact on the resonant modes and TZ, as shown in Figure 13. When X_1 was changed from 0.5 to 3.5 mm, the TZ shifted from 8.7 GHz to 8.45 GHz.

3.3. Proposed Filter: Wideband Filter with Enhanced Selectivity

In the third iteration, two L-shaped open-circuit stub resonators were connected to both the source and the load in order to further enhance selectivity, as shown in Figure 3c. As a result, two TZs were generated on the lower-frequency side of the passband because of the additional source-load coupling caused by the open-circuited stubs. The $|S_{11}|$ and $|S_{21}|$ of the proposed filter are shown in Figure 14. The proposed filter has three TZs: two on the lower-frequency side of the passband and one on its higher-frequency side. The TZs are located at 2.09, 2.72, and 6.82 GHz with respective attenuation levels of −33, −32.4, and −22.1 dB. The proposed filter has three poles, which are located at 3.3, 4.29, and 5.75 GHz. Moreover, the filter designed in the third iteration has a 3 dB fractional bandwidth of 69.31% at the center frequency of 4.67 GHz. The coupling topology of the filter is displayed in Figure 15. The simulated group delay varied from 0.31 to 0.8 ns in the passband (3.05–6.29 GHz), as shown in Figure 16. It can be observed that the filter had a small group delay in the passband; however, the group delay increased where the TZs were located. The large group delay values (peaks) in the group delay plots indicate the presence of the TZs.

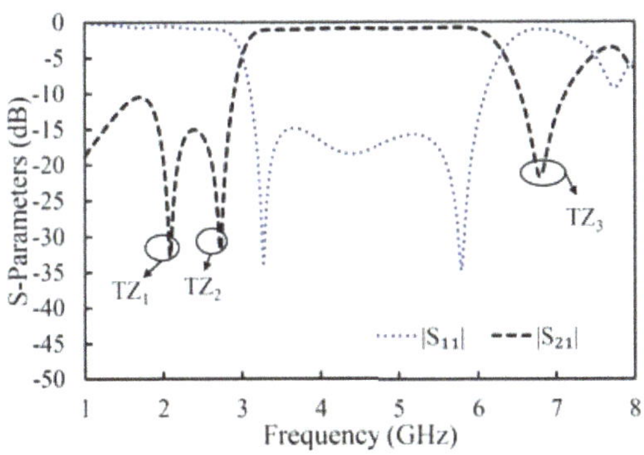

Figure 14. Simulated S-parameters (magnitude) of the filter in the proposed filter [1].

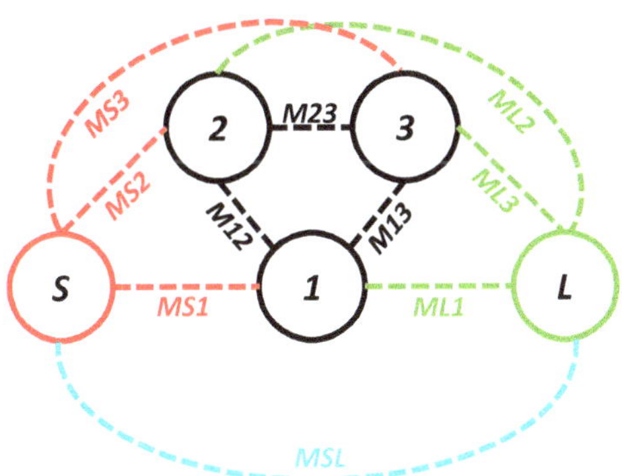

Figure 15. Coupling topology of the proposed filter.

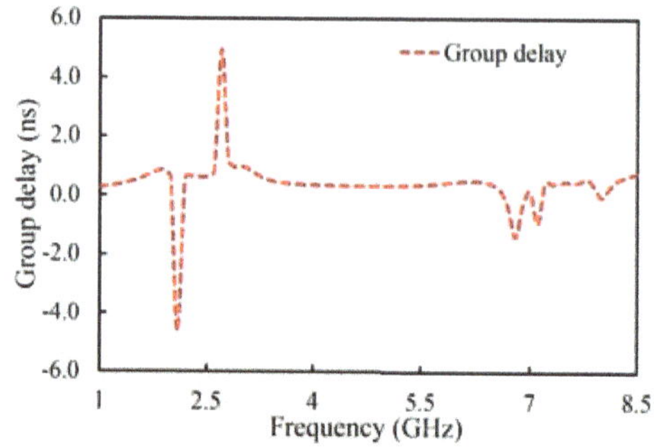

Figure 16. Simulated group delay of the filter in the proposed filter.

Figures 17 and 18 display the impact of varying parameters L_{s2} and L_{s1} on the first three resonant modes and the TZs of the filter. By increasing the length of L_{s2}, the coupling between the open-circuited stubs increased; hence, the first two TZs moved further apart, as illustrated in Figure 17. The position of the third TZ shifted to the lower-frequency side when L_{s2} was increased. The resonant frequency of the second mode shifted to the lower-frequency side by increasing the L_{s2}, while the first and third modes were resistant to any change in L_{s2}. The parameter L_{s1} had a negligible effect on the first three resonant modes and the first two TZs, but it had an inverse relationship with the third TZ. Therefore, L_{s2} and L_{s1} can be adjusted to position the TZs of the filter.

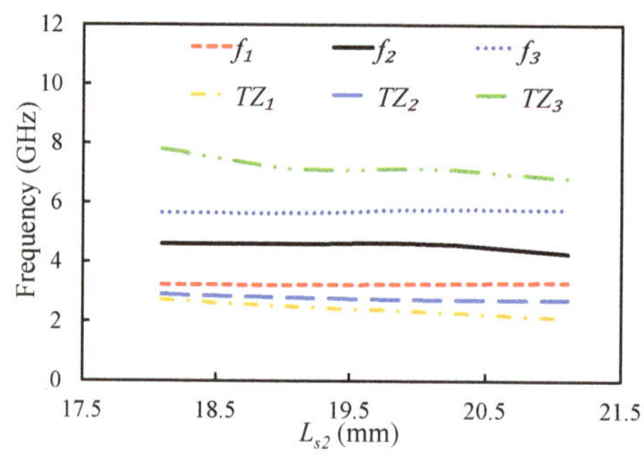

Figure 17. Mode chart for the first three resonant modes and transmission zeros against L_{s2}.

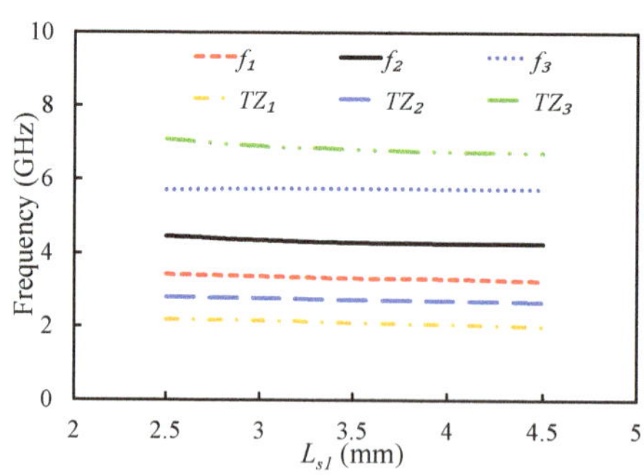

Figure 18. Mode chart for the first three resonant modes and transmission zeros against L_{s1}.

4. Results and Discussion

The fabricated prototype of the proposed filter is shown in Figure 19. RT/duroid 5880 (ϵ_r = 2.2, loss tangent of 0.0009, and thickness of 1.575 mm) was used as a substrate. The performance of the fabricated filter was measured with the Network Analyzer. The S-parameter results of the fabricated filter were compared with the simulation ones, as shown in Figure 19. The measured center frequency was 4.66 GHz with a fractional bandwidth of 67.8%. It can be seen that the insertion loss was better than 1 dB in the whole passband. The TZs were located at 2.09, 2.72, and 6.82 GHz, as shown in Figure 19. The measured group delay of the filter varied from 0.09 ns to 0.81 ns, as shown in Figure 20. The group delay of the filter is plotted over a 3D Smith chart [23]. The above region from the surface is a positive peak of the group delay, and the interior region from the surface is a negative peak, as shown in Figure 21. The comparison between simulation and measurement results shows good agreement, with differences on the order of 1 dB, on average, in the case of S_{11} parameters. The observed deviations are given mainly by connector effects, specifically to losses as well as reactive loading inherent to the connector soldering process. Radiation loss may result from etching the U-shaped slot on the upper surface of the HMSIW resonator. Therefore, the forward radiation loss (RL_{fl}) was calculated using Equation (3) [6]. The radiation loss was <10% in the whole operating band, as illustrated in Figure 22.

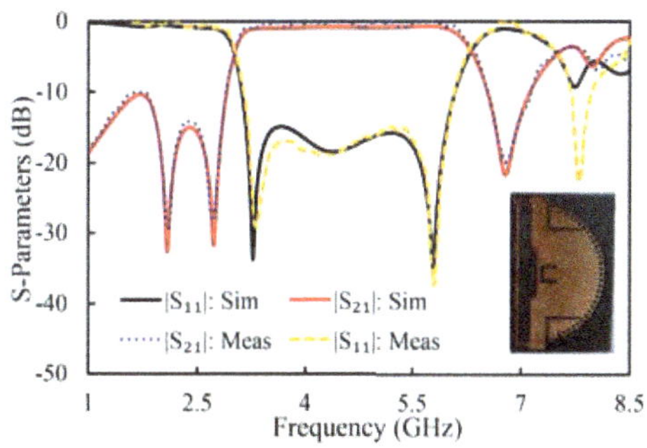

Figure 19. Simulated and measured S-parameters (magnitude) of the proposed filter.

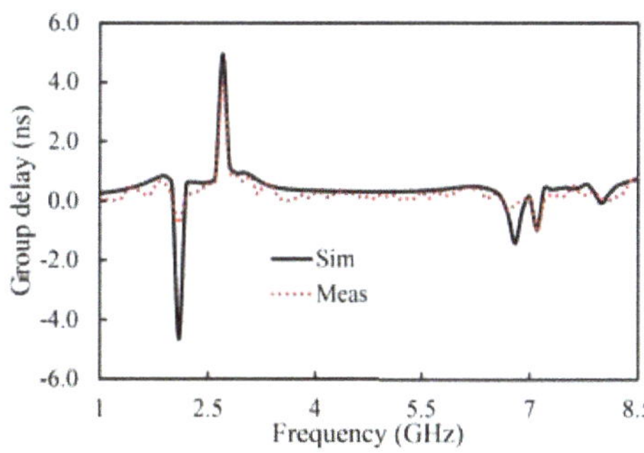

Figure 20. Simulated and measured group delay of the filter.

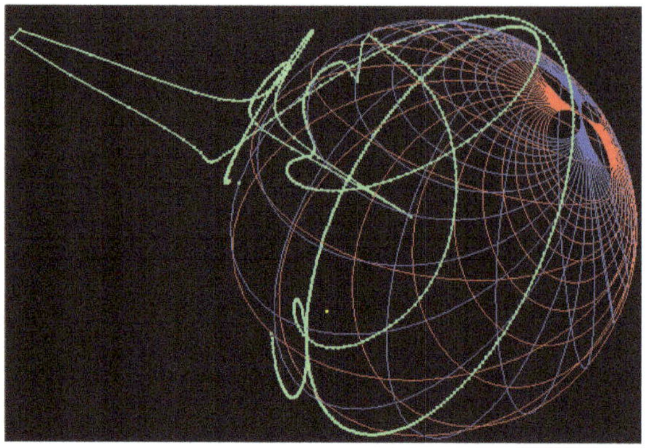

Figure 21. Group delay of the proposed filter over a 3D Smith chart.

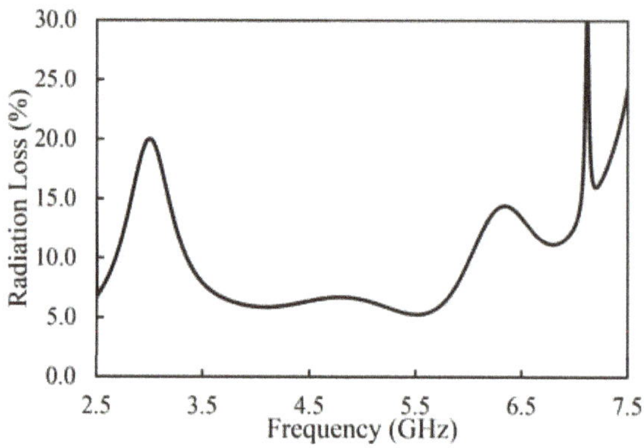

Figure 22. Simulated radiation loss of the filter.

$$RL_{fl} = 1 - |S_{11}|^2 - |S_{21}|^2 \quad (3)$$

Table 1 compares the proposed filter with the state-of-the-art wideband filters. The proposed filter has a higher fractional bandwidth than any other filter. The insertion loss of the filter is better than the other filters, except [11]. However, [11] has a lower fractional bandwidth and fewer TZs.

Table 1. Comparison with the published wideband filters.

Ref.	Size ($\lambda_o \times \lambda_o$)	FBW (%)	f_o (GHz)	IL (dB)	TZs
This Work	**0.69×1.03**	**69.31**	**4.67**	**0.9**	**3**
[6]	0.22×1.2	29	3.45	1.5	0
[7]	Not Given	10.34	5.8	1.1	3
[9]	≈0.58×0.91	69.1	4.5	1.4	2
[10]	0.63×1.25	42	8.5	1.1	3
[11]	Not Given	38	5.2	0.74	2
[14]	1.48×0.39	27.2	1	1.62	4

λ_o = free space wavelength at the center frequency (f_o), IL = insertion loss; FBW = fractional bandwidth

Based on the above studies, the following guidelines are suggested.

1. Select the dimensions of the cavity, based on the design specifications of the filter, using the following equation:
$$f_r = \frac{k_{mnp}c}{2\pi r\sqrt{\epsilon_{reff}}}$$
2. Select the radius of the vias and gap between them as suggested by [22]. The suggested guidelines are $d/a \geq 0.5$ and $d/\lambda_\circ \leq 0.1$ (where $\lambda_\circ$ is the wavelength at the center frequency, d is the diameter of the vias, and a is the gap between the vias).
3. Symmetrically cut the circular SIW cavity resonator into two portions. Each part is the HMSIW cavity resonator.
4. Adjust the dimensions of the open-ended slot (L_c and W_c) to position the resonant modes of the cavity.
5. Design a U-shaped slot in the center of the cavity resonator to generate a TZ on the higher-frequency side of the passband. The following equation can be used to estimate the dimensions of the slot:
$$f_z = \frac{c}{2L\sqrt{\epsilon_{reff}}}$$
6. Adjust the dimensions of the slot (L_1, and L_2) and position of the slot (X_1) to control the location of the TZ. Moreover, these parameters are also vital for positioning the resonant modes.
7. To further improve the selectivity, add L-shaped open-circuited stubs to generate the TZs on the lower-frequency side of the passband.
8. Adjust the dimensions of the L-shaped open-circuited stubs (L_{s1}, and L_{s2}) to control the location of the resonant modes and TZs.
9. Optimize the overall filter parameters to get the desired results.

5. Conclusions

A multimode wideband HMSIW filter with a fractional bandwidth of 69.31% at a center frequency of 4.67 GHz, return loss better than 15 dB, and insertion loss of <0.9 dB was designed and experimentally validated in this paper. The proposed filter was fabricated on the RT/duroid 5880 (ϵ_r = 2.2, loss tangent of 0.0009, and thickness of 1.575 mm) substrate. The measured results show good agreement with the simulation in terms of S-parameters and group delay. The proposed HMSIW bandpass filter can be integrated with any planar wideband communication system circuit, thanks to its planar structure. It covers the majority of fifth-generation (5G) sub-6 GHz bands (USA: 3.3–3.8 GHz, 4–5 GHz; Japan: 3.6–5 GHz; and South Korea, Taiwan, China, Russia, India, Australia, and EMEA: 3.3–3.8 GHz). The proposed filter is promising choice for next-generation wireless technologies (5G) due to its low insertion loss, competitive in band return loss, wide bandwidth, and good selectivity.

Author Contributions: Design and concept, A.I.; methodology, A.I.; investigation, A.I.; resources, A.I., J.J.T., S.K.W.; writing—original draft preparation, A.I.; writing—review and editing, A.I., J.J.T. and S.K.W.; validation, A.I., M.A., F.F., and E.L.; supervision, J.J.T. and S.K.W.; project administration, J.J.T., M.A., F.F., and E.L. All authors have read and agreed to the published version of the manuscript.

Funding: This work is partially supported by RTI2018-095499-B-C31, Funded by Ministerio de Ciencia, Innovación y Universidades, Gobierno de España (MCIU/AEI/FEDER, UE).

Acknowledgments: The authors would like to thank Andrei A. Muller, Alin Moldoveanu, Victor Asavei, and their whole team for providing us a student version of 3D Smith Chart tool (http://www.3dsmithchart.com).

Conflicts of Interest: The authors declare no conflict of interest.

References

1. Iqbal, A.; Tiang, J.J.; Kwang Lee, C.; Muhammad, S. Wideband Half Mode Substrate Integrated Waveguide (HMSIW) Bandpass Filter. In Proceedings of the 2020 17th International Conference on Electrical Engineering/Electronics, Computer, Telecommunications and Information Technology (ECTI-CON), Phuket, Thailand, 24–27 June 2020; pp. 710–712. [CrossRef]
2. Feng, W.; Gao, X.; Che, W.; Yang, W.; Xue, Q. High selectivity wideband balanced filters with multiple transmission zeros. *IEEE Trans. Circuits Syst. II Express Briefs* **2015**, *64*, 1182–1186. [CrossRef]
3. Iqbal, A.; Tiang, J.J.; Lee, C.K.; Mallat, N.K.; Wong, S.W. Dual-band half mode substrate integrated waveguide filter with independently tunable bands. *IEEE Trans. Circuits Syst. II Express Briefs* **2019**, *67*, 285–289. [CrossRef]
4. Iqbal, A.; Tiang, J.J.; Lee, C.K.; Lee, B.M. Tunable Substrate Integrated Waveguide Diplexer with High Isolation and Wide Stopband. *IEEE Microw. Wirel. Components Lett.* **2019**, *29*, 456–458. [CrossRef]
5. Iqbal, A.; Ahmad, A.W.; Smida, A.; Mallat, N.K. Tunable SIW Bandpass Filters with Improved Upper Stopband Performance. *IEEE Trans. Circuits Syst. II Express Briefs* **2020**, *67*, 1194–1198. [CrossRef]
6. Máximo-Gutiérrez, C.; Hinojosa, J.; Alvarez-Melcon, A. Design of wide band-pass substrate integrated waveguide (SIW) filters based on stepped impedances. *AEU-Int. J. Electron. Commun.* **2019**, *100*, 1–8. [CrossRef]
7. Lugo, C.; Papapolymerou, J. Planar Realization of a Triple-Mode Bandpass Filter Using a Multilayer Configuration. *IEEE Trans. Microw. Theory Tech.* **2007**, *55*, 296–301. [CrossRef]
8. Jia, D.; Feng, Q.; Xiang, Q.; Wu, K. Multilayer substrate integrated waveguide (SIW) filters with higher-order mode suppression. *IEEE Microw. Wirel. Compon. Lett.* **2016**, *26*, 678–680. [CrossRef]
9. Zhang, Z.C.; Liu, H.W.; Wong, S.W. Compact wideband bandpass filter based on double-T-shaped stub loaded resonator and loading technique for zero-voltage point. *Int. J. RF Microw. Comput. Aided Eng.* **2018**, *28*, e21197. [CrossRef]
10. Chen, R.S.; Wong, S.W.; Zhu, L.; Chu, Q.X. Wideband bandpass filter using U-slotted substrate integrated waveguide (SIW) cavities. *IEEE Microw. Wirel. Compon. Lett.* **2014**, *25*, 1–3. [CrossRef]
11. Jin, C.; Shen, Z. Compact triple-mode filter based on quarter-mode substrate integrated waveguide. *IEEE Trans. Microw. Theory Tech.* **2013**, *62*, 37–45. [CrossRef]
12. Wong, S.W.; Wang, K.; Chen, Z.N.; Chu, Q.X. Design of millimeter-wave bandpass filter using electric coupling of substrate integrated waveguide (SIW). *IEEE Microw. Wirel. Compon. Lett.* **2013**, *24*, 26–28. [CrossRef]
13. Zhu, X.C.; Hong, W.; Wu, K.; Tang, H.J.; Hao, Z.C.; Chen, J.X.; Chu, P. Design and implementation of a triple-mode planar filter. *IEEE Microw. Wirel. Compon. Lett.* **2013**, *23*, 243–245. [CrossRef]
14. Srisathit, K.; Worapishet, A.; Surakampontorn, W. Design of Triple-Mode Ring Resonator for Wideband Microstrip Bandpass Filters. *IEEE Trans. Microw. Theory Tech.* **2010**, *58*, 2867–2877. [CrossRef]
15. Liu, J.; Ding, W.; Chen, J.; Zhang, A. New ultra-wideband filter with sharp notched band using defected ground structure. *Prog. Electromagn. Res.* **2019**, *83*, 99–105. [CrossRef]
16. Ji, X.C.; Ji, W.S.; Feng, L.Y.; Tong, Y.Y.; Zhang, Z.Y. Design of a novel multi-layer wideband bandpass filter with a notched band. *Prog. Electromagn. Res.* **2019**, *82*, 9–16. [CrossRef]
17. Yeo, K.S. Wideband Grounded Coplanar Waveguide Bandpass Filter with Novel Coupling Structures. In Proceedings of the 2019 IEEE Asia-Pacific Microwave Conference (APMC), Singapore, 10–13 December 2019; pp. 1038–1040.
18. Hou, Z.; Liu, C.; Zhang, B.; Song, R.; Wu, Z.; Zhang, J.; He, D. Dual-/Tri-Wideband Bandpass Filter with High Selectivity and Adjustable Passband for 5G Mid-Band Mobile Communications. *Electronics* **2020**, *9*, 205. [CrossRef]
19. Zhang, P.; Liu, L.; Chen, D.; Weng, M.H.; Yang, R.Y. Application of a Stub-Loaded Square Ring Resonator for Wideband Bandpass Filter Design. *Electronics* **2020**, *9*, 176. [CrossRef]
20. Liang, G.Z.; Chen, F.C. A compact dual-wideband bandpass filter based on open-/short-circuited stubs. *IEEE Access* **2020**, *8*, 20488–20492. [CrossRef]
21. Pozar, D.M. *Microwave Engineering*; John Wiley & Sons: Hoboken, NJ, USA, 2011.

22. Xu, F.; Wu, K. Guided-wave and leakage characteristics of substrate integrated waveguide. *IEEE Trans. Microw. Theory Tech.* **2005**, *53*, 66–73.
23. Muller, A.A.; Sanabria-Codesal, E.; Moldoveanu, A.; Asavei, V.; Lucyszyn, S. Extended capabilities of the 3-D smith chart with group delay and resonator quality factor. *IEEE Trans. Microw. Theory Tech.* **2016**, *65*, 10–19. [CrossRef]

MDPI
St. Alban-Anlage 66
4052 Basel
Switzerland
Tel. +41 61 683 77 34
Fax +41 61 302 89 18
www.mdpi.com

Sensors Editorial Office
E-mail: sensors@mdpi.com
www.mdpi.com/journal/sensors

www.ingramcontent.com/pod-product-compliance
Ingram Content Group UK Ltd.
Pitfield, Milton Keynes, MK11 3LW, UK
UKHW060123010826
14089UKWH00006B/121

9 783036 535319